Physics and Applications of Pseudosparks

NATO ASI Series

Advanced Science Institutes Series

*A series presenting the results of activities sponsored by the NATO Science Committee,
which aims at the dissemination of advanced scientific and technological knowledge,
with a view to strengthening links between scientific communities.*

The series is published by an international board of publishers in conjunction with the
NATO Scientific Affairs Division

A	**Life Sciences**	Plenum Publishing Corporation
B	**Physics**	New York and London
C	**Mathematical and Physical Sciences**	Kluwer Academic Publishers
D	**Behavioral and Social Sciences**	Dordrecht, Boston, and London
E	**Applied Sciences**	
F	**Computer and Systems Sciences**	Springer-Verlag
G	**Ecological Sciences**	Berlin, Heidelberg, New York, London,
H	**Cell Biology**	Paris, and Tokyo

Recent Volumes in this Series

Volume 213—Interacting Electrons in Reduced Dimensions
edited by Dionys Baeriswyl and David K. Campbell

Volume 214—Science and Engineering of One- and Zero-Dimensional
Semiconductors
edited by Steven P. Beaumont and Clivia M. Sotomayor Torres

Volume 215—Soil Colloids and their Associations in Aggregates
edited by Marcel F. De Boodt, Michael H. B. Hayes,
and Adrien Herbillon

Volume 216A—The Nuclear Equation of State, Part A:
Discovery of Nuclear Shock Waves and the EOS
edited by Walter Greiner and Horst Stöcker

Volume 216B—The Nuclear Equation of State, Part B:
QCD and the Formation of the Quark-Gluon Plasma
edited by Walter Greiner and Horst Stöcker

Volume 217—Solid State Microbatteries
edited by James R. Akridge and Minko Balkanski

Volume 218—Applications of Statistical and Field Theory
Methods to Condensed Matter
edited by Dionys Baeriswyl, Alan R. Bishop, and José Carmelo

Volume 219—Physics and Applications of Pseudosparks
edited by Martin A. Gundersen and Gerhard Schaefer

Series B: Physics

Physics and Applications of Pseudosparks

Edited by

Martin A. Gundersen

University of Southern California
Los Angeles, California

and

Gerhard Schaefer

Late of Polytechnic University
Farmingdale, New York

Plenum Press
New York and London
Published in cooperation with NATO Scientific Affairs Division

Proceedings of a NATO Advanced Research Workshop on
Physics and Applications of Hollow Glow Switches,
held July 17-21, 1989,
in Lillehamma, Norway

Library of Congress Cataloging-in-Publication Data

NATO Advanced Research Workshop on Physics and Applications of Hollow
  Glow Switches (1989 : Lillehammer, Norway)
    Physics and applications of pseudosparks / edited by Martin A.
  Gundersen and Gerhard Schaefer.
        p.   cm. -- (NATO ASI series. Series B, Physics ; vol. 219)
    "Proceedings of a NATO Advanced Research Workshop on Physics and
  Applications of Hollow Glow Switches, held July 17-21, 1989, in
  Lillehammer, Norway"--T.p. verso.
    "Published in cooperation with NATO Scientific Affairs Division."
    Includes bibliographical references.
    ISBN 0-306-43539-X
    1. Glow discharges--Congresses. 2. Electric switches--Congresses.
  I. Gundersen, Martin A.   II. Schaefer, Gerhard, 1940-1989.
  III. North Atlantic Treaty Organization. Scientific Affairs
  Division. IV. Title. V. Series: NATO ASI series. Series B,
  Physics ; v. 219.
  QC711.8.G5N38   1989
  537.5--dc20   90-7212
                                                          CIP

This book is dedicated to Prof. Gerhard Schaefer

PREFACE

The purpose of the 1989 NATO ARW was to develop applications, and an improved understanding of the physics for high current emission and conduction observed in hollow cathode-hollow anode switches including the pseudospark and BLT. New applications include highly emissive cathodes for microwave devices, accelerators and free electron lasers, high power tubes, electron and ion beams, microlithography, accelerators, and other plasma devices.

Recent research has produced a new generation of gas-phase plasma switches that are characterized by very high current emission and conduction while operating in a glow mode. These switches include the pseudospark and the BLT, both of which have hollow electrodes, switch over 10 to 100 kA peak current, and have cathodes with emission $\geq$ 10,000 A/cm^2 over $\geq$ 1 cm^2 area. The cathode properties are especially remarkable - about 2 orders of magnitude larger emission than existing thermionic cathodes. Part of the meeting was devoted to understanding these properties, and exploiting applications of this cathode. The remarkable properties of these switches are very surprising in the light of considerable previous work in this area, and these results deserve study in order to understand the underlying physical mechanisms, and to develop ideas and insight into future applications, and foster coherent research in this area.

The operating cycle of pseudo-spark and BLT switches and related devices can be divided into four phases: hold-off, triggering, conduction, and recovery. There was very little discussion of the hold-off and recovery phases. Presumably the physics issues here are very similar to those in thyratrons. Most of the discussion centered around the triggering and conduction phase. It was generally agreed that triggering is accomplished through the formation of a hollow-cathode discharge which leads to the generation of an intense beam of positive ions in the inter-electrode region. It is thought that this beam heats the front surface of the cathode to form a super-emissive cathode which allows the very high current densities associated with the super-dense discharge. There is evidence that molybdenum and tungsten cathode surfaces are heated beyond the melting point during operation.

Most experimental studies of pseudo-spark discharges have centered around some type of optical spectroscopy or current and voltage measurements. It was generally agreed that there is a strong positive ion beam associated with the pseudo-spark discharge. There may be an electron beam as well, but it appears to be much weaker. There was a great deal of discussion about theoretical modeling of the discharge. The modeling efforts discussed dealt mainly with the early stages of triggering, and produced results in good agreement with what empirical information exists. The conduction (super-dense discharge) phase seems more difficult to model, and there were discussions about several points.

Advanced experimental and theoretical methods are now available to study these problems. These include laser induced fluorescence spectroscopy, highly time-resolved spectroscopic studies, surface studies, and sophisticated plasma simulation methods such as Monte-Carlo simulations and quantitative studies of microscopic processes in the bulk plasma and at the plasma-electrode interface. Powerful new technologies are enhancing our diagnostic capabilities -- including CCD detectors, tunable solid state lasers, T: sapphire, VUV laser spectroscopy, which will become routine -- enabling us to probe the resonance lines of many ions.

There are pressing engineering applications in the areas of lasers, particle beams, and high energy physics, where these principles should have significant impact. This meeting will bring together these different disciplines, and allow them to be focused on this problem area.

It was clear that fundamental data is needed on collisions involving **excited** atoms, molecules, radicals, and ions. All of the transport data we have is for unexcited gases -- and plasmas in high power devices have large fractional excitation and ionization. Furthermore, there was strong agreement that absolute measurements **are** worth the trouble. Agreement between a model and absolute measurements gives us much more confidence that we have the correct physics.

This meeting was also motivated by the fact that there are many new applications of these devices under consideration. These include new types of ion and electron beams for microelectronic technology, accelerators, other plasma loaded devices, plasma lenses for high energy physics, plasma accelerators, applications requiring very high cathode emission such as cathodes for pulsed accelerators and microwave sources. These Proceedings summarize results of a meeting between physicists, engineers, and applied scientists considering further developments in physics and engineering and industrial applications.

This introduction was written with the assistance of Dr. Klaus Frank, Prof. James Lawler and Prof. P. Frazer Williams, who provided summaries of the various sessions. Shirin Mistry, Sarah Novak and Roberta Gundersen provided considerable organizational and administrative skill.

This book is dedicated to Gerhard Schaefer. Prof. Schaefer provided the most important technical support in organizing this meeting. In the larger area of pulsed power research, Prof. Schaefer's overlapping expertise in quantum electronics and pulsed power was combined with exceptionally sound judgement -- both on a technical and personal level. He was invaluable, and he is missed.

Martin Gundersen
USC, Los Angeles

CONTENTS

Devices and Related Properties

The Properties of the Pseudospark Discharge.................................. 1
  J. Christiansen

Review of Superdense Glow Discharge....................................... 15
  K. Frank

Basic Mechanisms Contributing to the Hollow Cathode Effect.............. 55
  G. Schaefer and K. H. Schoenbach

Cathode-Related Processes in High-Current Density, Low Pressure
Glow Discharges.. 77
  W. Hartmann and M. A. Gundersen

Comparison of Electrode Effects in High-Pressure and Low-Pressure
Gas Discharges Like Spark-Gap and Pseudospark........................... 89
  J. Christiansen, K. Frank, W. Hartmann, C. Kozlik,
  W. Krauss-Vogt and R. Michal

Experimental Review

Mapping and Modeling of the Cathode Fall and Negative
Glow Regions 109
  J. E. Lawler, E. A. Den Hartog, W. N. G. Hitchon, T. R. O'Brian,
  and T. J. Sommerer

Emission Spectroscopy in Optically Thick Gas Discharges................. 133
  D. Karabourniotis

An Analysis of the High Current Glow Discharge Operation of the
BLT Switch... 155
  G. Kirkman-Amemiya, R. L. Liou, T. Y. Hsu, and M. A. Gundersen

Laser-Induced Fluorescence Measurements of Number Densities of
Neutral and Ionized Metal Atoms.. 167
  G. Lins

Streamers in Atmospheric Pressure N_2: Empirical Results 185
    P. F. Williams and F. E. Peterkin

Theoretical Modeling

The Solution of the Continuity Equations in Ionization and Plasma
Growth.. 197
    A. J. Davies and W. Niessen

Scaling Parameters for Optically Triggered Hollow Cathode Switches
Obtained by Computer Simulation 219
    H. Pak and M. J. Kushner

A Physical Model of Prebreakdown in the Hollow Cathode Pseudospark
Discharge Based on Numerical Simulations................................. 233
    K. Mittag

Self-Consistent Models of DC and Transient Glow Discharges 255
    J. P. Boeuf

Weak Collisions in Strong Double Layers 277
    H. Schamel, P. Hatjiimanolaki and P. Nicoletopoulos

The Effect of Pendel Electrons on Breakdown and Sustainment
of a Hollow Cathode Discharge 293
    K. H. Schoenbach, L. L. Vahala, G. A. Gerdin, N. Homayoun,
  F. Loke and G. Schaefer

A Two-Electron-Group Model for a High Current Pseudospark or
Back-Lighted Thyratron Plasma.. 303
    H. Bauer, G. Kirkman and M. A. Gundersen

Electron Ionization Rate Coefficients at Very High E/N..................... 319
    L. C. Pitchford

New Applications

Plasma-Based Device Concepts Based on the Pseudospark and BLT........ 331
    M. A. Gundersen

Emittance Measurement of a Pseudospark-Produced Electron
Beam 343
    M. J. Rhee and E. Boggasch

New Ways of Electron Emission for Power Switching and Electron
Beam Generation 349
    H. Riege

Index 359

x

THE PROPERTIES OF THE PSEUDOSPARK DISCHARGE

J. Christiansen

Department of Physics
University of Erlangen/Nürnberg
Erwin Rommel Str. 1
8520 Erlangen   FRG

INTRODUCTION

The contribution gives an overview on the state of research work done on
a new kind of gas discharge in the low pressure regime, which is based on the
principles of a hollow cathode discharge. The main property of the self-sustained
discharge is that high currents resp. high current densities are achieved within
very short time. This led to the name pseudospark discharge. The phenomenon
was discovered 12 years ago within the frame of a research work looking for the
possibility to accelerate particles in a space charge neutralized plasma [1].

The pseudospark is a self-sustained discharge, which is normally run
periodically (i. e. the voltage applied to the electrodes is saw-tooth like). Besides
running in a repetitive selfbreakdown mode, the pseudospark can also be initiated
from a stand by mode by different trigger methodes [2], [3].

The characteristics of the voltage breakdown is determined in both cases
by a rise of the breakdown voltage with falling gas pressure. Hence it shows a
behaviour similiar to the Paschen-law of a parallel plate system.

Research and development work done on modified discharge systems within
the last years led essentially to the following excellent applications:
1. a new kind of high power switch [4], [5], [6], [7], [8], [9]
2. a source of an intense, medium energy electron beam which is emitted at the
   anode side [10], [11], [12].

IGNITION AND TEMPORAL DEVELOPMENT OF THE PSEUDOSPARK DISCHARGE

The main property of the pseudospark - to establish a high current in a
gas discharge within very short time - proves to be an universal, i.e. gas

independent phenomenon; it is realized by a circular central hole in the cathode and an almost screened cavity behind the cathode hole. It turns out, that a symmetrical set up (fig. 1), where the anode also has a central hole, is adequate for almost all applications. It is expedient to separate the temporal evolution of the discharge into four different parts:
1. predischarge (ignition of the pseudospark)
2. development of a hollow cathode discharge
3. high current main discharge
4 decay of the discharge plasma

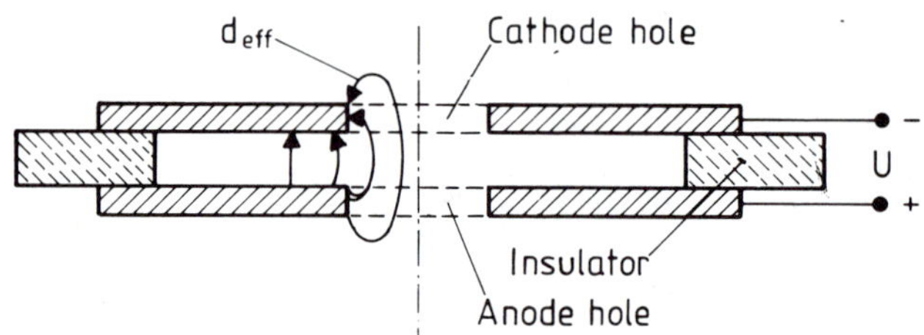

Figure 1. Basic scheme of a single gap pseudospark chamber

Phase 3 arises only, if a suitable amount of energy is delivered to the electrode system. Otherwise the pseudospark recovers its voltage hold off capability at the end of phase 2 (this happens e.g. if the self capacitance of the electrode system is discharged).

With progress of the second phase electrons are accelerated between cathode and anode, and leave the device through the anode hole. A high current density electron beam with medium energies of the size of the break down voltage is emitted.

Phase 1: Ignition of the Pseudospark

A systematic investigation of the predischarge mechanisms was carried out by Reichle [13] (experimental work) and by Mittag [14] (theoretical work). The results can be summarized as follows:

The electric field distribution in the discharge system causes typical E/n values of the order of 10^{-14} $V \cdot cm^2$ within the cathode backspace , whereas E/n amounts to 10^{-11} $V \cdot cm^2$ in the gap between the electrodes. Due to high α/p values there is a large charge carrier multiplication in the region of the cathode hole. On the other hand, electron impact ionization is less significant within the gap, i.e. the electrons run through the "accelerating gap" almost without collisions. Hence, electrons released from the walls of the cathode hole by particle impact or photo-effect cause a strong charge carrier multiplication resp. a rather high avalanche growth in the cathode backspace. Avalanche growth will be enhanced as soon as a positive space charge arises in the center of the cathode hole. The latter results from the fact, that electrons are extracted from the cathode backspace by the electric field which penetrates through the cathode hole, whereas the inert ions rest in the hollow cathode giving a virtual anode.

It is possible to estimate the time constant τ of charge carrier multiplication. Assuming that secondary electron release by ion impact on the walls of the hollow cathode is the dominating process, τ corresponds to the transit time of ions from the center of the cathode hole to the "edge" of the hole, or, more precisely $\tau \approx (\alpha_{axis} \cdot v_{+\ axis})^{-1}$ with

α_{axis} : Townsend-α coefficient and

$v_{+\ axis}$: radial ion drift velocity. τ amounts to $10^{-7} - 10^{-8}$ s for a space charge potential of several 100 V.

Hence a dense plasma emerges at the center of the cathode hole within very short time, in unison with theoretical calculations of Mittag. This behaviour was also studied by means of fast photographical methods. Furtheron, the measurements yielded, that, at the same time, the cathode backspace gradually becomes filled with plasma. This is the transition to

<u>Phase 2: Hollow Cathode Discharge</u>

The plasma propagates isotropically, at a velocitiy of $2 \cdot 10^6$ m/s, into the hollow cathode as soon as the discharge current reaches a value of the order of 10 A. Likewise the plasma expands into the electrode gap with about the same velocity. Fig. 2 shows a streak and fig.3 a nanogate photography of the described evolution.

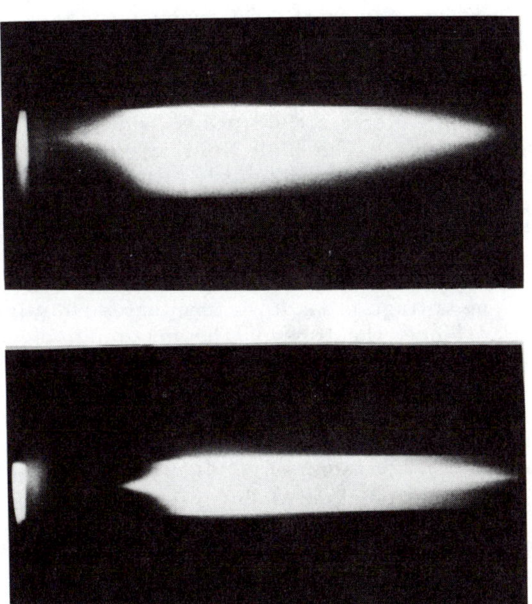

Figure 2. Streak photography of the cathode backspace
upper: slit near the cathode hole
lower: top of the cathode backspace
The total length of the time axis is 150 ns

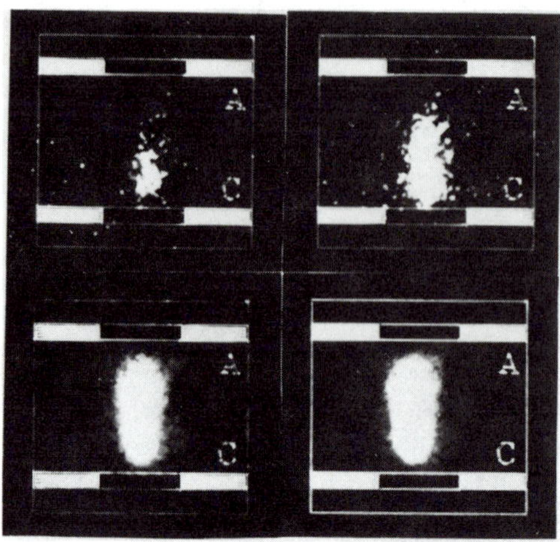

Figure.3 Nanogate photography of the beginning of the conducting phase of the pseudospark switch, registered between anode (A) and cathode (C). a) 70 ns, b) 68 ns, c) 62 ns, d) 54 ns before the onset of steep current rise

In this phase an electron beam is emitted at the anode hole. The development of the discharge may be interpreted in the following way:

The hollow cathode, being filled with plasma, is working like a virtual cathode for the electron extraction from the cathode backspace. The electrons are accelerated by the high electric field across the electrode gap. α/n values are small in this region due to the high field strength. Hence, electrons pass through the gap with almost no collisions as long as the impedance of the voltage source is small compared to the resistance of the accelerating gap. Ionizing processes by primary and secondary electrons cause the propagation of plasma into and the shortening of the gap. The voltage breakdown can easily be observed by electrical measurement of the gap voltage, and of the electron beam current, respectively. The voltage breakdown terminates emission of fast electrons from the pseudospark. Hence, the time of the maximum electron beam current (several 100 A) characterizes the end of phase 2, which typically lasts 10 to 30 ns.

The intense, space-charge neutralized electron beam with a current densitiy of 10^4-10^5 A/cm^2 offers a large range of applications. A remarkable property of the beam is its small diameter of 0,1 - 1 mm at the anode hole. The narrowing of the electron beam is mainly caused by magnetic compression due to the pinch effect, and by radial components of the electric field which are defined by the electrode geometry.

For the technical realization of a suitable electron beam source, it is useful to insert several isolated, floating electrodes between anode and cathode, as shown in fig. 4. By this means the voltage hold-off capability of the system is enhanced for constant gas pressure, giving a maximum densitiy of neutrals at the beginning of the discharge.

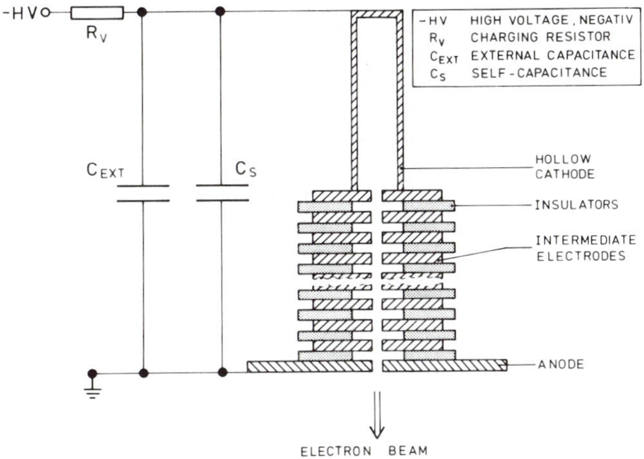

Figure 4. Pseudospark chamber for generating pulsed electron beams

Fig.5 shows a streak photography of the plasma of a high current discharge. The temporal expansion of the electron beam, respectively of the plasma surrounding the beam, is obvious. The picture shows impressively the transition of the hollow cathode discharge to the third phase, which describes the high current behaviour.

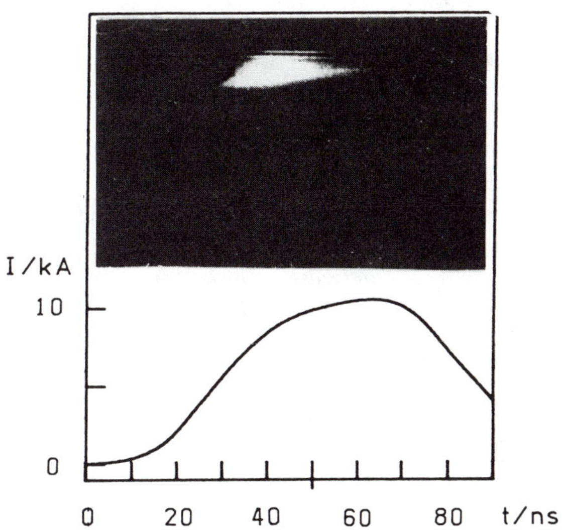

Figure 5. The discharge current and its radial and axial development in the main gap.
Upper photograph: Slit parallel to axis
Lower photograph: Slit parallel to electrodes

High Current Main Discharge resp. Superdense Glow Discharge

The third phase begins, as soon as the plasma diameter becomes equal to or greater than the diameter of the cathode hole. Phase 2 is terminated, because electron emission from the cathode backspace is limited by a screening of the hollow cathode due to the dense plasma filling the cathode hole. It is then possible to accelerate positive ions towards the cathode surface. The accelerating electric field is of the order of $k \cdot T_e / \lambda_D \cdot e$, if we assume a dense plasma. The current density is limited by the ion drift velocity, and can be described by Bohm's formula:

$$j_+ = 0.5 \cdot n_+ \cdot e \cdot v_+ \quad \text{with} \quad v_+ \approx \sqrt{2 \cdot k \cdot T_e / m_+} \; .$$

(This estimation is valid only for field-free plasmas, however; if there is any electric field gradient within the plasma, the limited ion current density can be considerably higher.)

It results that, assuming an electron temperature of the order of eV , the current densitiy is of the order of some 100 A/cm^2 , even if the Townsend coefficient γ_i approaches values of 1. De facto current densities greater than 10^4 A/cm^2 were observed at the time, when the discharge plasma, corresponding to fig. 4, has reached the "edge" of the cathode hole. In this case, we assume that the main part of the electrons has its origin at the electrode surface. We define this part of the discharge as the start of the high current discharge. Following the work of Gundersen and Hartmann [15], [16], who give an interpretation of this phenomenon, we will call it superdense glow discharge. The interpretation is based on the following assumption:

Gas particles are released from the cathode surface by ion impact or by thermal desorption. After ionization, the desorbed particles give a remarkable contribution of the order of 10^{16} cm^{-2} to the total numberdensity of ions which are available for the discharge. These ions are accelerated in the cathode fall voltage drop of about 10^6 V/cm. Hence, the electrode surface is heated within a thin surface layer by ion impact to a temperature of 3000–4000 K within 30–100 ns. An estimation of field-enhanced thermionic electron emission (Schottky-emission) by Hartmann and Gundersen using the above data yields that this is a possible mechanism to deliver the measured high current densities. In fact a melting of electrodes in the region of the central hole is observed. A more detailed discussion of this phenomenon is given in [15] and in another contribution. With this phenomenon in mind, further investigations were carried out to get information about the gas consumption of the pseudospark switch. The result is, that a high current discharge deposits a considerable part of the filling gas (hydrogen) in the electrodes.

The high current discharge is mainly determined by the external circuit. Considering an LC-circuit, the current of the circuit resp. the magnetic field due to the current will reach its maximum at the current maximum. With decreasing magnetic field there will be a current reversal at the outer shell of the current carrying plasma, as shown by Boggasch [17]. This sheeth is accelerated radially outwards, due to the magnetic pressure which could be shown in experiment. The plasma–wall interaction causes a distruction of the insulator. Therefore, the insulator should be shielded from the plasma, as it is shown in fig.6. For the high power pseudospark switch this modification of the plane-parallel plate system is extremely important. Generally one obtains many periods of current oscillations if a LC-circuit is discharged by a symmetric built pseudospark switch. The resistance of the plasma may be estimated by the damping of the oscillation, if the commutation losses are taken into account properly [4]. Fig.7 represents

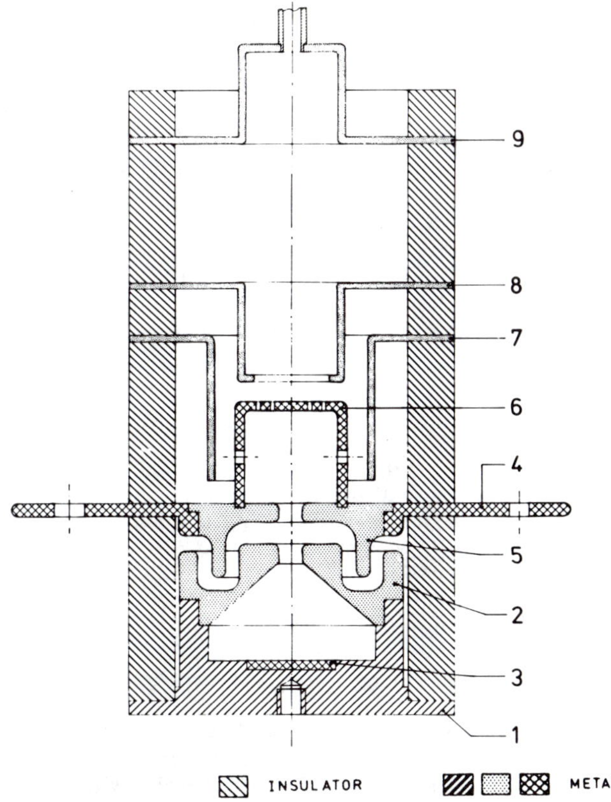

Figure 6. Cross section of a prototype pseudospark switch.
1,2,3: anode; 4,5,: cathode; 6: cathode backspace;
7,8,9: auxiliary trigger electrodes

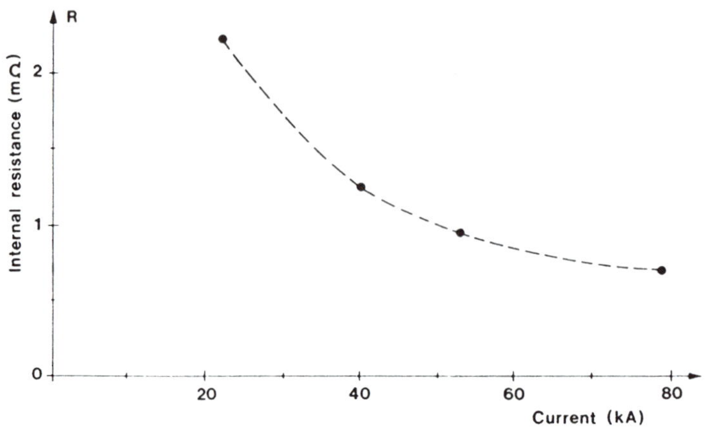

Figure 7. Internal resistance of a high current pseudospark switch.

the total equivalent series resistance of a switch carrying a current of 10^5 A. As can be seen the voltage drop across the switch during the high current phase is of the order of several 100 V, most of which can be attributed to the cathode fall voltage drop.

Decay of the Plasma

The duration of the high current discharge is determined by the parameters of the external circuit. In particuliar it is of interest t>>o study the long time behaviour of the system. For this reason ringing discharges with a period duration (full sine wave) of 10^{-4}– 10^{-3} s were performed [18]. The switch was operated at voltages between 5 and 10 kV, and the total charge transfer per discharge amounted to $\approx$ 1 Coulomb. The main results of these experiments are:
1. There appears no quenching phenomenon during the current carrying phase , i.e. the plasma channel does not become unstable.
2. The switch may regain its voltage hold off capability during current zero.

The second property shows the following systematic: If the switch operates within a high frequency circuit (cycle times of less than 10^{-4} s) no recovery phenomena are observed.. With prolongation of the sine wave duration a statistical current quenching at current zero is observed after several periodes. An earlier switch recovery is achieved if the pressure within the switch is decreased. Up to now the earliest reproducible quenching of the switch appeared at current zero at the end of the first full sine wave of the circuit. As this investigations still are at their beginning, only some rough conclusions are possible. During commutation the electron emitting cathode surface is cooled down, especially at long cycle times and the Schottky emission ceases. Due to wall and plasma bulk recombination, the number of charge carriers becomes insufficient to reinitiate the discharge, although there is a high electric field across the gap. This experiment shows the possibility to create a bipolar switch, which recovers during current zero. The corresponding current and voltage wave forms can be seen in fig. 8. As dI/dt is at its maximum when the current stops, rather steep voltage rise appears across the switch.

On the other hand there is no indication that the pseudospark switch, in contrary to a thyratron, will quench during the high current phase . Hence the pseudospark switch, which becomes known for its excellent short time switching behaviour, may also be used for applications, where a long lasting current shall be switched. Further investigations will improve the already wide range of applications of the switch.

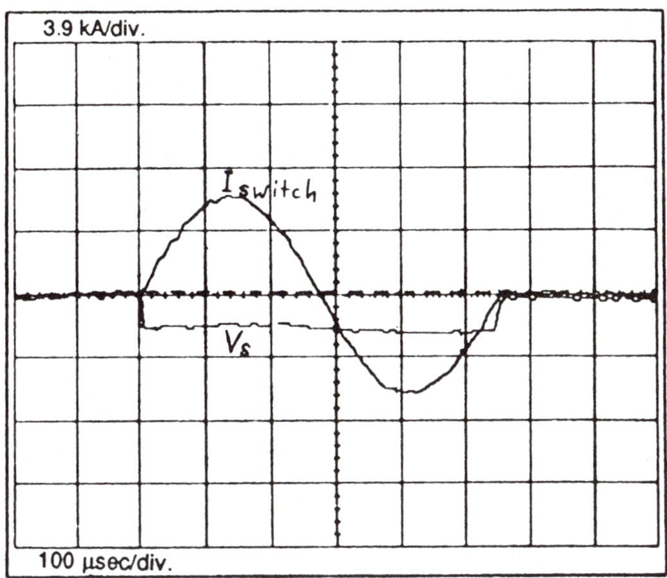

Figure 8. Voltage and current waveforms from a discharge circuit with a sine wave of 560 µs period and ~ 6 kA current amplitude

PROPERTIES OF THE PSEUDOSPARK SWITCH

The ability of the plasma originated by a pseudospark discharge to carry high currents at a low discharge resistance qualifies the pseudospark as a candidate for high power switches. The breakdown voltage at a given gas pressure, which is defined by the Paschen law, corresponds to the maximum operating voltage of the switch. Using hydrogen as filling gas an upper limit of the hold-off voltage for an one-gap system is given to be approximately 40 kV. If one accepts to work at very low pressure, it is possible to use even higher voltages. But one has to accept a reduced lifetime of the switch, because erosion of the electrodes is enhanced by particle impact at the cathode and by runaway electrons at the anode.

The development of the pseudospark discharge described in the second chapter offers the possibility to trigger the system by providing charge carriers to the sensible region of the cathode hole. It is possible to initiate the pseudospark discharge even with a low power trigger system [9].

One of the frequently used trigger mechanisms is based on the pulsed injection of charge carriers of an external glow discharge into the cathode. Further mechanisms are:

- a pulsed or dc corona discharge within the hollow cathode

- a surface discharge within the cathode hole, initiated by a steep high voltage (≤ 5kV) pulse across an insulator surface

- release of photoelectrons by an appropiate UV light source.

The latter method is used by the **b**ack-of-the-cathode **l**ighted **t**hyratron (BLT), as described by Gundersen. Details of the different trigger methods and their jitter- and delay properties will not be represented within this report. But it should be mentioned that the typical jitter of a discharge in hydrogen amounts to 3-10 ns and the delay to 50 - 500 ns independent of the used gas. If only a weak plasma is provided by the trigger mechanism the jitter is increased by one order of magnitude.

As the switch offers a wide range of applications, an adequate classification obviously is necessary for the discussion of switch design for specific applications.

High Current Switch

This type if switch is made of bulk tungsten electrodes with a borehole diameter of 10 mm [4]. The hold off voltage amounts to 7-18 kV, and maximum currents of 150 kA can be switched repeatedly. Working with puls energies of 2-3 kJ a lifetime of 10^6 discharges was achieved. Therefore, the switch may be used instead of an ignitron, having the same efficiency, but offering the advantages of a precise trigger with a jitter of less than 20 ns, a faster recovery, longer lifetime, and the independance of the mounting position.

High Current Rise Time Switch (MUPS)

A further interesting application in the high power regime offers the research work of Mechtersheimer and Kohler [19]. They used an elongated plane-parallel plate system with up to 20 discharge channels in a linear arrangement. By pulsing a common glow discharge in the common cathode backspace of this multichannel system it is possible to ignite all discharge channels simultanously. A trigger current of several Amperes is sufficient to get an excellent switch performance. The limits of this system, called MUPS (**Mu**ltichannel **P**seudospark **S**witch), were investigated by Brückner et al. [29]. They operated a switch with 20 parallel discharge channels at hold off voltages up to 40 kV and achieved a rate of current rise of $4 \cdot 10^{12}$ A/s . From a technical point of view, however, a disadvantage of this system is the lack of sealed-off-technologies suitable for the production of these switches. But this would be necessary to have a system with high life expectancy and without maintenance work.

Coaxial MUPS

The principle of a multi-channel switch is also applicable to a coaxial system. Fig. 9 shows an open-shutter photography of the plasma of a switch with three channels, which are arranged at the corners of a equilateral triangle. The borehole diameters are 3 mm, the borehole distance is 20 mm, and the electrode seperation is 3 mm. If the parameters of the trigger circuit are well choosen, it is possible to initiate the discharge channels simultanously. Hence the electrode erosion is reduced by a factor of three for a given switching power. This favours the application of this system for high power and high current applications, especially as a sealed-off switch is feasible as has been shown for single channel switches.

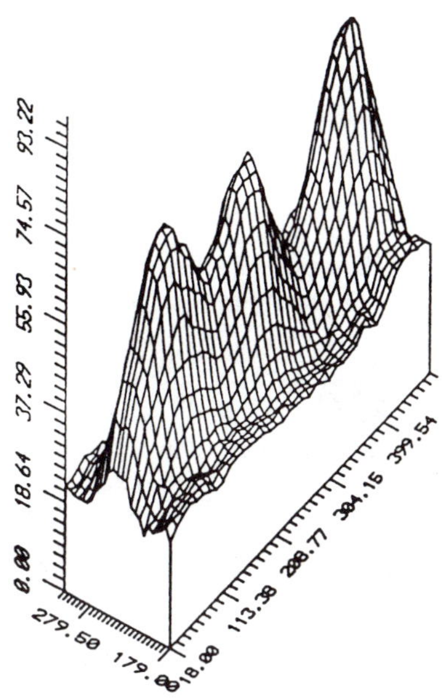

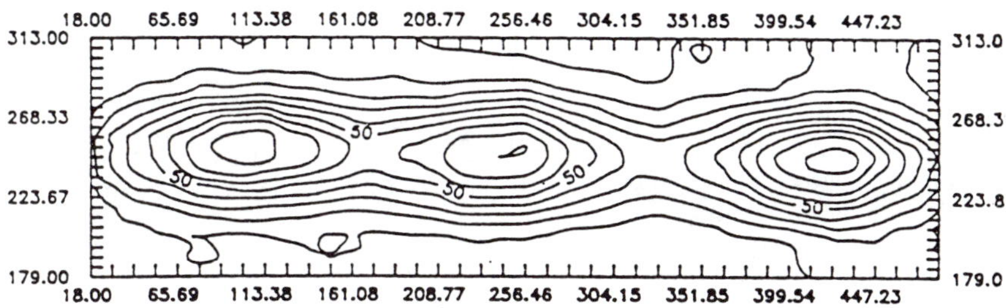

Figure 9. 3-D-diagram, and isolines of an open-shutter photography of a coaxial MUPS with three discharge channels. (rectangular current pulse with a length of 50 ns)

Opening Switch

The principles and the mechanims of an opening switch are - as far as they are known resp. understood - already described in chapter II.d . In order to exploit such a switch to close and open circuits with a cycle time of ms, some further remarks will be done.

Using a switch of the type seen in fig.5, and operating the switch at a pressure which corresponds to about 50% of the value given by the Paschenlaw, leads to a complete switch recovery during the second current zero of a ringing circuit. A further pressure reduction is not possbile, as the switch will not be triggerable anymore. The upper frequency for switch recovery of about 10^4 Hz is dependent on the time derivative of the current at current zero, and is therefore dependent on the external circuit. Hence it is possible to use this property of the pseudospark switch in a 50 Hz network as well as for capacitor discharges for flashlamp pumped solid state lasers.

It is proofen to discharge a capacitorbank step by step by several successive discharges, if the above mentioned principle is applied. If the switch is triggered soon after its recovery, a quasi continuous ringing current with short interrupts is achieved. This offers a closing and opening switch system for high average power applications.

The earlier mentioned method of a multi-channel switch is also applicable for an opening switch, giving the advantage of an enhanced life time. The multichannel principle is of general interest for the high power opening switch as well as for the conventional fast high power pseudospark switch.

Comparison of the Pseudospark Switch with Thyratrons and Spark Gaps

The advantage of the pseudospark switch compared to the thyratron is mainly a high rate of current rise, a 100% current reversal capability, and eventually a higher repetition frequency at a high average power level. It may further have a higher lifetime expectancy than a thyratron, a fact which has to be proofen by a careful investigation and reduction of electrode erosion.

In comparison to a conventional spark gap, the pseudospark offers a higher repetition rate capability and a distinct greater life time. Although the rate of current rise of a spark gap is not yet completely achieved, mutichannel devices could be suitable to overcome this disadvantage in the near future.

References

[1] J. Christiansen, C. Schultheiß, 1979,
     Z. Phys. A290; 35-41.
[2] D. Bloess et al., 1983,
     Nucl. Insrum. 205, 173-184.
[3] J. Christiansen, K. Frank, H. Riege, R. Seeböck,
     1983,
     Proc. XVIth ICPIG, Vol.2, 160-161.
[4] G. Mechtersheimer, R. Kohler, T. Lasser, R. Mayer,
     1986,
     J. Phys. E: Sci. Instrum. 19, 466-470.
[5] P. Billaut et al., 1987,
     Yellow Report CERN, 87-13.
[6] K. Frank et al., 1987,
     Proc. of SPIE, Vol.735, 74-81.
[7] G. Kirkman, W. Hartmann, and M. A. Gundersen,
     1988,
     Appl. Phys. Lett. 52(8), 613-615.
[8] K. Frank et al., 1988,
     Proc. of SPIE, Vol.871, 173-180.
[9] K. Frank et al., 1988,
     IEEE Trans. Plasma Sci. Vol.16, No.2, 317-323.
[10] W.Benker et al., 1988,
     Proc. SPIE, VOL.873, 249-254.
[11] W. Benker et al.,1989,
     Technica 2, 35-38.
[12] H.P. Schölch et al., 1989,
     Appl. Phys. A 48, 397-400.
[13] Roger Reichle, 1989,
     Ph. D. Thesis, Univ. Erlangen, FRG.

[14]   K. Mittag, W. Niessen, 1987,
       Int. AMSE Conf. on Modeling and Simulation,
       Karlsruhe.
[15]   W. Hartmann, M. A. Gundersen, 1988,
       Phys. Rev. Lett. $\underline{60}$ $\underline{(23)}$, 2371-2374.
[16]   W. Hartmann, V.Dominic, G. F. Kirkman, and
       M. A. Gundersen, 1988,
       Appl. Phys. Lett. $\underline{53}$ $\underline{(18)}$, 1699-1701.
[17]   Ekkehard Boggasch, 1987,
       Ph. D. Thesis, Univ. Erlangen, FRG.
[18]   W. Hartmann et al., 1989,
       Proc. IEEE Pulse Power Conf. Monterey,
       to be published.
[19]   G. Mechtersheimer and R. Kohler, 1985,
       J. Phys. E: Sci. Instrum. $\underline{20}$, 270-273.
[20]   V. Brückner, R. Baumgartl, R. Müller, 1988,
       Opt. Soc. Am., Technical Digest Series
       $\underline{Vol.7, TU\ U6}$

REVIEW OF SUPERDENSE GLOW DISCHARGE

Klaus Frank

Physikalisches Institut
Universität Erlangen-Nürnberg
D - 8520 Erlangen, FRG

ABSTRACT

The superdense glow, a diffuse, high-current density low-pressure gas discharge, has found application as the conduction mechanism in high-currrent closing switches and rectifiers, especially in the Russian research into pulsed power generation.
A characterization of the superdense glow will be given in comparison to the well-known types of glow discharges, including the hollow-cathode glow discharge.
The main fields of application of the superdense glow are the repetitive switching of high-voltage, high-current pulses, and as an intense source of charged particles. Some of the most important developments are presented. A critical assessment of the superdense glow is made as far as possible.

INTRODUCTION

Recently by W. Hartmann and M.A. Gundersen /1/ a possible explanation for the anomalous cathode emission in a superdense glow discharge was published. This special type of a glow discharge was first described by Russian authors in 1966, called "Superdense Hollow-Cathode Glow Discharge". The development of a new class of high-power switches, like the pseudospark switches or the Back-Lighted Thyratrons /1/, draw again attention to this discharge. Therefore it seems to be useful to look for the basic mechanism of the superdense hollow-cathode glow discharge as far as the literature was available.

A.   DIFFERENT TYPES OF GLOW DISCHARGES

1.   The Normal Glow Discharge

What is a glow discharge? Following A. von Engel /2/ this discharge derives its name from a luminous region near the cathode, the visual extent of which depends on the gas pressure and the current density. Fig. 1 shows the well known picture of the dark and luminous spaces (in a system with plane parallel electrodes). Apart from structured details, of main interest are the cathode and FARADAY dark spaces, separated by the negative glow. If a gas discharge is started by increasing the external voltage, different modes of glow discharge can be produced (see Fig. 2).

Physics and Applications of Pseudosparks
Edited by M. A. Gundersen and G. Schaefer
Plenum Press, New York, 1990

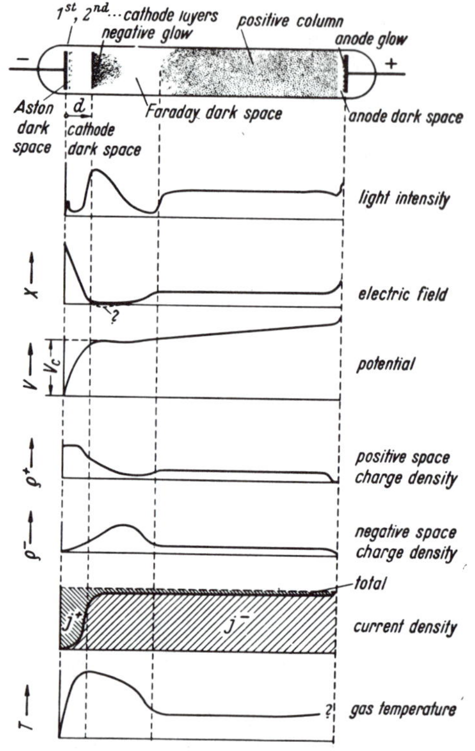

Fig. 1. $V_x, E_x, \rho^+, \rho e, j^+, je = f(x)$ in a cold cathode at low pressure and with normal flow discharge

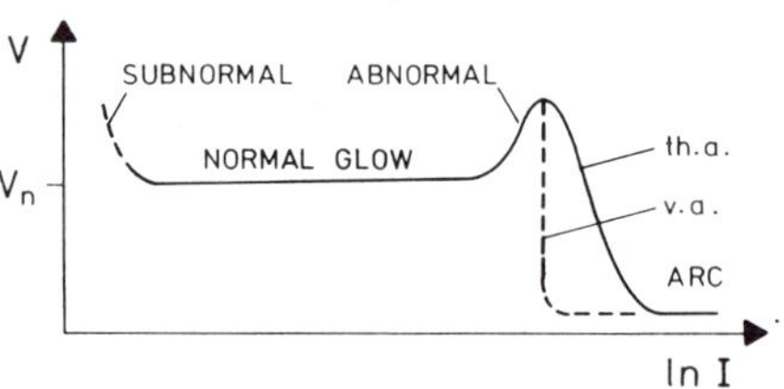

Fig. 2 . Three different modes of glow discharge dependent on the current i; th.a. = thermal arc; v.a. = vapour arc;

The normal glow discharge is characterized by a constant burning voltage, the 'normal cathode fall V_n'. By varying the dimensions of discharge tubes furthermore it was found that j/p^2 is a similarity parameter of the cathode. When the current is increased beyond the value at which the glow covers the whole cathode surface, then a larger current requires a higher cathode fall potential V_c ($V_c > V_n$). At the same time the 'reduced' dark space value ($p \cdot d_c$) begins to decrease, i.e. the dark space shrinks axially, whereby the region where most of the positive ions are produced moves from the dark zone near the boundary into the negative glow. The glow-to-arc transition occurs around 0.1 A, but these values may vary considerably. What happens physically during transition to an arc is that ion bombardment heats up a spot on the cathode enough that it emits electrons thermionically. The thermionic emission facilitates a great increase in discharge current with a low voltage across the tube. An arc may also be initiated by field emission from a rough or sharp cathode surface. Typical voltage drops and currents for arcs range from 10 - 100 V and 1 - 100 A, respectively, and depend on such facts as gas type, gas pressure, electrode material, surface conditions, etc.

2.   The Cold Hollow-Cathode Discharge

An important other form of a glow discharge is the hollow cathode discharge. Fig. 3 shows the typical configuratoin of a hollow-cathode discharge. This discharge is mainly determined by the so-called 'hollow-cathode effect' /3,4,5,6/. The investigations of this effect have delivered three important correlations between the external parameters current density j_c, cathode fall potential U_c, pressure p and cathode distance D (see Fig. 3) /2/:

-   U_c increases only weakly with j_c.
-   With U_c = const., j_c strongly increases with falling values of $(D \cdot p)$ /7/.
-   For given j_c and p there exists an optimum D with a corresponding
    minimum of U_c.

The appearance of much higher current densities compared to linear discharges makes the hollow-cathode discharge a fascinating device for different applications like lasers, particle beam sources, light sources, and high power switches etc. Therefore it seems to be worthwhile to have a better insight into this discharge, the boundary conditions, the 'hollow-cathode effect' mechanisms, and the influence of crossed electric and magnetic fields.

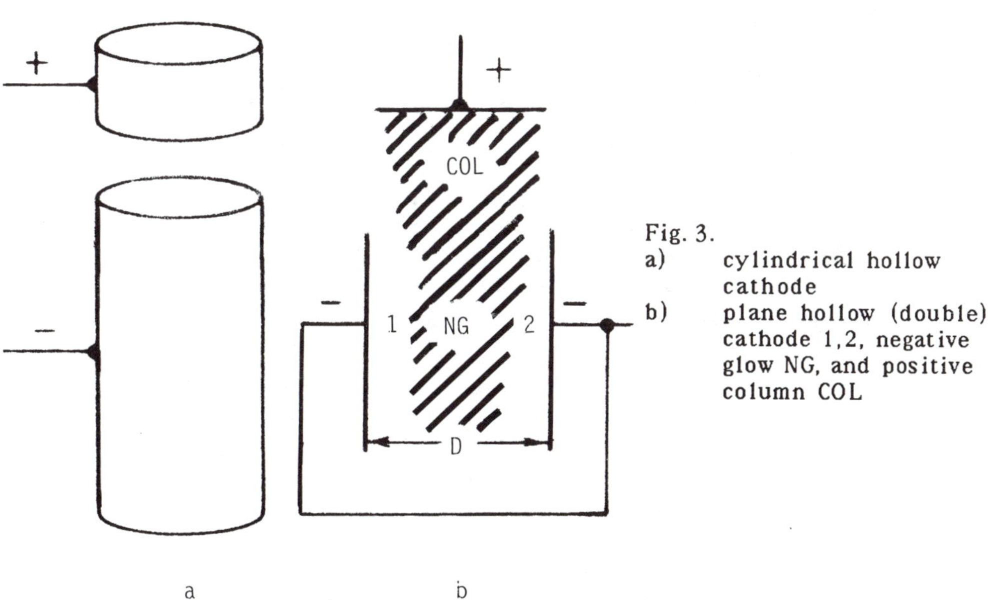

Fig. 3.
a) cylindrical hollow cathode
b) plane hollow (double) cathode 1,2, negative glow NG, and positive column COL

The 'hollow-cathode effect'.   The 'hollow-cathode effect' for a given type of gas is observed only within some range of the product $(D \cdot p)$, where D is the width of the cathode cavity and p is the pressure. The upper bound of $(D \cdot p)$ for which the 'hollow-cathode effect' can be observed occurs when D is approximately ten times greater than the free path length of the primary electrons. For a long cavity the lower pressure cutoff for the 'hollow-cathode effect' occurs when λ exceeds D.

If the output aperture of the cavity is comparable in area to the working surface of the cathode, the discharge characteristics are very sensitive to the boundary conditions, which determine, how many fast electrons are lost through the output aperture. As the length of the cylindrical cavity decreases, the lower bound of the pressure increases, resulting in a smaller range of $(D \cdot p)$ in which the 'hollow-cathode effect' is observed. This indicates that electron losses affect the discharge characteristics even when $\lambda < D$, and that the effects become more important as the pressure is reduced.

Experiments studying the boundary conditions and their effects on discharge characteristics toward the low-pressure end of the parameter region for which the 'hollow-cathode effect' is observed are described in /8,9,10,11/. The most important results are:

- The discharge characteristics depend strongly on the ratio
$$\alpha = 2\,D/A = \frac{\text{area of output aperture}}{\text{working surface of the cathode}}$$
  Normally the maximum fast electron energies usually exceed the energy at which the ionization cross section has its maximum, however, the probability of ionization is approximately ten times greater than the probability of inelastic processes. Therefore a hollow-cathode discharge is more efficient than a linear glow discharge.
  In the center of the cavity the fastest electrons oscillate perpendicular to the cathode surface. Due to repeated collisions with gas molecules the tangential velocity component of a fast electron increases steadily, and escape from the cavity becomes more probable. The probability of escape is maximal for electrons which form near the boundary of the cavity. A consequence of this fact is that the plasma and current density at the cathode surface are less than in the center of the cavity. This results in twisting of the plasma surface and formation of a peculiar electrostatic 'plug' on the boundary of the cavity.

- Fast electron losses through the output aperture of a cathode cavity increase the lower pressure bound of the region in which 'hollow-cathode effects' are observed. These losses result in formation of an electrostatic trap which blocks a further escape of the fastest electrons. A superimposed transverse magnetic field has a similar effect ($p \neq 1$ Pa).

- There exists an 'optimum pressure range' which corresponds to the appearance of the 'hollow-cathode effect'. The upper boundary of the 'optimum pressure range' corresponds to the maximum plasma density at the cavity axis. The lower boundary corresponds to minimum discharge voltage. It depends on the nature of the gas and the diameter and length of the cathode.

<u>Inversion of the cathode cavity of a glow discharge in a magnetic field</u>
<u>/12/.</u>   When the 'hollow-cathode effect' operates the current increase is accompanied by an increase in intensity of the glow in the cavity. Hence the question of the discharge mechanism has evoked discussion of the relative role of UV quanta, which may cause additional electron emission from the cathode. An investigation of a discharge with a cylindrical hollow cathode in a transverse magnetic field showed that increasing the field strength increases the radius of the central zone of the cavity free from fast electrons. If the cavity diameter greatly exceeds the width of the drop region the trajectories of the fast electrons can be represented by arcs of circles whose radii are determined by its energy and the magnetic field strength. It is obvious that the motion of fast electrons will be similar if the working surface of the cathode is not the inner surface of a cylindrical cavity, but the outer surface of a rod of the same diameter (corresponds to an inversion of the cathode cavity) (see Fig. 4). That means: The fraction of UV quanta and ions formed in the plasma which reaches the cathode is 100 % with a hollow cathode and 50 % replacing the cavity by a rod. The main results are:

- The rapid increase in current with increase in cathode drop is due to the current of ions produced by fast secondary electrons entering the plasma from the drop region.

- The efficiency of fast secondary electron production depends on the depth of penetration of oscillating primary electrons into the drop region and

increases with pressure reduction, increase in cathode drop, and increase
in ratio of the length of the primary electron trajectory in the drop
region to its total length.

Emission of electrons from the cathode due to the photo effect can play
only a minor role in the development of the 'hollow-cathode effect'.

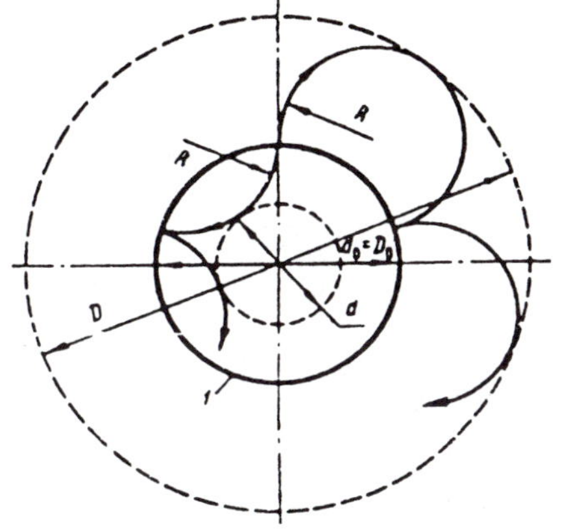

Fig. 4. Paths traced out by primary
electrons in the magnetic field
1)   cathode
D_0)   diameter of the rod cathode
d_0)   diameter of the cavity
R)   Larmor radius
D)   diameter of the envelope of
        the trajectories in a discharge
        with a rod
d)   diameter of the envelope of
        the trajectories in a discharge
        with a cavity

<u>Hollow-cathode glow discharge with vacuum operation of the cathode
hollow /13,14/.</u>   As already shown the voltage and current of a hollow
cathode discharge with p_0 about 1 Pa are very sensitive to the factor ($\alpha = 2 D/A$).
The existence of a relatively high lower pressure bound p is a disadvantage, since
it limits the electrical strength of the discharge devices and the range of
practical applications. Although the characteristics are insensitive to pressure
within the working range, for $p < p_0$, the discharge voltage U is observced to rise
rapidly as p decreases and the discharge either cuts off or goes into a low-current
high-voltage regime. The threshold pressure p_0 is normally greater or equal 1
Pa. The experimental set-up is shown in Fig. 5. The main results are:

-    The escape of fast electrons through the output aperture of the cathode
     cavity is the primary reason for the lower working pressure limit p_0 in a
     hollow-cathode glow discharge.

-    The pressure p_0 is proportional to α. For $\alpha \lesssim 10^{-2}$ the discharge is stable for
     $p \gtrsim p_0 = 10^{-2}$ Pa.

-    For $p \lesssim 10^{-2}$ Pa ('vacuum regime') cathode sputtering changes
     substantially. The composition of the gaseous mixture and the distribution
     of the plasma parameters in the cathode hollow.

-    For $\alpha < (m/M)^{1/2}$, where m and M are the electron and ion masses, a double
     layer forms near the output aperture, across which the potential drop
     exceeds the ionization potential of the working gas.

-    The size of the double layer is not known, only the potential drop (see Fig.
     6).

-    The double layer has a significant effect on the plasma outside the cathode
     hollow.

For more information about the formation of a double layer see /15,16,17/. For
more information about hollow-cathode discharges look into
/18,19,20,21,22,23,24,25/.

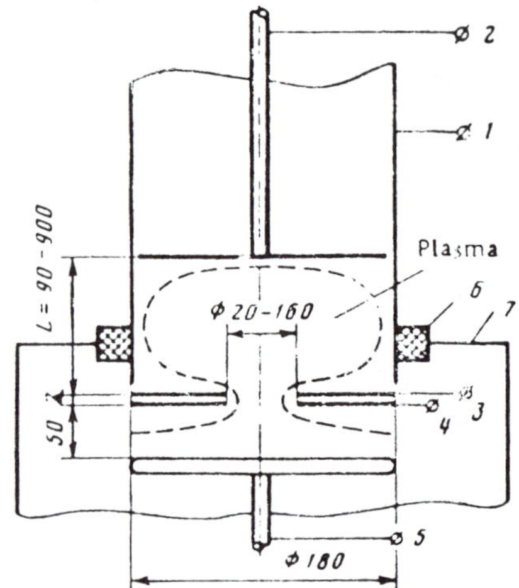

Fig. 5. Structural diagram of the
discharge apparatus:
1)   cylindrical cathode hollow
2)   movable end flange
3,4)   inner, outer cathode diagrams
5)   anode
6)   insulator
7)   vacuum chamber

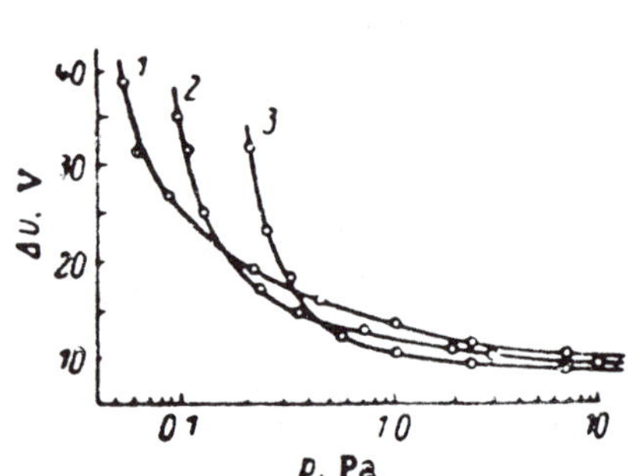

Fig. 6. Potential drop   U across the
double layer as a function of
pressure p for I = const.
1)   I = 0.2 A
2)   I = 0.5 A
3)   I = 2 A

3.   Superdense Cold Hollow-Cathode Glow Discharge

Dense glow discharges.   Sometimes in literature for the same type of
discharge the expressions 'overdense' or 'ultra-dense' or 'ultra-high' are used.
The paper, titled 'Superdense Hollow-Cathode Glow Discharge' was published in
1966 by L.Yu. Abramovich et al. /26/. Based on preliminary experiments which
gave evidence for violation of PASCHEN's law /27,28/ the authors B.N. Klyarfeld et
al. /29,30/ presented a new classification of glow discharges, in general D.C.
discharges (with plan-parallel electrodes)
-        the elementary glow discharge
-        the normal glow discharge
-        the dense glow discharge

Under pulsed conditions peak currents of more than 1000 A can be observed.
Then the cathode is strongly heated and has a strong electric field at its surface,
and the dense glow discharge becomes an arc. This transition is preceeded by
another type of glow discharge, the so-called 'superdense glow discharge' with a
low burning voltage, distinguished by this particular high current. From /27/ is
known that often the transition of a high vacuum discharge into the PASCHEN
region with the corresponding relationship between U_b and (p · d) is prevented
by an intermediate discharge region. This can be clearly seen in Fig. 7 where in
this region when U_b still depends on p, the smaller gap still resists lower
breakdown voltages. An explanation of this fact is not given. (The curves of Fig. 7

were obtained after an additional treatment of the electrodes with a glow
discharge in helium.) The second and most important example for violation of
PASCHEN's law is described in /31/ under the title 'Ignition of the high voltage
discharge in hydrogen at low pressures .
In uniform electric fields the striking voltage for a discharge satisfies the
similarity rule. It is a function of the product (p · d). A number of investigations
/28,32/ have shown noticeable deviations from the similarity rule when the
electric field at the cathode begins to exceed 10^6 V/cm, and spontaneous emission
of electrons begins. This experiment served to investigate hydrogen discharge
striking voltages to study these deviations from similarity in the left-hand
branch of the PASCHEN curve which were not of the nature of spontaneous
electron emission. In the second place these particular forms of discharge which
had a high voltage drop at the electrode'were studied.

Applicability of PASCHEN's law to the ignition of hydrogen discharges.

The tube intended for the study of discharges under the conditions of the left-
hand branch of the PASCHEN curve was constructed to prevent any discharge by
way of the 'long path'. A construction was used in which the electrodes could be
kept fixed, or one electrode cold be moved so as to vary the separation d between 4
and 32 mm (electrode diameter D = 80 mm, U ≤ 40 kV). Repeated measurements of
striking potentials $U_b = f (p · d)$ for different values of d are given in Fig. 8. These
curves are quite widely separated. Such effects can be repeated reproducibly in
tubes of different constructions, and also in tubes where both the distance d and
the hydrogen pressure p can be varied. Repeated experiments have also shown
the absence of local effects at the electrodes, and that the results are the same
when the electrode polarities are interchanged (this in contradiction to other
publications).
All this proves that the deviations from the similarity law are not experimental
errors or pecularities of the apparatus, but are inherent in the nature of the
hydrogen discharge. Plotting log p vs log d (for constant U_b) one obtains a family
of straight lines intersecting the coordinate axis at a very nearly constant angle,
differing from 45°. This implies that the striking potential is a function of the
derived quantity $p · d^k$. Then U_b follows the relation

$$U_b = A/(p · d^k)B$$

or for U_b in kilovolts:

$$U_b = 46 · 10^{-3}/(p · d^{0.58})6$$

This formula has been verified up to 30 kV, but only up to current densities of 0.1
A/cm^2. It shows that U_b is more sensitive to variations in pressure than to

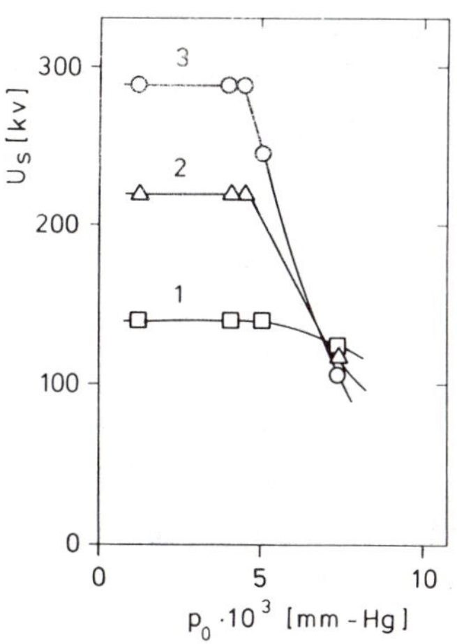

Fig. 7. U_b (p) for various gap
distances d with a positive
anode
1) d = 14 cm
2) d = 8 cm
3) d = 3 cm

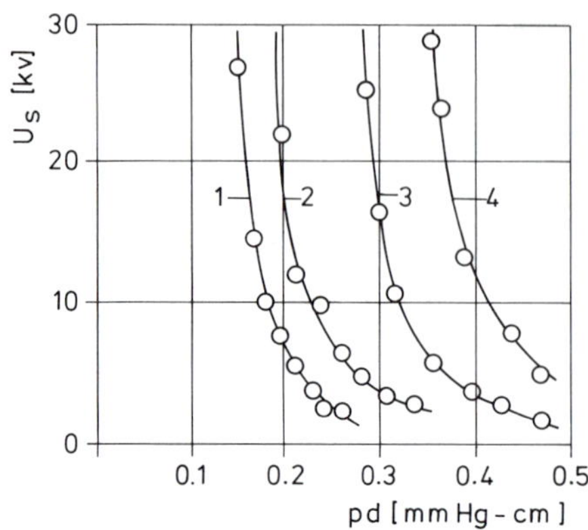

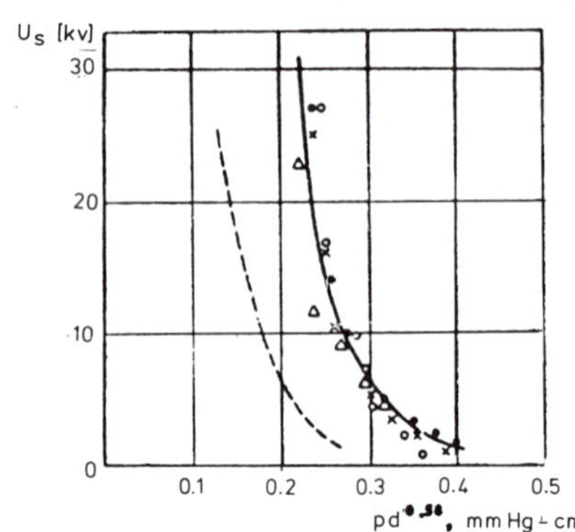

Fig. 8. Deviations from the PASCHEN law in hydrogen. Curves of $U_b = f(p \cdot d)$ at T = 20 °C
1)   d = 4
2)   d = 8
3)   d = 16
4)   d = 28 mm

Fig. 9. Striking voltage of hydrogen discharges as function of $p \cdot d**0.58$
• d = 2.8 mm, x d = 16 mm
▽ d = 8 mm,   o d = 4 mm
Dashed curve is from data by Quinn /33/.

variations in the distance between electrodes. The data obtained at different values of d agree well with the relation $U_b = f (p \cdot d**0.58)$ as shown in Fig. 9.

Incidental remark: Properly speaking the term 'superdense glow' up to now was not used in literature. But there are hints for the existence of an intermediate discharge mode preceeding the transition of a dense glow discharge into an arc. Referring to the authors of /26/ the 'overdense' or 'superdense' glow discharge runs only with a hollow cathode under pulsed conditions. This fact is in a certain contradiction to an older paper /31/ (see page 817 there). Therefore it is justified, with respect to the new devices to apply the expression 'superdense glow' without specifying geometry, gas type and circuit. The 'superdense glow' classifies pulsed low-pressure discharges with current densities of kA/cm^2 or tens of it.

The reasons why hydrogen discharges deviate from the similarity rule can be stated only in relatively general terms. Field emission cannot play any part here, since the deviations are observed even when the field strength at the cathode is below 10^4 V/cm. Some reasons for the duration might be:

- ionization processes in the gas space, which may arise in various ways in molecular gases
- cathode surface effects

High voltage form of the discharge at very low pressures.   (D.C. voltage)
On the left-hand branch of the PASCHEN curve the normal glow discharge cannot in general occur, since the thickness of the normal region of the cathode potential drop is greater than the distance between the electrodes. It has the characteristics of a silent discharge (there is only a weak space charge between the electrodes). It also has the characteristics of a hindered discharge (there is the cathode drop potential thickness comparable or greater the electrode gap). Fig. 10 shows U-I characteristics of the high-voltage form of hydrogen discharges.

A D.C. high-voltage discharge does not turn into an arc until the discharge current reaches 1 A. In pulsed discharges, transition to the arc occurs for currents of 1000 A only at pressures above 0.2 mm Hg. At lower hydrogen pressures, transition to the arc does not occur until the total discharge current is above 1000 A ($j = 20$ A/cm^2). This result indicates that the high voltage form of the discharge can be supported for short periods of time without going over into an arc, at extremely high current densities, probably considerably in excess of 100 A/cm^2. For a comparison of the nature of discharge effects on the left-hand and right-hand branches of the PASCHEN curve, Fig. 11 shows schematically the U-I characteristics of a high-voltage discharge (curve 1) and a discharge on the right-hand branch (curve 2).

Summary: This paper the first time mentioned the possibility of high-current densities of 100 A/cm^2 or more with hydrogen discharges under pulsed conditions. If such gas discharges with a hollow cathode transform under pulsed conditions into a high-current hollow-cathode low-pressure discharge, then this is called a superdense hollow-cathode glow discharge.

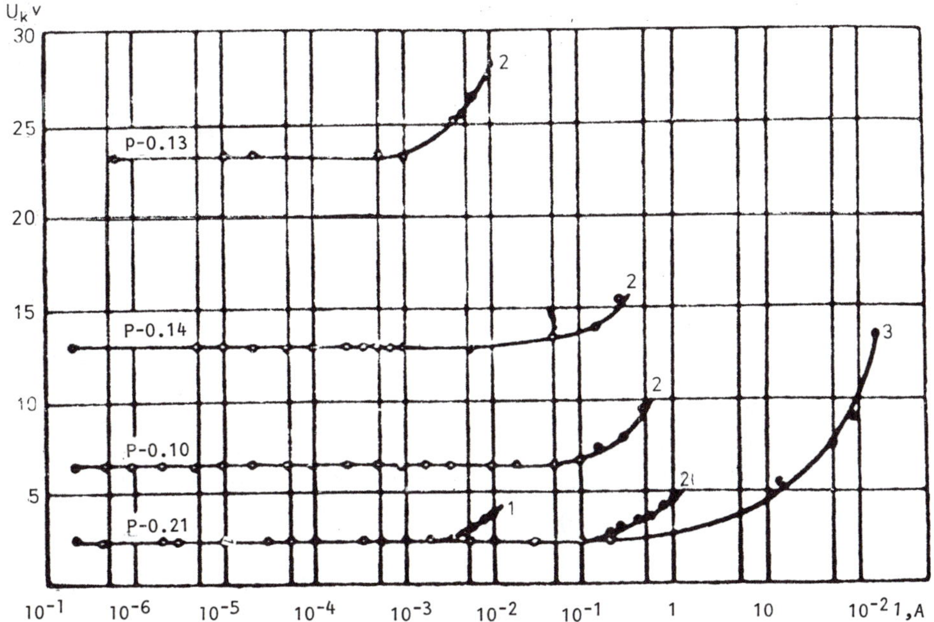

Fig. 10. U-I characteristics for
high voltage form of a
hydrogen discharge at
constant field strength
d - 16 mm
cathode surface - 50 cm^2
pressure p in mm Hg
1)   thin nickel foil
    electrodes
2)   massive copper
    electrodes
3)   massive copper
    electrodes with pulsed
    operatio

        <u>Superdense hollow-cathode glow discharge.</u>   This discharge is
characterized by a high current density on the cathode and a relatively low
maintaining voltage. As shown above the appearance of a superdense glow
discharge is preceeded by a dense glow discharge. The dense glow discharge
converts to a superdense glow discharge, when the discharge current increases
and exceeds some critical value. There is a rapid fall of voltage on the electrodes
and the current begins to increase, in some cases up to values of several kA.
The shape of the cathode plays a significant role in the transition. It has been
impossible to obtain a superdense glow discharge between plan-parallel
electrodes. The electrode configurations where a superdense glow discharge is
easily produced shows Fig. 12. The electrical circuit produces single unipolar
current pulses of sinusoidal or rectangular form. The pulse length is 80 - 100 μsec
in both cases. Measurements in hydrogen, helium, neon, and argon were
performed and show that the pressures are higher, the smaller the molecular
weight of the gas is. Fig. 13 shows one of the oscillograms obtained with a
rectangular current pulse in the superdense glow discharge stage. The U-I-
characteristics for the case of a cup-shaped cathode show (see Fig. 14) that the
maintaining voltage decreases with increasing current. Below 0.05 mm Hg the

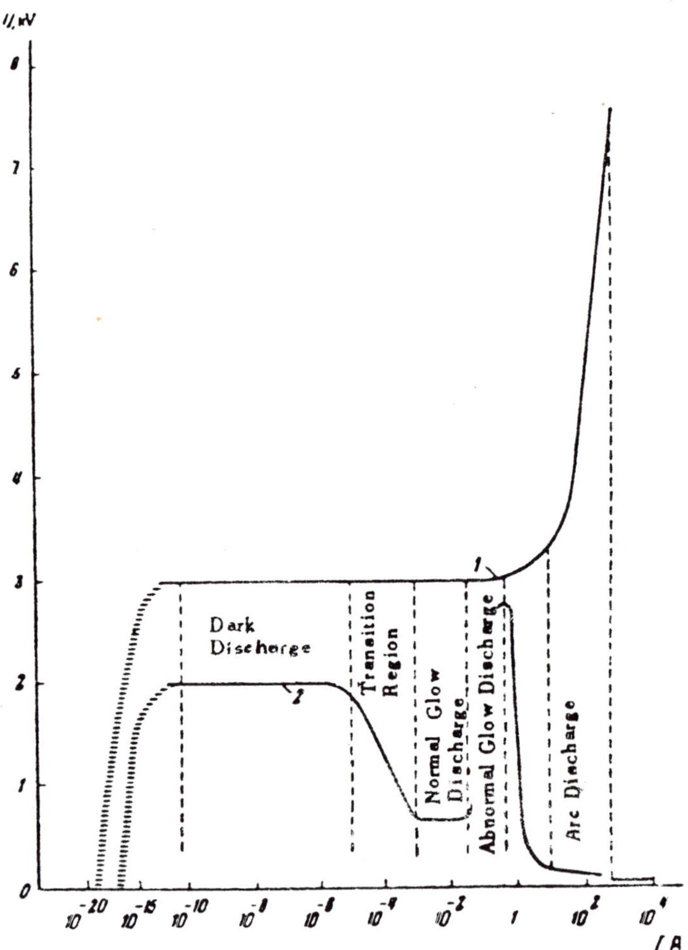

Fig. 11. Schematic form of the
U-I characteristics of a
discharge
1) for $(p \cdot d) < (p \cdot d)_{min}$
2) for $(p \cdot d) > (p \cdot d)_{min}$
(ten times enlarged)

starting gap voltage, however, rises steeply up to 30 kV. The gap voltage also strongly depends on the cathode configuration, i.e. for a cylindrical cathode the maintaining voltage is several times higher.
The maximum peak current for the cup-shaped cathode was 2500 A, corresponding a current density at the bottom of the cathode of 50 A/cm^2.

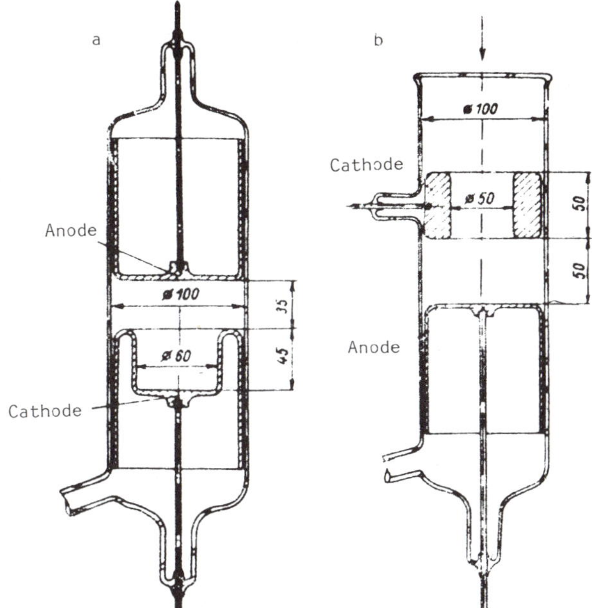

Fig. 12. Schematics of discharge tubes
a) tube with cup-shaped cathode
b) tube with cylindrical cathode

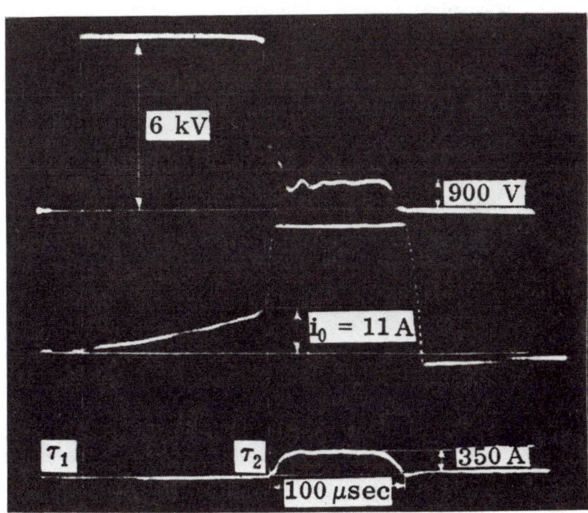

Fig. 13. Oscillograms of voltage and discharge current. The gas is helium, pressure: 0.2 mm Hg. The top oscillogram shows the maintaining voltage, the middle oscillogram shows the discharge current (x 50), the bottom oscillogram shows the discharge current not amplified.

The presence of an annular glow in the superdense glow discharge stage
indicates that beyond the limits of the very thin layer of positive space charge at
the cathode surface the electrons dissipate the energy obtained in the cathode-
fall space on ionization and excitation of the gas. The appearance of a high
concentration of ions and excited atoms with a much greater collision cross
section than normal atoms leads to the transfer of the electron energy to the gas
within a layer of a few millimeters thickness. Assuming electron emission is
mainly caused by ion impact, the current carried by positive ions at the cathode
surface is

$$\frac{j_p}{j_p + j_e} = \frac{1}{1 + \gamma}$$

j_p - ion current density
j_e - electron current density
γ - coefficient for ion-electron
         emission

For current densities of 20 A/cm^2 and the corresponding voltages (see Fig.
14) then the thickness of the cathode fall d_c is about $1 \cdot 10^{-2}$ cm, i.e. this thickness
is many times smaller than the free path of the electrons and even of the ions.
This means that the potential-fall space transmits the counter streams of ions and
electrons practically unchanged, that means

$$\gamma \cdot \delta = 1$$

$$\delta = \frac{\text{number of ions leaving the annular glow}}{\text{number of electrons entering the layer}}$$

For smaller discharge currents this plasma efficiency coefficient δ was
determined in the range 1.3 to 2.8, that means at least one order of magnitude
higher than in a conventional glow discharge /34,35/.

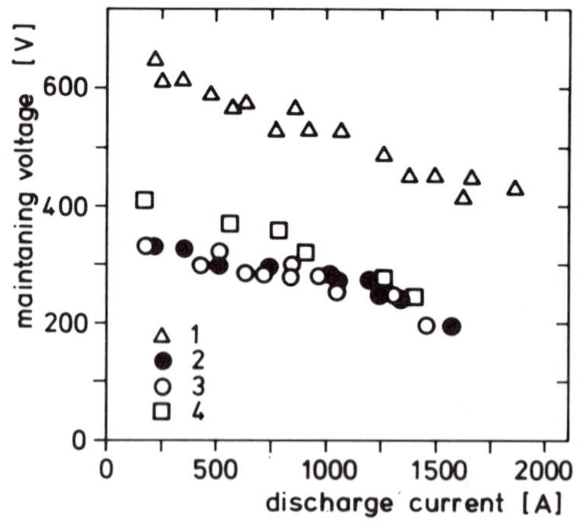

Fig. 14. U-I characteristics of
a superdense glow discharge
at different neon pressures
for a tube with cup-shaped
cathode
1) 0.05
2) 0.2
3) 0.5

Increasing the current density in a cup-shaped cathode beyond 50 A/cm^2 leads to the transition of superdense glow discharge to an arc (see next section).

4.   High-Current Hollow-Cathode Glow Discharge

The study of this type of glow discharge was performed by I.I. Aksenov and co-authors /36/. Apparently these more fundamental experiments were stimulated by a former work concerning the development of a low-pressure hollow-cathode spark gap (see part B.2. of this article). It is a more accurate investigation of the superdense glow discharge. The distinctive feature of a pulsed high-current hollow-cathode glow discharge (see /26,34,35/) is that it can exist without becoming an arc at currents of hundreds or even thousands of amperes and at relatively low voltages (several hundred volts). A better understanding both of high-current hollow-cathode, glow discharge switches and high-power arc switches depends on the mechanism of this type of glow discharge. In the first aspect, data which would provide the optimum conditions for minimizing the voltage U_b are important. As the precursor of an arc, in addition to reducing U_b, it is important to find the conditions for which the critical current for transformation to an arc is minimized.
Since the present current density is several orders of magnitude higher than in other variants of the glow discharge, it is to assume that cathode sputtering will have a determining role (see 'Cathode-Related Processes in High-Current Density, Low-Pressure Glow Discharge' by W. Hartmann). Due to reverse diffusion, metal atoms which have left the cathode are concentrated in a layer whose thickness at gas pressures of tenth of a torr does not exceed a few millimeters, and which lies near the cathode. Thus, the most significant changes in the volt-ampere characteristics may be expected when the width of the cathode cavity is reduced to this size.

<u>Experimental set-up.</u>   There have been studied cathodes in the form of a cup open at the anode end (Fig. 15 a) or separated from the anode by a perforated barrier (Fig. 15 b). An auxiliary electrode located in the cathode cavity is used to produce a preionized gas and to trigger the discharge at the required time by delivering a short control pulse of negative polarity. It should be noted that the discharge characteristics are stabilized only after several tens of hours of operation. The systems have to be baked out and be conditioned similar to 'sealed-off 'devices. The gas pressures were stabilized by heated gas reservoirs. With respect to these requirements most of the experiments were performed with the arrangement shown in Fig. 15 c. It is a hollow cathode system consisting of two parallel disks, 110 m in diameter and 12 mm thick. There are 12 radial slits 3 mm in width in each disk. A plane anode is located parallel to the cathode, 4 mm from the nearest disk. The second cathode disk can be moved relative to the first by means of a bellow and screws to change the width of the cathode cavity from 0.5 mm to 40 mm. The switched main current pulse has the form of half a sinusoid lasting 18 μsec, the repetition rate is 50 sec^{-1}. The working gases are hydrogen, deuterium, and oxygen.

<u>Results.</u>   Using devices shown in Fig. 15 a and Fig. 15 b it is established that U_b has a dependence on the cathode diameter d typical of a hollow cathode discharge (monotonically decreasing with decreasing diameter) over the entire range of currents investigated. As the cathode diameter is increased the

preferred sputtering of metal into angles towards the region between the side
walls and bottom becomes more noticeable. The conditions for development of the
hollow-cathode effect are optimal, and it is to this part of the cathode that the
majority of the discharge current flows. By preventing that high-energy γ-
electrons can move directly to the anode (see Fig. 15 b with a diaphragm) the
breakdown voltage U_b is lowered by a factor of two. A qualitative explanation is
that one has a more effective utilization of these electrons for ionization.
Now back to the arrangement of Fig. 15 c. Fig. 16 shows the U-I characteristics for
discharges in such a system. The value of the differential resistance decreases as
the current increases within the limits of each curve. Fig. 17 shows the
dependence of U_b on gas pressure and the cavity width. It also follows from Fig.
16 and 17 that in the given range of gas pressures the voltage decreases as the
width of the cathode cavity is decreased, but there is a limit beyond which a
further reduction in d leads to a growth of U_b. At larger pressures the minimum
in the $U_b(d)$ curve is lower and sharper and shifts somewhat to the left. This
agrees with the theory of a hollow-cathode discharge that the greatest reduction
of U_b is observed when the distance between the cathode plates is such that the
negative glow regions of the two cathodes merge into one. The width of this
region decreases with increased j and p.
Defining two coefficients q_u and q_i it is easy to describe th enhancement of the
hollow-cathode effect.

$$q_u = (U_{bo} - U_{bi})/U_{bo}$$

$$q_i = (I_i - I_0)/I_0$$

with U_{bo} and I_0 the voltage for a given current and the current for a given
voltage in a system with d = 40 mm, U_{bi} and I_i are the same quantities with d less
than 40 mm (see Fig. 18 and Fig. 19).

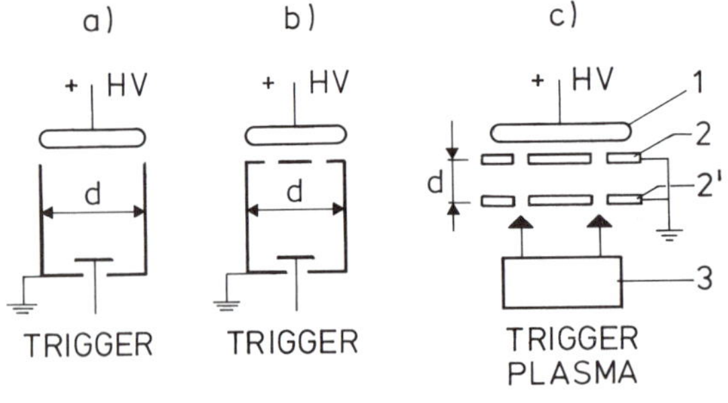

Fig. 15. Schematic drawing of the
experimental devices:
1)   anode
2)   fixed cathode disk
2')   moveable cathode disk
3)   pre-ignition discharge

Besides the voltag another important practical parameter is the critical current
at which the discharge becomes an arc, I_{cr}. Since the probability of creating a
cathode spot increases with the density of a glow discharge, it is possible to expect
a reduction in I_{cr} as d is decreased. The results of some measurements of I_{cr} as a
function of d are given in Fig. 20. The left branch of the curve corresponds to
conditions when the discharge shifts from the inner surface of the cathode to the
outer surfaces. Besides this it is obvious that the transition current changes
comparatively little with pressure.

28

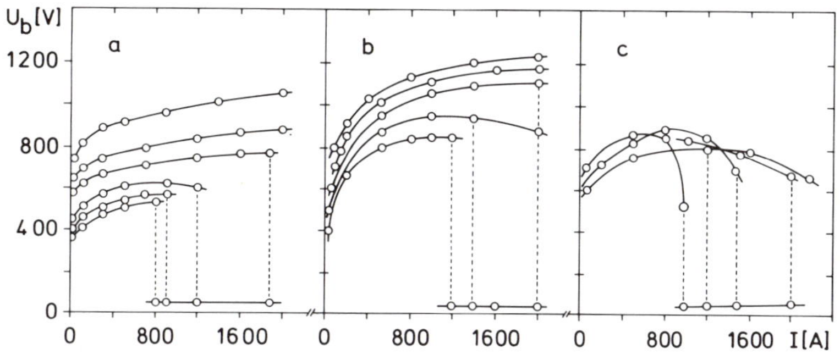

Fig. 16. Volt-ampere characteristics of the discharge. The parameter is the width of the cavity d in mm.
a)   p = 0.6 torr H_2
b)   0.6 torr O_2
c)   0.17 torr O_2
Stainless steel cathodes

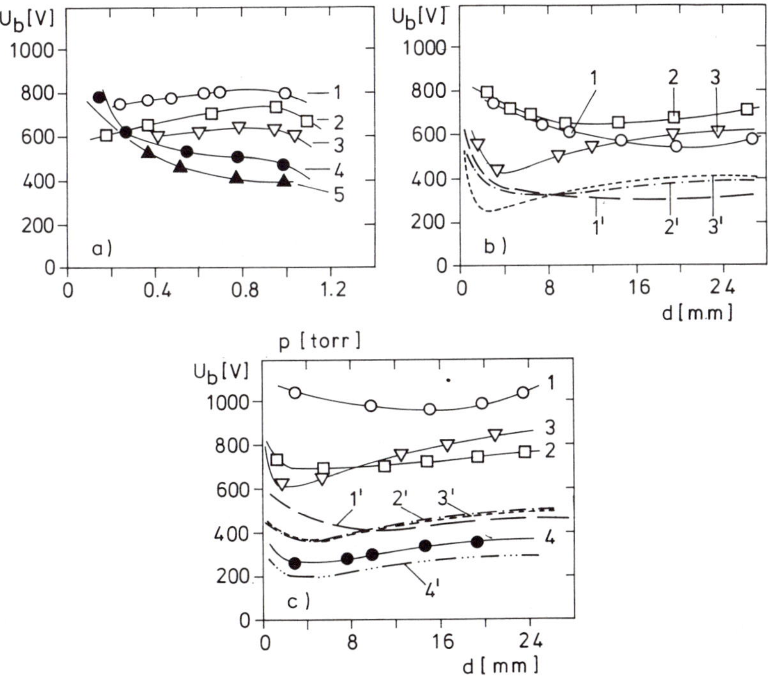

Fig. 17. Effect of the gas pressure and width of the cavity on the voltage of the discharge.
1) d = 40, 2) d = 20, 3) d = 9, 4) d = 5, 5) d = 2 mm, H_2 stainless steel, I_a = 300 A,
b: 1-3) I_a = 750 A, 1'-3') 7A, 1,1') p = 0.2, 2,2') 0.4, 3,3') 0.6 torr, H_2, stainless steel,
c: 1-3) I_a = 750 A, 4) 400 A, 1'-4') 7A, 1,1') p = 0.2, 2,2') 0.4, 3,3') 0.6, 4,4') 0.4 torr, O_2,
aluminium

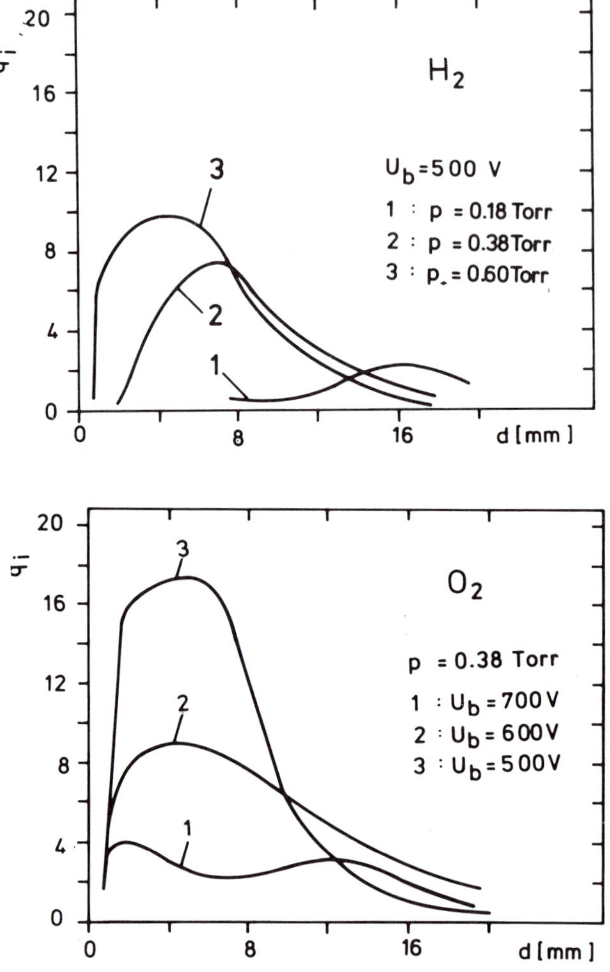

Fig. 18. Effect of a hollow cathode with respect to current

Further results:

The voltage of the dischrge in deuterium is significantly higher compared with hydrogen. From /37/ it is known that for the generalized secondary electron emission coefficients γ the following is valid:

$\gamma_{H_2} \approx \gamma_{D_2}$ and $\gamma_H \approx \gamma_D$

From inceasing the ion mass follows a corresponding increase in velocity which needs a higher U_c (cathode fall potential) which itself is approximatly equal to U_b.

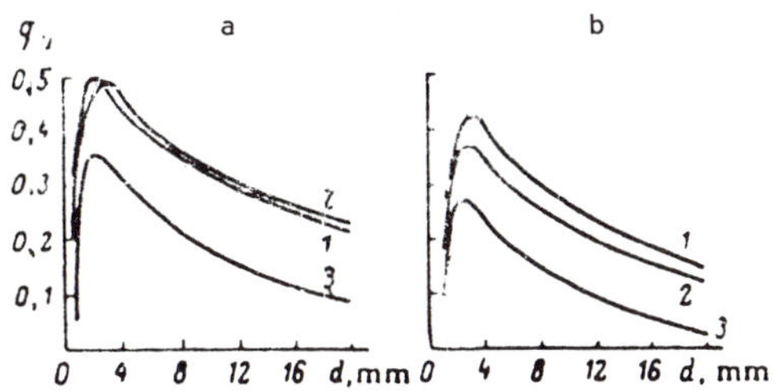

Fig. 19. Efficiency of a hollow cathode
with respect to voltage
a)   I = 50
b)   I = 500 A
1)   stainless steel, H$_2$
2)   stainless steel, O$_2$
3)   aluminium, O$_2$, p = 0.6 torr

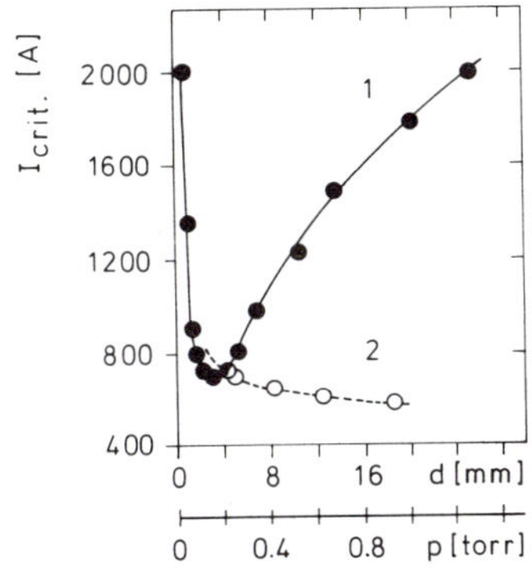

Fig. 20. Dependence of the current at which a glow
discharge becomes an arc on the width of
the cathode cavity and on hydrogen
pressure.

curve 1:   p = 0.35 torr
curve 2:   d = 4 mm

- For similar discharges the density of the metal vapors, ρ_m, in a hollow
  cathode transforms as a^{-2} on going from one cavity width d_1 to another d_2
  (a: = $d_1/d_2 < 1$). At the same time the gas density, ρ_g, tranforms as a^{-1}.
  Therefore the ratio ρ_m/ρ_g varies as a^{-1}. One expects, if d is decreased a
  reduction in the effective ionization potential U_i . This is correlated to a
  transition to a negative slope in the U-I characteristic.

- The appearance of such negative slope sections are due to changes in the
  coefficient γ and an increased role of the metal vapors in the production of
  charge carriers.

- Nevertheless the sharp differences in U_b for H$_2$ and O$_2$ as well as for H$_2$
  and D$_2$ can be taken as evidence that the gas type still have an important
  role in the maintenance of the high-voltage glow discharge.

- Considering the different U-I characteristics one has to pay attention to
  the way they have been produced (influence by the pulse length, the
  device preparation and the choice of reference points).

- Several authors (see /12,38/) investigating the hollow-cathode effect were
  neglecting the influence of photons to increase the generalized secondary
  electron emission coefficient.

But these experiments which cast doubt on the importance of the photo-effect
have been carried out at relatively low values of γ. If in a hollow cathode the
intensity of emission from metal atoms, E_r, is proportional to $j^{3.3}$ (plane cathode:

$E_r \sim j^{1.7}$) then for j values typical for these experiments one might expect a noticeable increase in the role of the photo-effect.

__Summary.__   This paper is very important. It clearly shows the experimental problems investigating high-current, hollow-cathode discharges. Beyond this it delivers in a certain scale possibilities to separate the both discharge modes.

5.   Other Phenomena Related To Dense Discharges

Nonliner Plasma Effects

__Magnetization by self-fields.__   The electron component of the plasma of a hight-current discharge differs from that of a discharge with a low current density in that it is highly magnetized by the self-magnetic field of the current. An increase of the discharge current is accompanied by an increase in the number of fast superthermal electrons, and it results in a displacement of the ohmic regime by DREICER runaway of electrons. In the axial region of a high-current discharge, where the fast electrons are concentrated the DREICER current may turn to be substantially higher than the ohmic current. The resulting magnetic field configuration in the plasma is governed by these currents. Even if most of the current is carried by thermal electrons in the ohmic regime, the pronounced DREICER increment can sharply change the plasma characteristics. In particular, the radial profile of the electron density under such conditions may differ radically from the canonical profile, with a maximum at the center of the plasma column and a monotonic decrease towards its edge /39/.

__Fast-electron current in hollow-cathode glow discharges /11/.__   The negative glow of a hollow-cathode discharge contains a substantial number of fast electrons with energies near U_c. The measurement of the current density and energy spectrum of these electrons as a function of the discharge current and the pressure yield important information on the hollow-cathode effect. The results are that an increse in the discharge current in a hollow-cathode glow discharge is due to a larger oscillating-electron current and to the broadening of the fast-electron energy spectrum caused by collisionless loss in the negative glow. As the pressure in the discharge is reduced, at an essentially constant anode voltage, the oscillating-electron current increases, as does the width of the fast-electron energy spectrum.

__Electrode phenomena.__   The mechanism for the formation of the dense cathode plasma is not well understood. Furthermore it is not clear what mechanism is responsible for the sharp increase in the conductivity of the cathode-plasma gap when there is a high-current electron beam. It is shown experimentally that the density of the cathode plasma increases when a critical charge density $q_+ = (1 - 5) \cdot 10^{-6}$ Cb/cm^2 is reached at the surface of the dielectric film covering the cathode /40/. Antonov et al. /41/ believe that the plasma forms as a result of the desorption and ionization of gas on the surface of these films caused by electrons which have tunneled through the film. In contradiction is the more popular version: Explosive destruction of dielectric films and the appearance of effective sources of electrons and plasma: cathode spots. A more detailed presentation of electrode phenomena accompanying superdense glow is given later on by W. Hartmann ('Cathode-Related Processes in High-Current Density, Low-Pressure Glow Discharges').

6. Hollow-Anode Configurations

These are devices with a plane cathode. Two examples are given.

The hollow-anode thyratron /42/. These new types of thyratrons were developed in order to improve the capability to withstand reverse currents up to 50 % or more.

High-voltage hollow-anode gap discharge /43/. The discharge works at pressures between 10^{-5} and 10^{-3} torr. It is a focusing of the produced electron and ion beams observed. This behaviour is preferred by the presence of a cavity in the anode. Under pulsed conditions starting with current densities of 10 A/cm^2 the electron beam density within a spot diameter of 1 mm on axis is 100 times higher than elsewhere in the gap. This stimulates the development of breakdown at the center of the cavity. Increasing the initial pressure leads to a decrease in the height of the current pulse and to an increase in the delay time before breakdown occurs.

B. APPLICATIONS

1. Production Of Intense Particle Beams

This especially means production of intense electron beams. Besides the Russian literature and the papers dealing with glow discharge electron beams for material processing, welding, melting etc. it seems worthwhile to have a look on two recent experiments using hollow cathodes. The first device was described in a paper by the authors J.J. Rocca et al. in 1982 (see /44/). In /45,46/ there is a more detailed description on glow-discharge created electron beams. Fig. 21 shows the experimental set-up. The hollow-cathode glow-discharge electron gun was designed for the excitation of CW lasers (see again /44/). In this type of electron gun the plasma that develops inside the hollow-cathode acts as a source of electrons for the electron beam. These hollow-cathode electron guns have two modes of operation. One mode is a high impedance one in which the electron beam is produced, and the other one is a low impedance mode where no electron beam is produced. Operation in the beam mode occurs over a limited range of current and pressure. This statement is in good agreement with the ideas of a dense glow discharge (in the pulsed mode: superdense glow discharge). In the high impedance mode the majority of the discharge voltage drop occurs in the external plasma. The pressure p, at which the hollow-cathode electron gun will operate, is determined by the internal cathode diameter D. Consequently the scaling gas discharge law

$$p \cdot D = C$$

can be used to design such electron guns, the constant C depends on the nature of the gas and the cathode surface. It is possible to produce in this way multikilowatt electron beams with generation efficiency values of 50 % to 80 %.
In 1979 Christiansen and Schultheiss /47/ published an intense particle beam source, based upon the pseudospark discharge. The simplest design for the pseudospark electron beam source (see Fig. 22) consists of two plane-parallel electrodes, with a center bore hole. The cathode is designed as a hollow-cathode. The most important data of the beam are:

- Pulse duration up to some 10 ns
- Beam diameter less than 0.5 mm
- Mean electron energy about 2/3 of U_b
- Current density of about 10^5 A/cm^2
- Energy per pulse up to 0.5 J

- Current rise time of 10^{11} A/sec
- Power density up to 10^9 W/cm^2
- Peak currents up to 10^3 A
- Repetition rate up to 10^3 Hz
- Life time greater 10^6 discharges

For more information see /48,49,50,51, 52, 53, 54, 55, 56/.

2.   High Power Closing Switches

High Peak Currents And Long Pulse Operation

High-current switching device with a hollow cold cathode. The rectification effect in a system of electrodes with a hollow cold cathode is described later. In course of prolonged investigations of arc rectifiers it was noted that their electrodes rarely fail and the devices have a long service life which is determined by the low electrode erosion and the absence of damage to the anode insulator. A study of the current distribution over the surface of the electrodes and the dynamics of the discharge motion during various stages of a pulse showed that the current is distributed over a large area of the surface of the electrodes and moves over it during the pulse, which explains the low electrode erosion. On the basis of arc rectifiers (see later) another class of devices was designed to operate in circuits in which onesided conduction is not required. Such a device is called high-current plasma commutator /57/. The main features of design and the dynamics of motion of the current layer are shown in Fig. 23 and 24.

For currents $I_p \gtrsim 250$ kA the discharge forms like in Fig. 24 b with plasma foci at both the cathode and the anode. At the same time intensive erosion and formation of craters began on the electrodes. To prevent this at higher peak currents so-called antipinch loops were installed. Table 1 presents the essential data of plasma commutators.

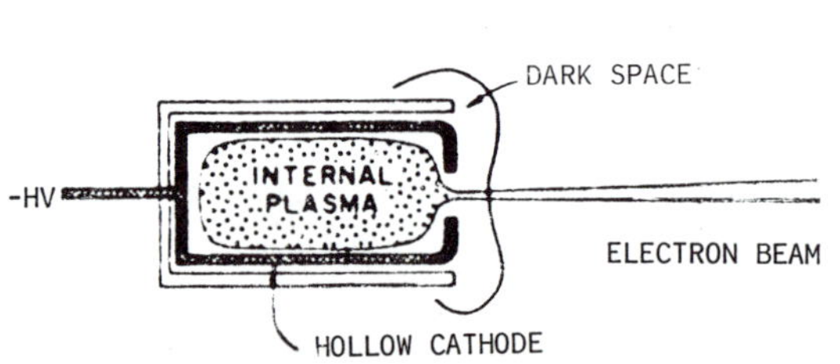

Fig. 21. Schematic representation of a hollow-cathode electron beam source

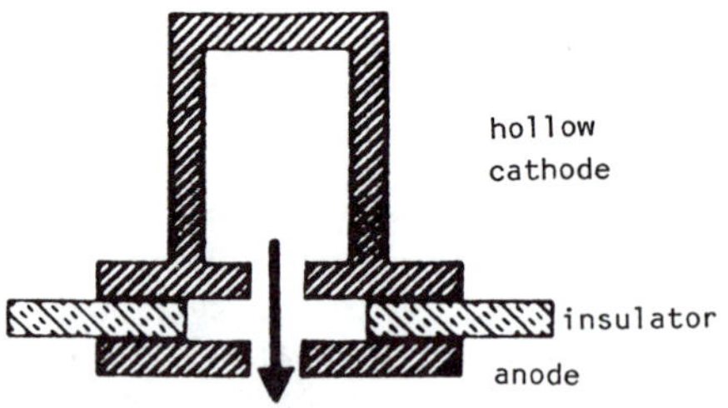

Fig. 22. Scheme of a pseudospark electron beam source

Table 1. Parameters of Plasma Commutators

working gas	air, different other gases
gas pressure	$10^{-2} - 10^{-3}$ torr
hold-off voltage	$\leqslant 25$ kV
peak current (one device)	$\leqslant 300$ kA
cavity diameter	$\leqslant 300$ mm
electrodes	water cooled (at high peak currents)
with antipinch loops	$I_p \geqslant 250$ kA (up to 500 kA possible)
trigger methods	pulsed magnetic field or plasma injection
jitter	10^{-7} sec
voltage drop (for I = 500 kA and U = 20 kV)	300 – 400 V
parallel operation	up to six commutators
max. peak current (with six in parallel)	1.5 MA
repetition rate (at 1.5 MA)	0.1 Hz
lifetime	10^6 discharges

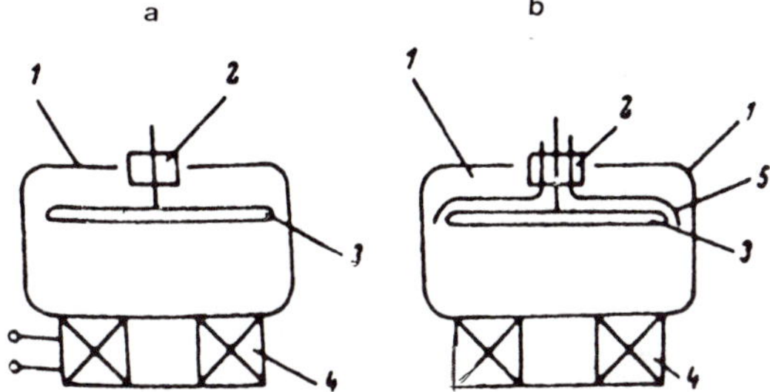

Fig. 23. Schematic diagram of plasma commutators
a) U = 15
b) U = 25 kV
1) hollow cathode
2) insulator
3) anode
4) solenoid
5) metal screen

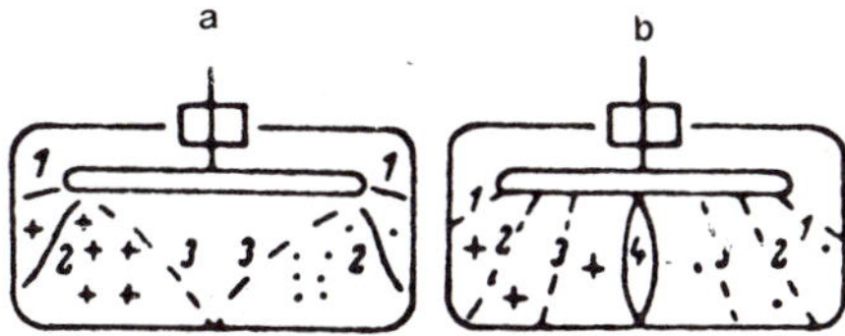

24. Dynamics of motion of the current layer during a quarter period of current a) $I_p \leqslant 200$, b) $I_p > 200$ kA. The numbers indicate the position of the current layer at various times.

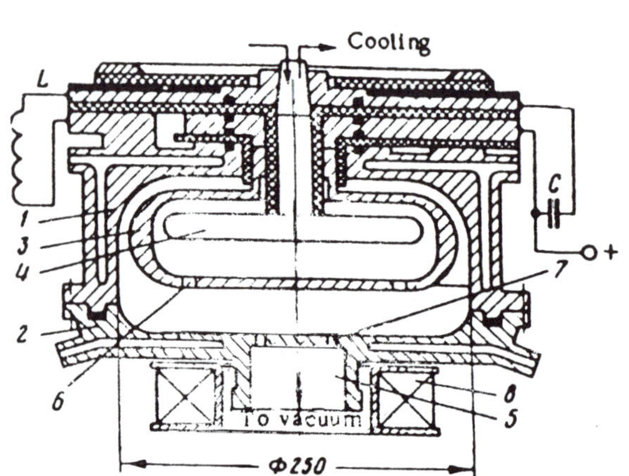

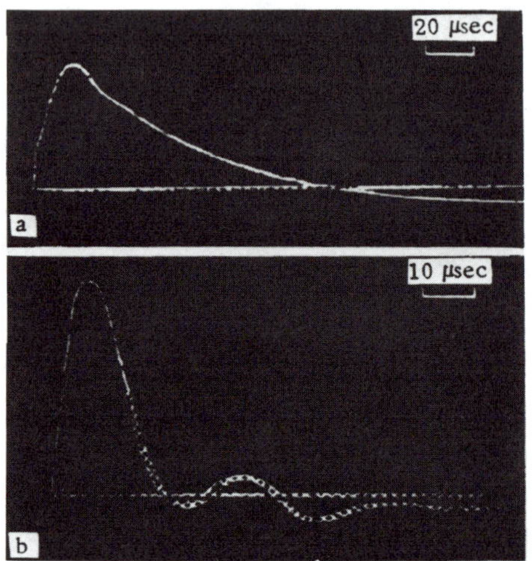

Fig. 25. Construction of commutator unit /58/

Fig. 26. Shape of current pulse with a circuit inductance L = 0.7 µH (b)

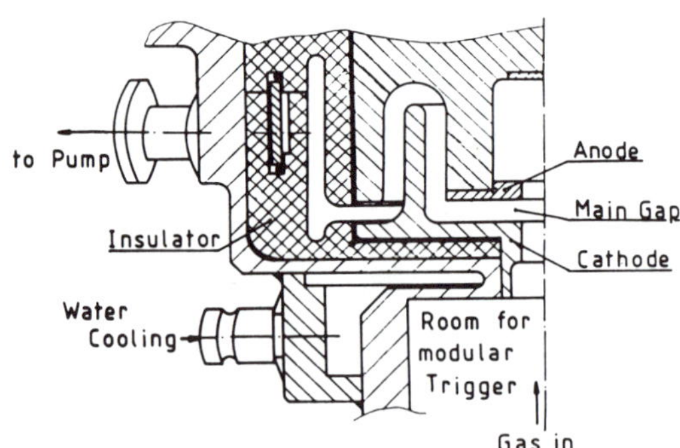

Fig. 27. Mechanical layout of the CERN high-current pseudospark switch

For more information see Figs. 25, 26, /58,59/.

Higher-power pseudospark switch /60,61,62,63/.   Based on the principles of
the pseudospark discharge /47,49/ applications have been developed in the range
of thyratron-like medium-power switches at typically 20 to 40 kV and up to 30 kA
(see B.2.). High-current pseudospark switches have been built for a high-power
20 kJ pulse generator which is being used for longterm tests of plasma lenses
developed for the CERN Antiproton Collector (ACOL). These switches have operated
for several hundred thousands shots, 20 to 50 ns jitter at 16 kV charging voltage
and more than 100 kA peak current amplitude. There have been tested three
trigger possibilities, the surface discharge trigger, the pulsed glow discharge
trigger and the electron beam trigger. In general the surface discharge trigger
was used /64,65/. Fig. 27 shows the mechanical layout of the high-current
pseudospark switch and table 2 lists the main characteristics of the high-current
pseudospark switches. The typical average resistance at 100 kA is less than 1 mΩ
(see Fig. 28). An essential requirement for their application in the plasma lense
pulse generator is the simultaneous switching of four switches. This parallel
operation worked in an excellent manner.

Table 2. High-current pseudospark
swtich features

Maximum hold-off voltage	(kV)	> 25
Maximum peak current	(kA)	200
Nominal hold-off voltage	(kV)	16
Nomonal peak current	(kA)	100
Nominal charge transfer per pulse	(As)	0,400
Reverse current	(%)	95
Maximum dI/dt at 16 kV	(A/s)	1.5×10^{11}
Switching delay	(ns)	100 - 600
Jitter (pulse energy = 1 J)	(ns)	+ 10
Jitter (pulse energy = 3.5 kJ)	(ns)	$< \pm$ 100
Inductance	(nH)	40
Repetition period	(s)	3
Filling gas		Helium
Gas pressure	(Pa)	1 - 10
Electrode material		Densimet 18
Insulator material		Araldite
Lifetime (shots at 3.5 kJ)		> 400000

Medium Power Devices With Variable Pulse Durations

Low-pressure hollow-cathode spark gaps.   This type of a controllable
spark gap with a hollow cathode and a trigger electrode in the cathode cavity has
been constructed by I.I. Aksenov and co-authors /66/. The motivation of this
paper was devoted to an investigation of the trigger delay and anode voltage
decay, the determining merits of the spark gap as a controlled switch. In the
following the most important results are presented.

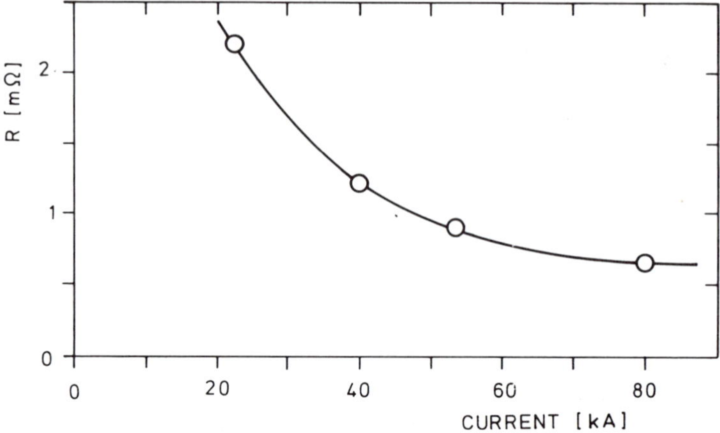

Fig. 28. Internal resistance of the
high-current switch at peak current
(dI/dt = 0)

   <u>Experimental set-up and principles of operation.</u>   A diagram of the spark
gap is shown in Fig. 29. The electrodes are mounted in a steel cover 3. The anode 2
is a disc and is located in the anode chamber bounded by the walls and the cathode
4. The trigger electrode, consisting of two parallel discs 6 and 7 joined by
conducting supports, is located in the cavity 5 on the opposite side of the cathode.
Cavity 5 functions as the trigger chamber and is coupled to the main spark gap 1
through slits in the cathode 8. Like the cover all electrodes are fabricated from
stainless steel. The insulators are made of glass. The width of gap 1 is 5 mm. At
hydrogen pressure of the order of a few tenths of a torr the breakdown voltage of
the anode chamber U_b is determined by the left-hand branch of the PASCHEN —
curve. At these pressures the dimensions of the trigger chamber (height: 50 mm,
diameter: 100 mm) correspond roughly to the minimum of this curve.

The gaps between the anode and cover walls, and also between the leads and the
insulator walls, do not exceed 4 - 5 mm, which preludes the possibility of
breakdown over an extended path. There is no heated cathode. Its absence leads,
under certain circumstances, to substantial deviations from the exponential
decay of anode potential. The appearance of fluctuations prolongs the period for
establishing the discharge between anode and cathode.

When the device is in operation, a low-current pre-ignition glow discharge is
maintained in the trigger chamber. The main purpose of this discharge is to
reduce and stabilize the time lapse between the trigger pulse and breakdown of
the spark gap. With small cathode slits (width about 2 mm) and pre-ignition

currents I_p of some milliamperes the breakdown voltage of the device is almost
unaffected. This changes daramatically by increasing the slit width to 3 or 4 mm.
The magnitude of U_b also falls sharply with increasing pre-ignition current I_p,
if this current is negative (- Ip) in direction. That means, the trigger electrode
being negative with respect to the cathode. Vice-versa a pre-ignition discharge
with the trigger electrode positive (+ I_p) causes a substantial increase in U_b and
makes this quantity more stable.

The spark gap is triggered by feeding a trigger pulse to the discs 6 and 7.
Triggering takes place in several stages. The duration of each of these depends on
a number of conditions, in particular, on the magnitude and polarity of the
trigger pulse. Currents and voltages at the spark gap electrodes during the firing
period are shown in Fig. 30 a,b.
With the arrival of a negative pulse at the trigger electrode (Fig. 29 a) the
current (i_t) in the corresponding circuit begins to rise, at first slowly (t_1), but
then rapidly (t_2), until it reaches a maximum value at which a high-voltage mode
of glow discharge is established in cavity 5. The initial burst of current is a
displacement current.
The current in the anode circuit develops later, some time after arrival of the
trigger pulse. The anode voltage drops from U_a to U_s, the voltage of a self-
maintaining discharge. The decay time t_f is, by convention, usually regarded as

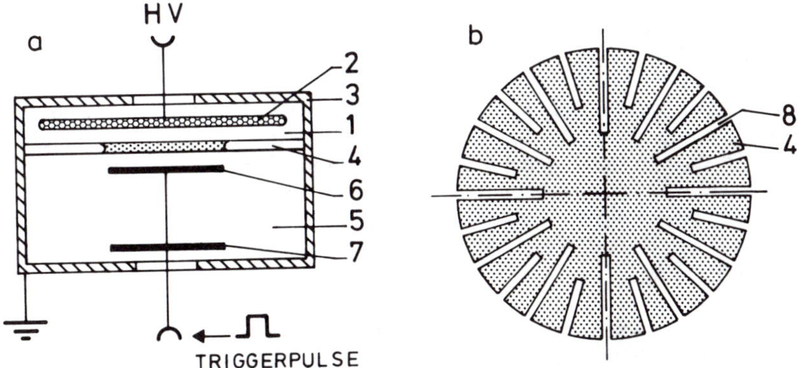

Fig. 29. a) Diagram of the spark gap
b) the cathode (top view)

the interval between the points corresponding to $0.9 \cdot (U_a - U_s)$ and $0.1 \cdot (U_a - U_s)$.
In this experiment the total trigger delay t_d is taken as the interval between the
time at which the pulse voltage is applied to the trigger gap and the point on the
$U_a(t)$ curve corresponding to termination of the period t_f. When the trigger pulse
is positive the current and voltage curves initially have the same form as in the
previous case. The distinction lies in the fact, in addition to t_t $(= t_1 + t_2)$, the
voltage decay period is preceeded by a further interval t_a, during the course of
which the current i_t in the trigger electrode circuit and the voltage at the
trigger electrode drop to values several times smaller than their maxima. The
delay in breakdown of the main gap and the appearance of t_a is explained by the
retarding effect which the trigger electrode field has on the electrons. The
ordered motion of the electrons in the trigger chamber is directed away from the
cathode slits.

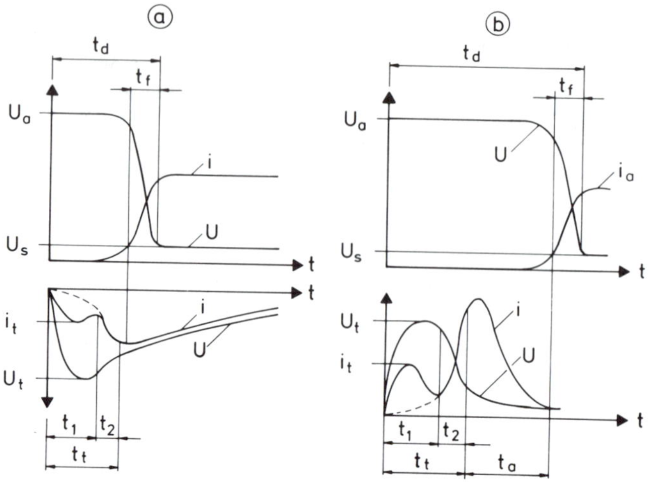

Fig. 30. Currents and voltages during triggering

<u>Results</u>

<u>Trigger Delay Time</u>

The dependence of the delay time t_d on the amplitude and polarity of the trigger pulse, on the magnitude and direction of current in the pre-ignition discharge, and on the hydrogen pressure in the switch is shown in Fig. 31, 32 and 33.

The main results are:

- The $t_d(U_t)$ curves for triggering pulses of negative polarity correspond qualitatively to the analogous curves for a pulsed thyratron with a positive grid. In fact, when there is a negative voltage on the trigger electrode, the latter can be regarded as a cathode, the role of grid then being played by electrode 4. After breakdown of the main gap 1, either electrode 4 or the whole of chamber 5 function as the cathode.

- The magnitude of t_d is markedly affected by the direction of the current in the pre-ignition discharge (see Fig. 31).

- Switching on the pre-ignition discharge substantially reduces the delay time. A rapid decrease of t_d takes place as I_p increases, until this reaches about 4 mA (see Fig. 32).

- When I_p is positive then, irrespective to the polarity of U_t, the spark gap triggering process is distinguished by stability ($U_t = 3$ kV, the spread in t_d is a few nanoseconds).

- To the left of a minimum value of p the spark gap does not trigger, while to the right the discharge is continuous (see Fig. 33).

- With a cathode slit width of 2 mm, the time t_d is independent of U_a (from 3 to 30 kV)

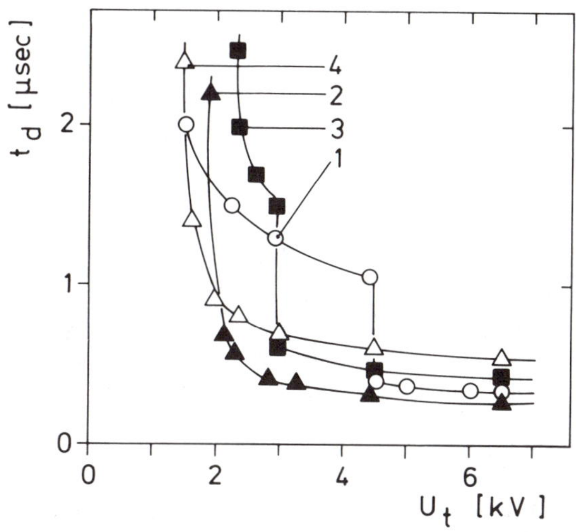

Fig. 31. Dependence of delay time on
the trigger amplitude
p   =   0.29 torr H_2
U_a =   16 kV
I_p (in mA): 1 and 4: +4 mA,
2 and 3: −4 mA
The continuous and dashed lines
correspond to positive and negative
pulses respectively

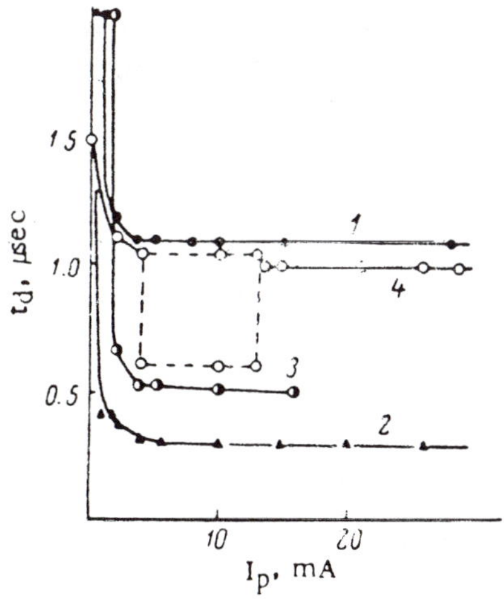

Fig. 32. Dependence of delay time on
current in the pre-ignition
discharge
p   =   0.29 torr H_2
U_a =   16 kV
U_t =   3.5 kV
1) $(+U_t, +I_p)$
2) $(-U_t, +I_p)$
3) $(-U_t, -I_p)$
4) $(U_t, -I_p)$

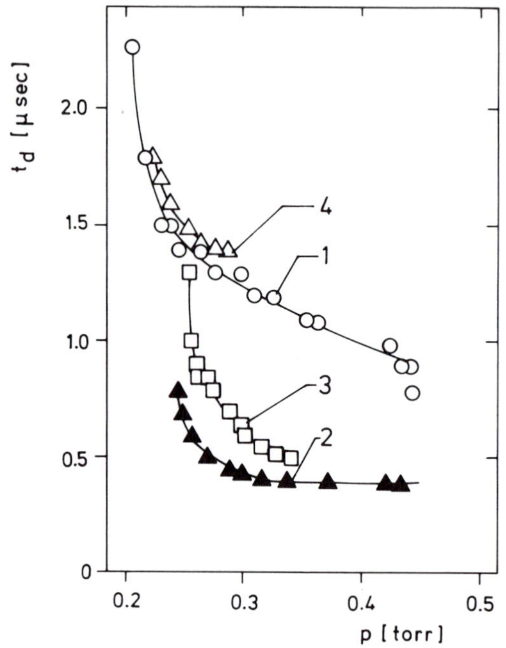

Fig. 33. Dependence of delay time pressure

U_a -   16 kV
I_p -   4 mA
U_t -   3.5 kV

1) $(+U_t, +I_p)$
2) $(-U_t, +I_p)$
3) $(-U_t, -I_p)$
4) $(+U_t, -I_p)$

(The dashed lines denote the limits within which the device remains controllable.)

Discharge Formation Time

The quantity t_f is one of the most important parameters of switch, although it is an insignificant fraction of the total delay time t_d. It is this quantity that determines the effective rise time of the anode current pulse and the losses during the formation of the main discharge. It is evident from Fig. 34 that the switching time decreases with increasing pressure up to 0.35 torr, it then remains almost constant. An explanation for this fact is not given.

Summary.   In table 3 the most important data of this device are summarized.

During the investigations the spark gap had the function of a switch in a line-type modulator. The wave impedance of the modulator shaping line is $10\,\Omega$, the load on the line is matched, and resonance charging was used.

The device has shown that it can be triggered over a wide range of anode voltages by pulses of either sign. However, the most favourable conditions with regard to reduction of t_d, enhancement of the stability of this quantity and extension of the range of working pressures are realized when the trigger pulse is of negative polarity and the pre-ignition discharge current is positive.

Table 3. Low-pressure spark gap
data

anode voltag U	:	≤ 30 kV
peak current I	:	≤ 1.5 kA
repetition rate f	:	50 sec^{-1}
pulse length T	:	3 μsec
gas	:	H_2
gas pressure	:	some tenth of torr
gap width	:	5 mm
electrode material	:	stainless steel
jitter (best case)	:	some nsec
lifetime	:	unknown
conduction voltage U_S	:	unknown
rate of currennt	:	unknown

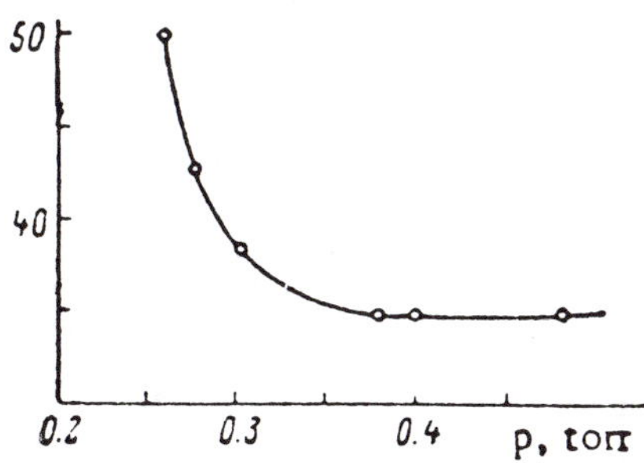

Fig. 34. Pressure dependence of the anode discharge formation time

U_a	=	16 kV
U_t	=	-3 kV
I_p	=	+4 mA

    <u>Medium-power pseudospark switch.</u>   The pseudospark switch and the BLT (Back-Lighted Thyratron), based on the pseudospark geometry, are presented during this workshop. Besides there is a manifold of publications on fundamentals and applications of these switches. Nevertheless it is useful to present the most important data of the pseudospark switch, because the question for the main discharge mechanisms provoked the explanation by a superdense glow. Fig. 35 shows the cross section of a prototype pseudospark switch. The discharge is initiated in a geometry consisting of two plane-parallel electrodes with a circular center hole each; with an electrode distance typically of the order of several millimeters. At gas pressures of 30 to 50 Pa of hydrogen (deuterium) the discharge can easily be triggered by the injection of electrons into the hollow cavity of the cathode. At voltages of 30 kV and more, and peak discharge currents of up to 30 kA, rates of current rise of up to 10^{12} A/sec have been achieved with single discharge channels. Triggering of the switch is most easily accomplished with the so-called 'charge injection trigger' or 'pulsed glow discharge trigger'. This trigger method is most commonly used. To improve the discharge characteristics sometimes a blocking electrode is used. This design of a pseudospark switch has now been realized in ceramic-metal brazing technique with an integrated gas resevoir. The state-of-the-art of this switch is comparable with a conventional thyratron (see table 4).

Table 4. Pseudospark switch
performance data

Typical ratings:		
hold-off voltage	:	32 kV
peak current	:	30 kA
current reversal	:	90 %
repetition rate	:	30 Hz
pulse length	:	100 ns
dI/dt	:	10^{12} A/sec
trigger discharge		
(power consumption)	:	1 W
delay	:	250 ns
jitter	:	4 ns
working gas	:	H_2 (D_2)
lifetime	:	$8 \cdot 10^6$
energy per pulse	:	25 J

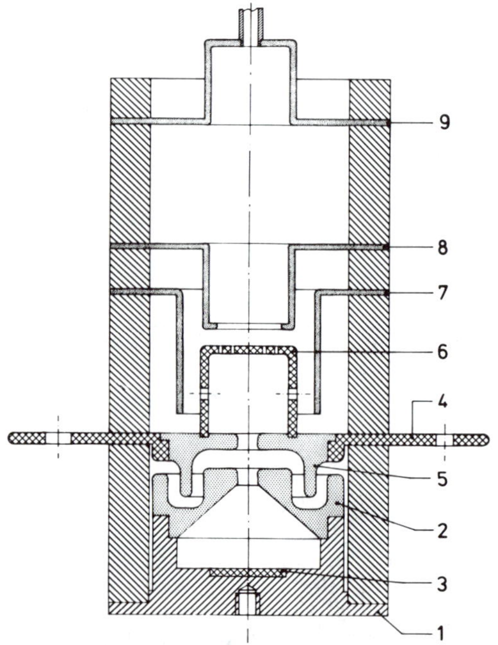

Fig. 35. Cross section of the
prototype pseudospark switch

For more information see /49,62,64,65,67/.

3.   High-Power Opening Switches (Hollow Cathode Devices)

Opening gas discharge device for interrupting voltages up to 50 kV /68/.
Fig. 36 shows the circuit with the gas discharge device. The circuit represents a
square-voltage pulse shaper with an inductive energy storage device.
Each of the two gas-discharge devices GD_1 and GD_2 has a hermetically sealed
space which contains two symmetric electrodes made in the form of cylindrical

cavities. Their insulation from each other withstands the total switching voltage.
Between the electrodes is a set of diaphragms insulated from each other. Each
device is placed in a constant longitudinal magnetic field, produced by a solenoid,
and is filled with hydrogen from the device P to a pressure $p_0 = 10^{-2}$ to 10^{-3} torr.
The device is put into conducting state by means of either of the two igniter
electrodes introduced into the space of the hollow electrodes, pulsed from the
ignition device ID with the appropriate polarity of voltage applied to the device.
The device has a thyratron switching-on characteristic. The process of current
passing through the gas discharge device is accompanied by intense absorption
of hydrogen by the intermediate electrodes of the device so that in a certain
interval of time, dependent on the value of the initial pressure p_0 and the
current, the density of the gas in the discharge zone falls below the critical value
and spontaneous cutoff of the current occurs.
The largest shaper parameters attained are: $I_0 = 500$ A, U = 50 kV, $\tau = 50$ msec.

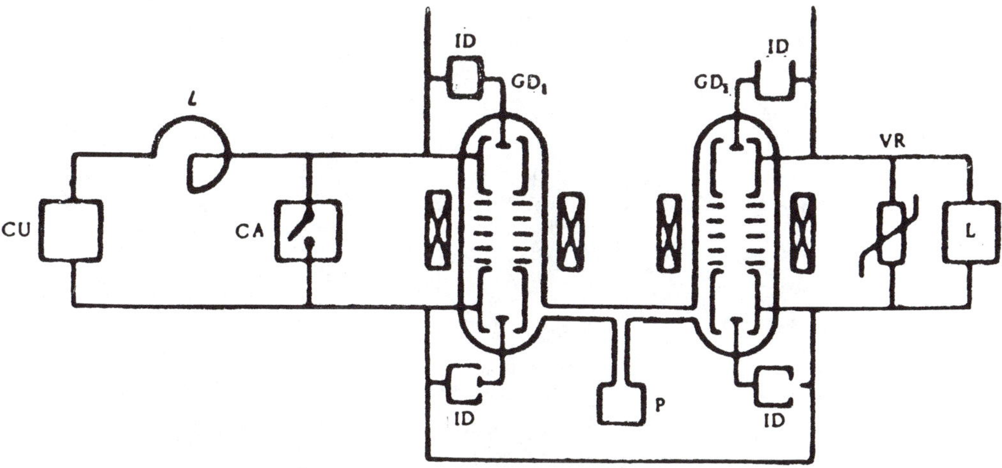

Fig. 36. Circuit of shaper and design of gas-discharge switching elements

Non-conducting of a pseudospark switch after zero-crossing of the current
pulse /69/.   Originally the medium power pseudospark switch (see B.2)
was developed for replacement of advanced thyratrons, but it has also shown its
feasibility for long pulse duration applications /69/. The pseudospark switch
utilized with this experiment was identical with the device described just before.
Hydrogen is used as the working gas throughout the experiments. The variation
of hydrogen pressure led to the discovery of a very rapid switch recovery during
current zero at the end of a full sine wave. Switch recovery was fast enough to
permanently stop conduction after conducting a full current sine wave of 560
μsec in duration and up to 6 kA in amplitude. This ability of opening after
conducting a certain number of sine waves depends critically on the hydrogen
pressure and the dI/dt at current zero (see Fig. 37). The maximum dI/dt at current
zero where the switch was still able to recover fast enough to stop conduction was
in the order of $0.6 \cdot 10^8$ A/sec.

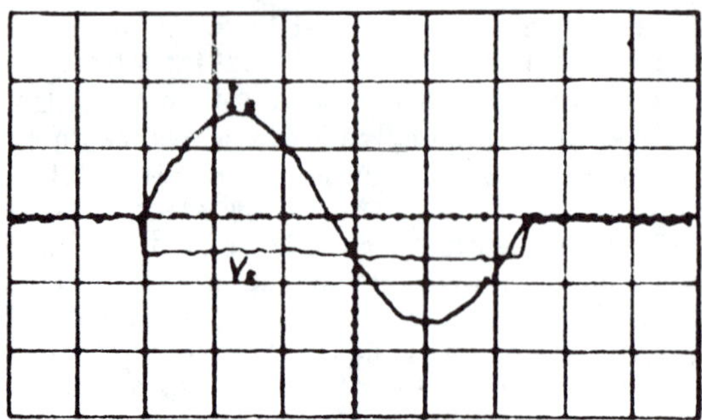

Fig. 37. Voltag and current waveforms
at a hydrogen pressure of 14.2 Pa;
100 μsec/div
3.9 kA/div (upper trace);
10 kV/div (lower trace)

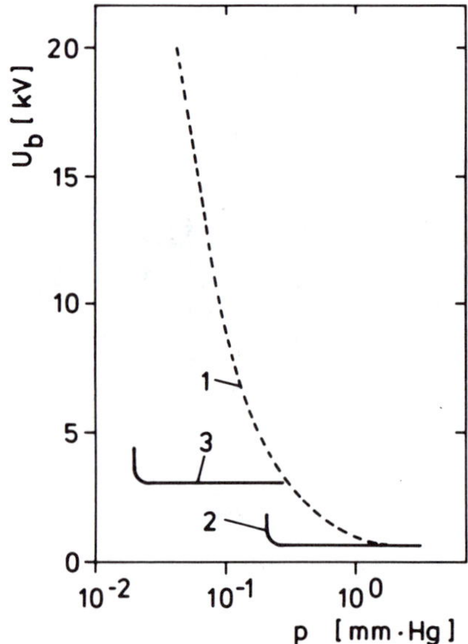

Fig. 38. Ignition potential as a function of air pressure for flat and hollow Al
cathodes. The horizontal curves pertain to the hollow cathode.

4.   High Power Rectifying Devices (Hollow-Cathode)

Rectification in a low-pressure arc with a hollow cold cathode.   This
paper was published by E.A. Koltypin and co-authors /70/. The hollow-cathode
device operates in the arc mode, not in the high voltage glow mode, although the
distinct difference between both is the principle of working. At first we have to
consider the anomalous stage of glow discharge immediately preceeding the arc.

46

At this stage the current can increase only as a result of rising density. This involves a rise of the voltage across the discharge and a contraction of the dark cathode space. The growth of the current density and voltage continue until the arcing potential is reached. Then the discharge voltage falls to several tens of volts, allowing an essential unlimited increase of the current. This discontinuity characterizes the superdense glow-to-arc transition /36,71,72/. The point corresponding to the discontinuity of the U-I characteristic of the discharge is called the arc ignition potential U_{ig}.

For a plane-parallel electrode system with a cold cathode the variation of U_{ig} with the gas pressure shows curve 1 in Fig. 38.

Qualitatively, the nature of this dependence can be understood on the basis of the following considerations: Since the anomalous glow discharge preserves the similarity relationship for a given voltage $p \cdot d_k$ = const. for the dark cathode space region (p - gas pressure, d_k - thickness of the cathode dark space), in addition, E_k/p also remains constant (E_k = electric field strength near the cathode). Thus, as the pressure drops, the size of the cathode dark space increases and the electric field decreases. As a result U_{ig} increases rapidly with decreasing pressure if the transition takes place at a definite gap voltage. It has been shown in /26,31/ (compare A.3.) that at low pressures, when (p · d) of the gas discharge gap is less than $(p \cdot d)_{min}$, i.e., lies on the left branch of the PASCHEN curve, the gas discharge does not pass through the normal glow discharge stage, but turns directly into the anomalous stage, which is characterized by a high burning voltage on the one hand, and by considerable difficulties in the transition to an arc discharge on the other.

Any change in cathode configuration involves a change in the width of the cathode space in the anomalous glow discharge stage as well as in the electric field intensity at the cathode. Thus, the dependece of U_{ig} on gas pressure will also be different. In the case of a hollow cathode the development of a discharge differs most strikingly from that with a flat one. It is well known /20,36/ that the so-called 'hollow-cathode effect', characterized by very high current densities, can take place in a discharge with a hollow or dual cathode under certain conditions. The condition for the 'hollow-cathode effect' can be written approximately

$$a \cdot p < 1 \text{ cm} \cdot \text{mm Hg}$$

a - diameter of a hollow cylindrical cathode
p - gas pressure in mm Hg

In the case of the 'hollow-cathode effect' the current density is much higher as mentioned before, this means d_k is considerably lower than these on the plane cathode, and d_k does not depend on the pressure within the whole region where the 'hollow-cathode effect' exists (see Fig. 40, curve 2 for a = 1 cm and curve 3 for a = 10 cm).

'Thus the arc discharge in the hollow cathode system is excited easily under low pressures when U_{ig} on the plane cathode reaches several tens of kilovolts .
It is not difficult to imagine electrode shapes when at one polarity the properties of the hollow cathode are realized while at the other polarity characteristics for the high-voltage discharge arise. Fig. 39 shows one of the simplest variants of the hollow cathode-plane anode system.
Fig. 40 presents the dependence of U_{gi} on the pressure for different polarities of the electrodes of the system shown in Fig. 39.

The properties of the hollow cathode manifest themselves most vivilly when it is combined with a magnetic field. If the system is placed in a magnetic field in such a manner that part of the magnetic lines of force (for example, lines I and II in Fig. 41) intersects twice the cathode C without passing through the anode A.

Thus the field configuration gives rise to an electron potential well in the region III. In fact, electrons in this region are prevented by the magnetic field from reaching the anode while the electric field does not allow them to approach the cathode. Electrons moving in crossed electric and magnetic fields drift around the system axis forming an ionization region in the interelectrode space. This phenomenon occurs if the anode is positive. If the anode is negative the electrons are rapidly captured by the positive cathode.

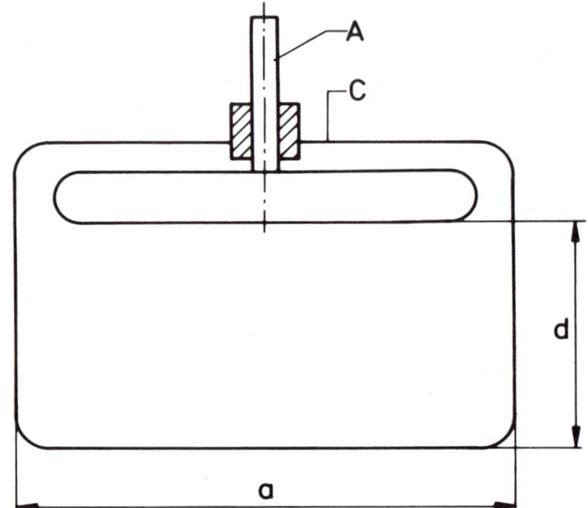

Fig. 39. Hollow cathode and flat anode system of electrodes

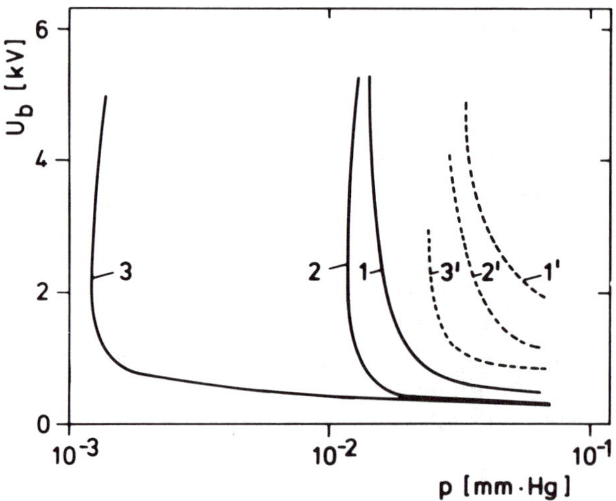

Fig. 40. Rectifier ingition potential as a function of pressure. Dashed and solid curves correspond to a positive and negative cathode, respectively. H = 0 (1 and 1'), 100 (2 and 2'), and 300 (3 and 3') Oe

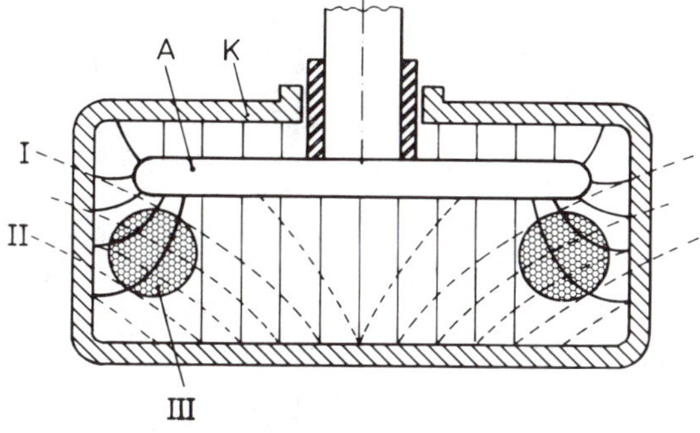

Fig. 41. Hollow cathode-flat anode electrode system in a non-uniform field

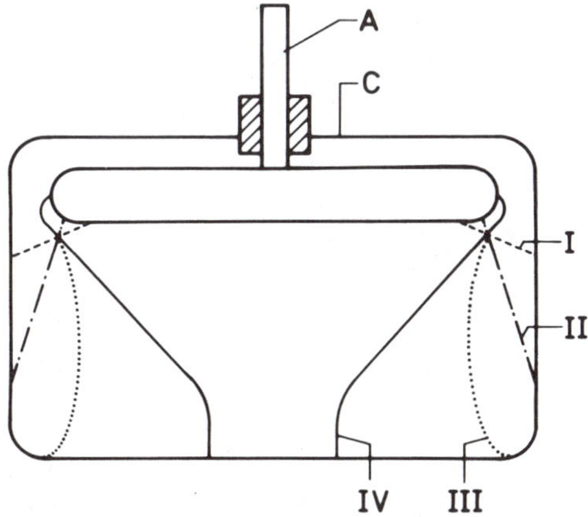

Fig. 42. Schematic diagram of a rectifier in which the dashed lines show the configuration of a self-pinched arc discharge at various times

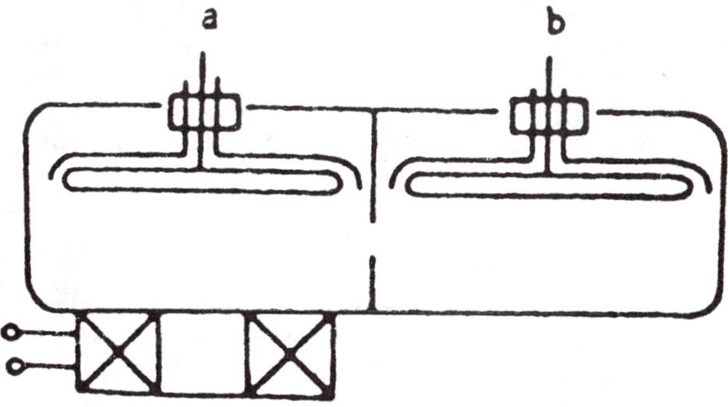

Fig. 43. Schematic drawing of two hollow-cathode arc rectifiers

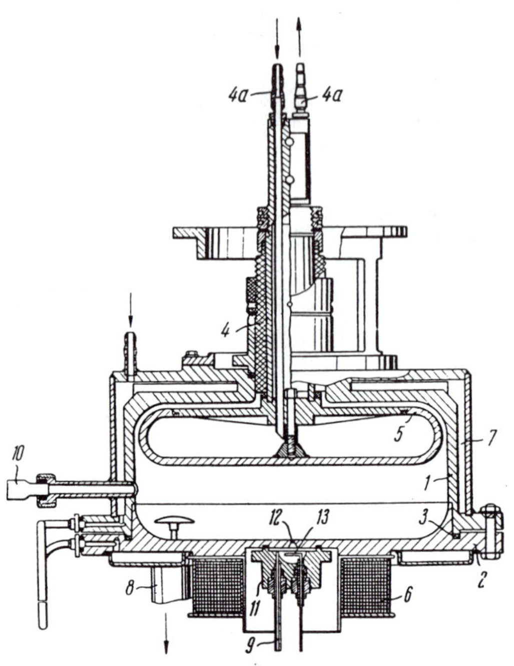

Fig. 44. A hollow-cathode arc
rectifier for 100 kA and 10 kV

Using an external trigger source (i.e. electron beam), the arc discharge can be
caused to ignite either at a given voltage or a given time, but always when the
anode is positive. A second important point for the rectification potential of such
a device is that voltage hold-off of the discharge gap to reverse voltages is
restored after the forward current pulse passes. This is possible only if the rate of
increase of the negative anode voltage is low compared to the rate of deionization
in the discharge gap.

With a = 10 cm the 'hollow-cathode effect' occurs at pressures of the order of 10^{-2}
mm Hg, the deionization rate seems to be fast enough

The behaviour of the discharge as current flows through the rectifier shows Fig.
42. A kind of pinched plasma column (stage IV) forms at the maximum current
region on the cathode bottom. This is a limiting effect concerning the maximum
peak current the rectifier can pass through, the aim is up to 100 kA at a pulse
length up to $(2 - 10) \cdot 10^{-4}$ sec.

Fig. 43 shows a two hollow-cathode arc rectifier /73/ constructed by the same
authors /74/. Fig. 44 and Fig. 45 show a practical device and the underlying
principle. Similar experiments were performed by M.Yu. Gel'tsel' and co-authors
/75/.

CONCLUSIONS

With the 'Superdense Cold Hollow-Cathode Glow Discharge' the range of
definition of glow discharges was extended by a completely new variant. It is to
think about carefully, whether the classification made by /30/ is to enlarge: by
the 'Superdense Glow Discharge'. Unfortunately this classification has not been
adopted in literature. The achievement and use of current densities up to 25
kA/cm^2 /18/ with hollow cathode discharges opened a new class of applications.

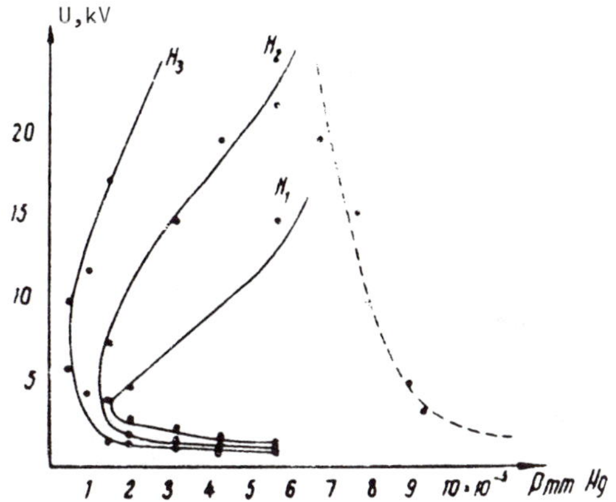

Fig. 45. Dependence of the starting
potential upon the pressure for
various magnetic fields

Originally this discharge was discovered with the investigations of high current
hydrogen discharges at low pressures as possible access to controlled nuclear
fusion. This is still obvious looking for the essential applications of such
discharges, i.e. commutating very high peak currents with long pulse durations
/57/.

Owing to these works described within the chapter, mainly performed in
Russian laboratories we have a quite better understanding of the Superdense
Glow Discharge. Nevertheless there is a lack of understanding concerning
important parameters like the mechanism within the cathode drop region or
emission processes at the electrodes. It is shown that there exists a clear
separation of the two discharge modes, the Superdense Glow and the transition to
the Hollow Cathode Arc. The experiments to the 'hollow cathode effect' not only
are of poor theoretical character, they deliver practical advices for construction
high duty practical devices.

Disadvantageous with the Russian papers is the fact that there is only an
analytical approach to understand and to describe the mechanism controlling the
Superdense Glow Discharge. It is to state an almost complete lack of attempts of
numerical simulations or models. Furthermore the main field of applications lies
in high current, long pulse devices; applications for replacement of thyratrons
are not found. This is an astonishing fact. In general, the available data are very
often of poor quality. A deepened judgement of these experiments is presented in
/76/.
In U.S.A. and Europe recently promising devices appeared like the pseudospark
/67,77,78/ or the BLT based upon the principle of a Superdense Glow Discharge.
Besides of these applications a lot of excellent work is done investigating its
phenomena without explicitly using the denomination 'Superdense Glow'. The
emphasis on using modern methods modeling gas discharges and its progress
promise a far better understanding of the discharge mechanism in the next
future. A critical point remains, there is still a lack of coordination in
experiments and comparison with numerical simulations. It seems to be very
important to improve as soon as possible the interaction of experiments with
theory. Furthermore it seems to be useful to take up the proposal of a new
classification of different glow discharge types (/29,30/) and to complete it by the
'superdense glow discharge' making a note of the only appearance under pulsed
conditions.

REFERENCES

1. W. Hartmann and M.A. Gundersen, Phys. Rev. Lett. $\underline{60}$, pp 2371 (1988)
2. A. von Engel, 'Electric Plasmas: Their Nature And Uses', Taylor and Francis Ltd., London and New York (1983)
3. F. Paschen, Ann. Physik $\underline{50}$, pp 901 (1916)
4. P.F. Little and A. von Engel, Proc. Roy. Soc. London $\underline{A224}$, pp 209 (1964)
5. D.J. Surges and H.H. Oskam, Physica $\underline{37}$, pp 457 (1976)
6. H. Helm, Z. Naturforsch., $\underline{27a}$, pp 1812 (1972)
7. A. Guentherschulze, Z. Physik $\underline{9}$, pp 313 (1923); Z. Tech. Phys. $\underline{11}$, pp 49 (1930)
8. V.G. Grechanyi and A.S. Metel, Sov. Phys. Tech. Phys. $\underline{27}$, pp 284 (1982)
9. V.I. Kirichenko, V.M. Tkachenko, V.B. Tyutyunnik, Sov. Phys. Tech. Phys. $\underline{21}$, pp 1080 (1976)
10. A.S. Metel, Sov. Phys. Tech. Phys. $\underline{30}$, pp 1133 (1985)
11. G.V. Okhmatovskii, Sov. Tech. Phys. $\underline{23}$, pp 552 (1978)
12. V.N. Glazunov and A.S. Metel, Sov. Phys. Tech. Phys. $\underline{26}$, pp 559 (1981)
13. V.G. Grechanyi and A.S. Metel, Teplofizika Vysokikh Temperatur $\underline{22}$, pp 444 (1984) (in Russian)
14. V.N. Glazunov and A.S. Metel, Sov. J. Plasma Phys. $\underline{8}$, pp 625 (1982)
15. V.A. Nikitinskii and B.I. Zhuravlev, Sov. Phys. Tech. Phys. $\underline{27}$, pp 564 (1982)
16. A.P. Semenov, Sov. Phys. Tech. Phys. $\underline{32}$, pp 1101 (1987)
17. A.S. Metel, Sov. Phys. Tech. Phys. $\underline{29}$, pp 141 (1984)
18. B.I. Moskalev, 'Hollow Cathode Discharges' (in Russian), Energiya, Moscow (1969)
19. V.N. Glazunov, V.G. Grechanyi, A.S. Metel, Sov. Phys. Tech. Phys. $\underline{27}$, pp 1084 (1982)
20. A.S. Metel, Sov. Phys. Tech. Phys. $\underline{31}$, pp 1396 (1986)
21. G.V. Grekova, E.L. Lapshin, G.V. Okhmatovskii, Sov. Tech. Phys. Lett. $\underline{1}$, pp 145 (1975)
22. V.S. Borodin and Yu. M. Kagan, Sov. Phys. Tech. Phys. $\underline{11}$, pp 131 (1966)
23. G. Schaefer and K.H. Schoenbach, AIAA 25th Aerospace Sciences Meeting, Reno, January 12-15 (1987)
24. G. Schaefer, P.O. Husoy, K.H. Schoenbach, H. Krompholz, IEEE Transactions of Plasma Science, PS-12, pp 271 (1984)
25. A.M. Zinin, N.P. Kozlov, V.I. Khvesyuk, Sov. Phys. Tech. Phys. $\underline{18}$, pp 1374 (1974)
26. L.Yu. Abramovich, B.N. Klyarfel'd, Yu.N. Nastich, Sov. Phys. Tech. Phys. $\underline{11}$,, pp 528 (1966)
27. D.D. Aleksandrov, N.F. Olendzkaia, S.V. Ptitsyn, Sov. Phys. Tech. Phys. $\underline{4}$, pp 836 (1958)
28. W.S. Boyle and P. Kisliuk, Phys. Rev. $\underline{97}$, pp 255 (1954)
29. B.N. Klyarfel'd, L.G. Guseva, A.S. Pokrovskaya, Soboleva, Sov. Phys. Tech. Phys. $\underline{11}$, pp 520 (1966)
30. B.N. Klyarfel'd and L.G. Guseva, Sov. Phys. Tech. Phys. $\underline{10}$, pp 244 (1965)
31. A.S. Pokrovskaia-Soboleva and B.N. Klyarfel'd, Soviet Physics JETP $\underline{5}$, pp 812 (1957)
32. E. Finkelmann, Arch. Eletrotech. $\underline{31}$, pp 282 (1937)
33. R.B. Quinn, Phys. Rev. $\underline{55}$, pp 482 (1935)
34. L.Yu. Abramovich, B.N. Klyarfel'd, Yu.N. Nastich, Sov. Phys. Tech. Phys. $\underline{14}$, pp 939 (1970)
35. Yu.N. Nastich and L.Yu. Abramovich, Sov. Phys. Tech. Phys. $\underline{17}$, pp 810 (1972)
36. I.I. Aksenov, V.A. Belous, S.A. Smirnov, Sov. Phys. Tech. Phys. $\underline{20}$, pp 1094 (1974)

37. U.A. Arifov, 'Interaction of Atomic Particles with Solid Surfaces' (in Russian), Nauka, Moscow (1968)

38. B.N. Klyarfel'd and B.I. Moskalev, Sov. Phys. Tech. Phys. 14, pp 800 (1969)

39. G.N. Alferov, V.P. Drachev, E.A. Kuznetsov, V.K. Mezentsev, G.I. Smirnov, Sov. J. Plasma Phys. 14, pp 356 (1988)

40. D.I. Proskurovskii and V.F. Puchkarev, Sov. Phys. Tech. Phys. 24, pp 1479 (1979)

41. M. Antonov, L.B. Gevorkyan, A.G. Ponomarenko, Sov. Phys. Phys. Lett. 4, pp 401 (19789

42. H. Menown and C. Neale, 'Thyratrons for Short Pulse Laser Circuits', IEEE Thirteenth Pulse Modulator Symposium, June (1978)

43. M.A. Vasilevskii, A.V. Demidov, I.M. Roife, V.I. Engel'ko, E.G. Yankin, Sov. Phys. Tech. Phys. 32, pp 769 (1987)

44. J.J. Rocca, J.D. Meyer, Z.Yu, M.R. Farrell, and G.J. Collins, Appl. Phys. Lett. 41, pp 811 (1982)

45. J.J. Rocca, J.D. Meyer, J.R. Farrell, and G.J. Collins, J. Appl. Phys. 56, pp 790 (1984)

46. Zeng qi Yu, J.J. Rocca, and G.J. Collins, J. Appl. Phys. 54, pp 131 (1983)

47. J. Christiansen and C. Schultheiss, Z. Phys. A290, pp 35 (1979)

48. W. Benker, J. Christiansen, K. Frank, H. Gundel, W. Hartmann, T. Redel, M. Stetter, 'Pulsed intense electron beams from pseudospark discharges', in Microwave and Particle Beam Sources and Propagation', SPIE Vol. 837, pp 249 (1988)

49. D. Bloess, I. Kamber, H. Riege, G. Bittner, V. Brueckner, J. Christiansen, K. Frank, W. Hartmann, N. Lieser, C. Schultheiss, R. Seeboeck, W. Steudtner, Nucl. Instr. and Meth. 205, pp 173 (1983)

50. J.T. Maskrey and R.A. Dugdale, Journal of Physics E: Scientific Instruments, Vol. 5, pp 881 (1972)

51. H.F. Ranea, Sandoval, N. Reesor, B.T. Szapiro, C. Murray, and J.J. Rocca, IEEE Transactions on Plasma Science, PS-15, pp 361 (1987)

52. X.L. Jiang, K.F. Chen, Y.B. Piao, Acta Physica Sinica 32, pp 1344 (1983)

53. P. Choi, H.H. Chuaqui, M. Favre, and E.S. Wyndham, IEEE Transactions on Plasma Science, PS-15, pp 428 (1987)

54. Yu. Kreindel, 'Plasma Sources of Electrons' (in Russian), Atomizdat (1977)

55. N. Ulyanov, High Temperature 16, pp 955 (1978) (Russian translation)

56. G.W. McClure and K.D. Granzov, Phys. Rev. 125, pp 919 (1962)

57. O.G. Bespalov, A.S. Knyazyatov, A.I. Nastyukhova, P.A. Smirnov, and A.N. Udovenko, Sov. Phys. Tech. Phys. 26, pp 1051 (1981)

58. O.G. Bespalov, V.V. Veselovskii, A.I. Nastyukha, and A.N. Udovenko, Pribory i Tekhnika Eksperimenta, No. 3, pp 94 (1982)

59. O.G. Bespalov, A.S. Knyzyatov, A.I. Nastyukha, P.A. Smirnov, and A.N. Udovenko, Pribory i Tekhnika Eksperimenta, No. 1, pp 149 (1979)

60. E. Boggasch, V. Brueckner, and H. Riege, in Proc. 5th Pulsed Power Conf., Arlington, VA, pp 820 (1985)

61. P. Billault, H. Riege, M. van Gulik, E. Boggasch, K. Frank, and R. Seeboeck, CERN Technical Report, CERN 87-13 (1987)

62. K. Frank, E. Boggasch, J. Christiansen, A. Goertler, W. Hartmann, C. Kozlik, G. Kirkman, C. Braun, V. Dominic, M.A. Gundersen, H. Riege, and G. Mechtersheimer, IEEE Transactions on Plasma Science, PS-16, pp 317 (1988)

63. I. Vitkovitsky, 'High Power Switching', Van Nostrand Reinhold Company, New York (1987)

64. K. Frank, E. Boggasch, J. Christiansen, A. Goertler, W. Hartmann, and C. Kozlik, SPIE Vol. 735, Pulse Power for Lasers, pp 74 (1987)

65. K. Frank, J. Christiansen, O. Almen, E. Boggasch, A. Goertler, W. Hartmann, C. Kozlik, and A. Tinschmann, SPIE Vol. 871, Space Structures, Power, and Power Conditioning, pp 173 (1988)

66.   I.J. Askenov, V.A. Belous, and S.A. Smirnov, Sov. Phys. Tech. Phys. <u>16</u>, pp 1119 (1972)

67.   K. Frank, J. Christiansen, W. Hartmann, O. Almen, A. Goertler, C. Kozlik, and A. Tinschmann, SPIE Vol. 1023, Excimer Lasers and Applications, pp 98 (1989)

68.   S.V. Agadzhanyan, I.N. Grigor'ev, E.I. Lapshin, and E.A. Polyakov, Pribory i Tekhnika Eksperimenta, No. 1, pp 80 (1983)

69.   W. Hartmann, O. Almen, K. Frank, R. Kowalewicz, A. Goertler, and J. Christiansen; J. Heuer, U. Braunsberger and J. Salge, in Proc. 7th Pulsed Power Conf., Monterey, CA, to be published

70.   E.A. Koltypin, A.I. Nastyukha, and P.A. Smirnov, Sov. Phys. Tech. Phys. <u>15</u>, pp 1703 (1971)

71.   W.S. Boyle and F.E. Haworth, Phys Rev. <u>101</u>, pp 935 (1956)

72.   A.V. Kozyrev, Yu.D. Korolev, and G.A. Mesyats, Sov. Phys. Tech. Phys. <u>32</u>, pp 34 (1987)

73.   E.A. Koltypin, A.I. Nastyukha, and P.A. Smirnov, Sov. Phys. Tech. Phys. <u>15</u>, pp 1710 (1971)

74.   A.I. Nastyukha, E.A. Koltypin, and P.A. Smirnov, U.K. Patents Nos. 1071133 and 1123739

75.   M.Yu. Gelt'tsel and A.A. Podminogin, Sov. Phys. Tech. Phys. <u>15</u>, pp 1327 (1971)

76.   T.R. Burkes, 'A Critical Analysis And Assessment Of High Power Switches', Electrical Engineering Department, Texas Tech University, Lubbock, TA 79409 (1978)

77.   G. Mechtersheimer, R. Kohler, T. Lasser, and E. Meyer, J. Phys. E: Sci. Instr. <u>19</u>, pp 466 (1986)

78.   G. Mechtersheimer and R. Kohler, J. Phys. E: Sci. Instr. <u>20</u>, pp 270 (1987)

BASIC MECHANISMS CONTRIBUTING TO THE

HOLLOW CATHODE EFFECT

G. Schaefer* and K.H. Schoenbach[†]

*Weber Research Institute, Polytechnic University
Farmingdale, N.Y. 11735, USA

[†]Dept. of Electrical Engineering, Old Dominion University
Norfolk, VA 23508, USA

INTRODUCTION

If a single plane cathode in a glow discharge is replaced by a cathode with some hollow structure such as a cylindrical or slit shaped hole, then, in a specific range of operating conditions the negative glow is found to be inside the hollow structure of the cathode. Under such conditions at a constant current the voltage is found to be lower and, at a constant voltage, the current is found to be orders of magnitude larger than for the plane cathode. This effect is called the hollow cathode effect (Pashen, 1916).

Many different geometric configurations have been investigated. Some examples are shown in Figure 1. Figure 2 shows the comparison of the discharge characteristics of a planar and a hollow cathode. Often the amplification factor, J/J_o, is used to express the hollow cathode effect, where J_o is the current density of the planar cathode and J the current density of the hollow cathode (Badareu and Waechter, 1958). Figure 3 shows measured amplification factors for two gases and two cathode materials.

Due to the well-defined plasma geometry and the intense emission hollow cathode discharges (HCD) have been used for a long time as spectral lamps (for reviews see S. Caroli, 1985; Mavrodineanu, 1984). For this application hollow cathode discharges are operated usually continuously with currents typically below 1A. Also pulsed hollow cathodes with currents up to kA have been used (Kielkopf, 1971). After the advent of lasers hollow cathodes have immediately been considered as suitable excitation sources for gas lasers (Chebotayev, 1963) again typically with continuous discharges and currents increasing into the 100A regime (for a review see Gerstenberger et al., 1980). Other hollow cathode discharge applications include ion sources (Kuen et al., 1978), plasma jets (Schaefer, 1986); hollow cathode electron beams (Rocca et al., 1987) and plasma contactors (Deininger et al., 1987).

Physics and Applications of Pseudosparks
Edited by M. A. Gundersen and G. Schaefer
Plenum Press, New York, 1990

More recently, interest in pulsed hollow cathode discharges has significantly increased. Three important applications are mentioned here.

1. Hollow cathode switches are used as high current closing switches (Koltypin et al., 1971). These switches allow high current densities with unheated cathodes without the usual erosion associated with an arc. They, therefore, have greater lifetimes than spark gaps under similar conditions. Hollow cathode switches are usually triggered with a magnetic field as Cross Field Switch Tubes. These devices show excellent operation conditions with respect to fast rep-rated, high current operation, but at this time lack short rise times. Hollow cathodes are utilized in Hollow Anode Thyratrons (Menown and Newton, 1973) which allow operating with high reverse

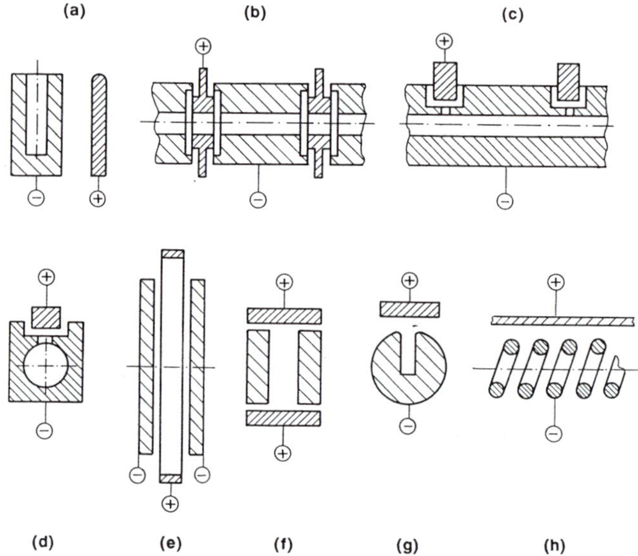

Figure 1.   Some typical hollow cathode geometries: a), b), c) cylindrical; d) spherical; e), f) parallel plate; g) slit; h) helical.

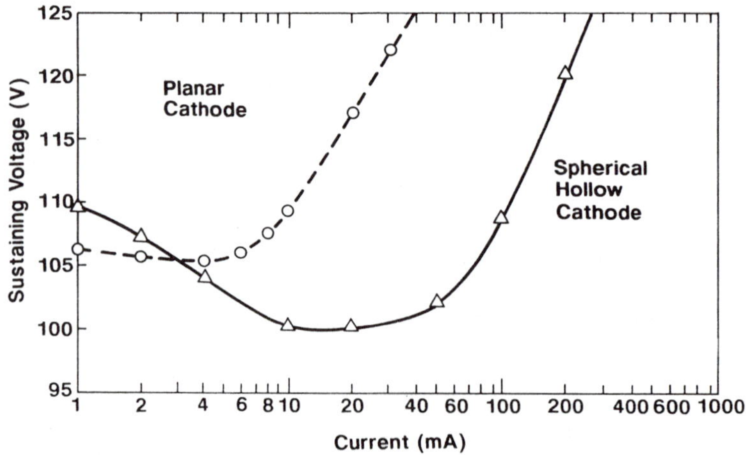

Figure 2.   Voltage - current characteristics of a spherical hollow cathode and a planar cathode discharge (Gewartowski and Watson, 1965).

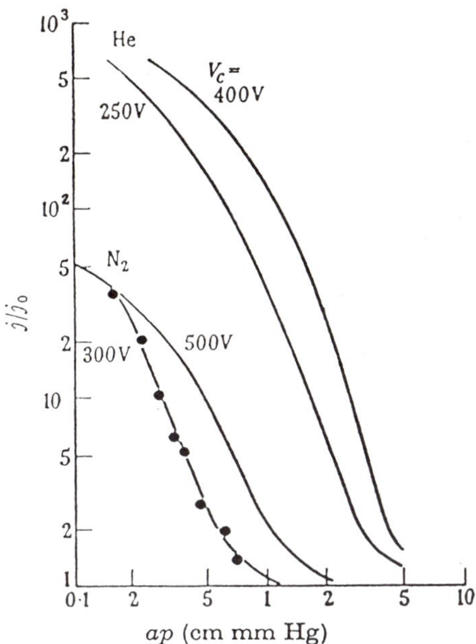

Figure 3.   Current amplification factor versus the product of cathode separation for a parallel plate hollow cathode, a, and discharge pressure, p, for Helium and Nitrogen. Lower curve with aluminum electrodes (Little and Engle, 1954), upper curve with iron electrodes (Guentherschulze, 1930).

currents.   More recently, the development of Pseudo Spark Switches (Christiansen and Schultheiss, 1979) has enhanced interest in the hollow cathode effect which is utilized at least during the discharge initiation period.   Short risetimes have been achieved with different trigger methods providing preionization (Mechtersheimer et al., 1986; Kirkman and Gunderson, 1986; Circel, 1987).

2.   Hollow cathodes are being considered as electrodes for atmospheric pressure diffuse discharge devices such as TEA lasers and diffuse discharge switches (Schaefer et al., 1984). It is assumed that the onset of discharge instabilities is shifted towards higher current densities. However, for these applications, the scaling of hollow cathode operation to high pressures has to be investigated first. For atmospheric pressure the holes should have diameters of the order of a few microns depending on the filling gas.

3.   Pulse hollow cathodes are being considered as plasma sources for gas laser systems operating via highly excited states (Falcone and Pedrotti, 1982).

In all of the mentioned applications of pulsed hollow cathodes, the device performance strongly depends on the discharge initiation mechanism and the rise time. In some of the hollow cathode applications as switches currents up to MA have been achieved (Koltypin et al., 1971). Also the transition into the "Superdense Hollow Cathode Glow Discharge" (Abramovich et al., 1966) was discovered and current densities at the cathode surface of up to 10^4 A/cm have been measured (Gunderson). At this time it is not clear whether the basic mechanisms contributing to the "low current density" hollow cathode effect still dominate or at least contribute to the characterization of these discharges or whether these effects play a role only in the initiation phase.

A systematic study and complete understanding of Hollow Cathode Discharges does not exist at this time although a very large number of papers has appeared in the open literature. One reason is that most experimental investigations have been carried out over a narrow range of operating parameters. Another reason is that it is difficult to compare the results obtained with different hollow cathode designs with respect to geometry, dimensions, materials etc., operated with different gases and in different pressure ranges (Kirichenko et al., 1976).

There are several mechanisms contributing to the hollow cathode effect:

1. Electrons emitted from the cathode surface inside the hollow structure which are accelerated in the cathode fall mainly contribute to ionization in the negative glow. Those electrons which have crossed the negative glow without significant energy loss will be reflected at the opposite cathode surface (pendular electrons) and change the potential distribution of the cathode fall. These electrons also contribute to a further enhanced ionization rate in the negative glow (Guentherschulze, 1923; Helm, 1972). This effect significantly influences the electron energy distribution function in the hollow cathode plasma (Borodin and Kagan, 1966; Gill and Webb, 1977).

2. The cathode fall in a hollow cathode under high current density conditions can be significantly thinner than for a plane cathode, reducing the probability for charge transfer collisions. Therefore the average ion velocity at the cathode surface is increased, causing an increased secondary electron emission rate (Babaderu et al., 1960).

3. Neutral, energetic particles (metastables and photons) generated in the negative glow inside the hollow cathode have a much higher probability of hitting the surface of the cathode due to the hollow geometry, increasing the electron emission rate of the cathode (Little and Engle, 1954). Also more positive ions are lost in the negative glow of a planar electrode since the negative glow is essentially field free and the ion transport is dominated by diffusion, (Sturges and Oskam, 1967).

4. The higher plasma density inside the hollow cathode makes multistep processes more likely (Sturges and Oskam, 1967; Witting, 1971).

5. The confined structure of the hollow cathode leads to a higher density of sputtered atoms of the cathode material with lower ionization potential and in rare gas discharges Penning ionization can occur (Musha, 1962).

All these mechanisms strongly depend on the operation conditions, the hollow cathode geometry, the fill gas, and the cathode material, and it is not clear which mechanism will, in a given device, dominate the hollow cathode effect (Kirichenko et al., 1976). Hollow cathode operation, in general, is restricted to a certain range of pD (1 torr cm $<$ pD $<$ 10 torr cm, for rare gases), where p is the gas pressure and D the diameter of the hollow cathode (Gerwatowski and Watson, 1965). This range is shifted to smaller values of pD if molecular gases are used.

In addition, the motion of the charged particle can be significantly altered using crossed magnetic field. Electrons emitted from the cathode which, without a magnetic field are accelerated away from the cathode surface, now move on cycloid type trajectories with an ExB drift parallel to the cathode surface (Metel and Nastyukha, 1981). This shifts the region with the highest ionization and excitation rates closer to the wall. As a consequence, the cathode fall thickness decreases and the impedance of the hollow cathode discharge decreases (Badareu et al., 1967; Tkachenko et al., 1972).

The purpose of this paper is to present an overview of the basic mechanisms contributing to the hollow cathode effect describe experiments which prove the existance and the importance of these effects, and to discuss the relation between these mechanisms and operating conditions and parameters of the discharge.

CURRENT CONTINUITY

Before we discuss the influence of the special geometry on the cathode fall and the negative glow we will briefly summarize the properties of these regions in the abnormal glow discharge. The major properties of these regions is to provide current continuity between cathode and the positive column (see Figure 4).

Electrons emitted from the cathode surface are accelerated in the thin cathode fall region and act as an electron beam. These beam electrons generate a plasma, the negative glow, similar to an electron beam sustained plasma. Typically, approximately half the energy of the initial beam electron is converted into ionization, generating positive ions and low energy thermal electrons, while the other half of the energy is used for various excitation processes. The negative glow is essentially field free which means that the transport of ions and thermalized electrons is dominated by diffusion. Ions which reach the boundary to the cathode fall are accelerated in the cathode fall and their energy at the cathode surface is dominated by the cathode fall voltage and charge exchange collisions with neutrals in the cathode fall region. These ions cause electron emission from the cathode surface. Other mechanisms contributing to electron emission are photoemission caused by UV-radiation emitted from the negative glow and collisions of metastables with the cathode surface. Any change of the current voltage characteristic as experienced when the cathode geometry is changed from a plane cathode to a hollow cathode must therefore change the balance of one or more of these effects. In the following we will discuss some special conditions which are characteristic for hollow cathodes.

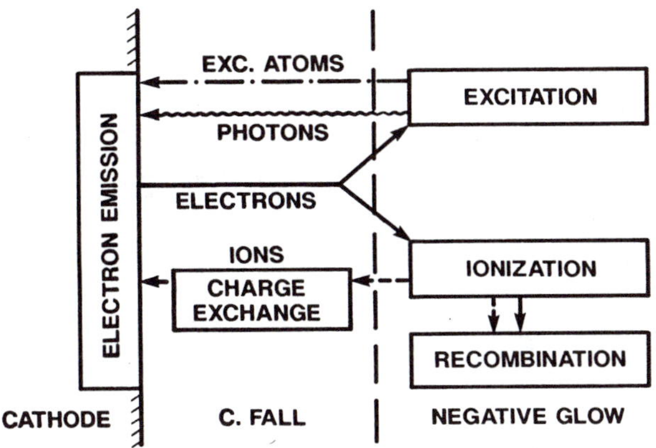

Figure 4.   Simplified carrier balance at the cathode

PENDULAR ELECTRONS

Electrons emitted from the cathode surface are accelerated in the cathode fall and loose their energy through exciting and ionizing collisions in the negative glow. If some of these electrons reach the opposite cathode fall region without significant energy loss they will reduce the positive space charge at the boundary of the negative glow, and as a consequence reduce the cathode fall potential. The influence of these electrons on the space charge and potential distribution is strongest inside the cathode fall itself where these electrons are decelerated down to nearly zero velocity (Schmidt, 1903; Guentherschulze, 1923; Schueler, 1926). This effect changes both, the electron energy distribution in the negative glow and the potential distribution in the cathode fall. It has also been discussed that the effect of the pendular electrons is enhanced through ionization inside the cathode fall. In this case secondary electrons are generated which fall through part of the cathode fall and become pendular electrons themselves.

The experimental verification of the pendular electrons effect was given by Helm (1972). He incorporated a small hole into the surface of a cylindrical hollow cathode and used an energy analyser to measure the energy distribution of the electrons penetrating through this hole. He confirmed that the pendular electrons contribute to the hollow cathode effect in a specific pD regime when p is the gas pressure and D the diameter of the cylindrical hollow cathode. At lower values of pD the electrons emitted from the cathode with some initial velocity have a high probability of crossing the hollow cathode and being absorbed at the opposite cathode surface, while at high pD values no electrons reach the opposite cathode fall region. Figure 5 shows the measured ratio of pendular electrons to discharge electrons versus pressure.

Another proof of the existence of the pendular electrons was given by Metel and Nastyukha (1981). They used a cylindrical hollow cathode with a cylindrical foreign body in the center of the cathode. This body blocked the motion of the beam electrons. With an axial magnetic field B the electron trajectories could be bent such that the electrons passed the foreign body.

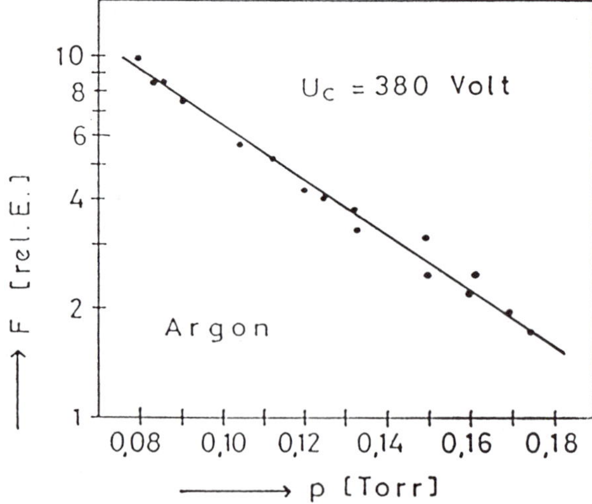

Figure 5.   Ratio, F, of pendular electrons to discharge electrons versus pressure in relative units (Helm, 1972).

Above this value of the magnetic field the discharge voltage at a constant current dropped significantly.

ELECTRON ENERGY DISTRIBUTION FUNCTION

The first experiments to determine the electron energy distribution were performed by Badareu and Popescu (1958) using a Langmuir Probe in a Double Cathode (Parallel Plate Cathode). The electron energy distribution measured in dry air was interpreted as the existence of two groups of electrons, with mean energies of ~ 0.6 and 5 eV. Similar experiments were performed by Borodin and Kagan (1966) in a cylindrical hollow cathode and the results were compared with measurements performed in the positive column of a glow discharge (Figure 6). These experiments show that the hollow cathode has more electrons in the high energy tail of the distribution (above ~ 16 eV) than the positive column, but also more electrons in the low energy regime (below ~ 4 eV). These results again suggest the existence of two electron groups. These probe measurements, however, did not have the necessary sensitivity to measure the low density of the beam electrons.

A full analysis of the electron energy distribution has been performed by Gill and Webb (1977) using a differentially pumped retarding field analyser. These measurements cover the full energy range from zero up to the maximum possible energy of the beam electrons. The high energy tail is shown in Figure 7. The energy distributions comprise three regions: (1) The peak with the highest energy in Figure 7 represents the beam electrons with the full cathode fall energy. (2) Following an empty gap we find electrons which have undergone one or more inelastic collisions and secondary electrons generated with high energies. (3) The low energy group consists of the electrons, after their energy relaxation.

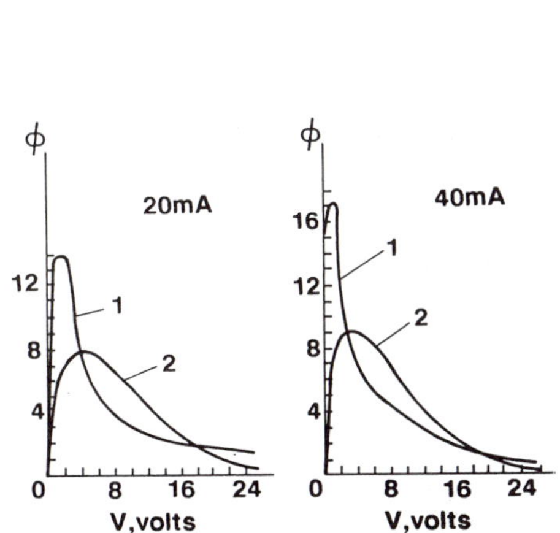

Figure 6. Comparison of the electron energy distribution functions in a hollow cathode and in a positive column (Borodin and Kagan, 1966).

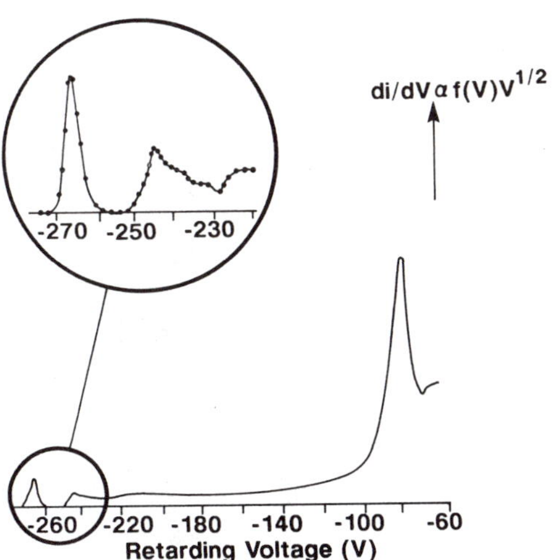

Figure 7. Measured electron energy distribution function close to the negative glow/cathode dark space boundary showing the cathode fall beam component and the first collision structure in detail (Gill and Webb, 1977).

Further studies have concentrated on the density of fast electrons and their dependence on HCD operating conditions (Okhmatovskii, 1978; Handle et al., 1984). Badareu et al. (1967) Popovici et al. (1968) and Ciobotaru (1972) suggested that an additional coupling exists between the different groups of the distribution function. The interaction of the beam electrons with the plasma of the negative glow generates waves from which the thermal electrons gain energy.

More recently so called Hollow Anode Cathode Discharges (K. Rosza, 1980) or High Voltage Hollow Cathode Discharges have been developed in which the anode is inserted inside the hollow cathode and the cathode is partially shielded from the plasma. Such discharges allow to increase the discharge voltage and as a result to raise the high energy tail of the electron energy distribution function (J. Mizeraczyk, 1984).

CATHODE FALL POTENTIAL DISTRIBUTION

As discussed before, the change of the space charge distribution will reduce the cathode fall potential if the cathode is operated at constant current density or increase the current density if operated at constant cathode fall voltage. In both cases the thickness of the cathode fall is reduced. This effect is easily demonstrated by monitoring the distance of the luminous zone (the negative glow) from the cathode surface (Badareu and Popescu, 1958) as shown in Figure 8.

The first field measurements in the fall region of a hollow cathode have been performed by Little and Engle (1954) using an electron beam. The experimental setup using a double cathode allowed to operate a discharge either with a single cathode or with a double cathode with varying separation of the

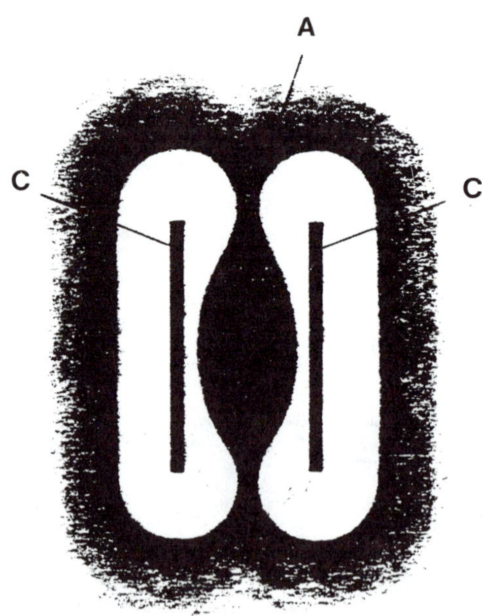

Figure 8.   Double cathode with annular anode, glow shown shaded (Badareu and Popescu, 1958).

two cathode plates. Figure 9 shows the result of a field measurement for different values of voltage and current. The resolution of this method has been improved by Warren (1955). Badareu et al. (1960) used probe measurements to compare the potential distribution in front of the hollow cathode with that at a plane cathode. Metel and Nastyukha (1981) have used Stark-splitting to measure the electric field distribution in front of a hollow cathode operated in Hydrogen. More recently techniques based on Rydberg state spectroscopy have been developed allowing field measurements in a variety of gases (Doughty and Lawler, 1984; Ganguly and Garscadden, 1985).

ION TRANSPORT THROUGH THE CATHODE FALL

It was first discussed by Sena (1946) that the transport of ions in its own background gas is governed by charge exchange collisions. This process slows down the ions, but at the same time generates fast atoms. Badareu et al. (1960) applied this concept to the transport of ions in the negative glow and through the cathode fall of a hollow cathode. They pointed out that the increased current density in a hollow cathode compared to a planar cathode is the consequence of both, the increased ion energy when they hit the cathode surface and also the increase of the ion current.

It should be mentioned here that the field in the negative glow of a hollow cathode is not zero but strongly depends on the operating conditions. Pahl et al. (1972) and Kirichenko et al. (1976) use probe measurements to determine the axial and radial potential distribution in the plasma of a cylindrical hollow cathode (s. Figure 10). These measurements show that at low pressures (low values of pD) the potential has its maximum in the center of the cylinder generating a force on the positive ions towards the cathode wall throughout most of the volume inside the hollow cathode. At high pressures (high values of pD) the potential maximum moves to some distance r/R from the center of the cathode, generating a slight depression in the center which acts as a weak trap for the positive ions.

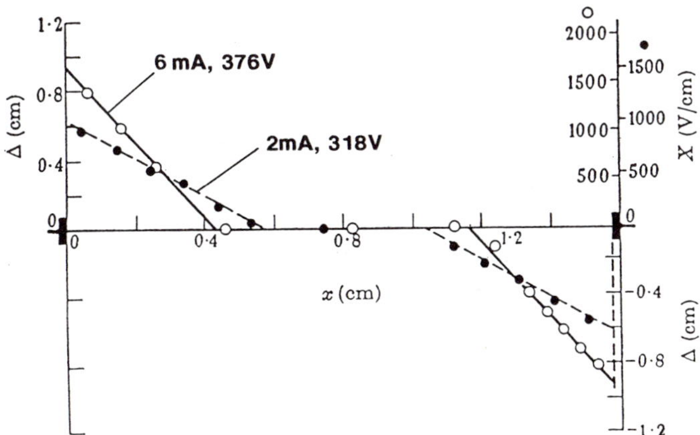

Figure 9.   Deflection of beam, Δ, and filed strength, X, as a function of position in a parallel plate hollow cathode (Little and Engle, 1954).

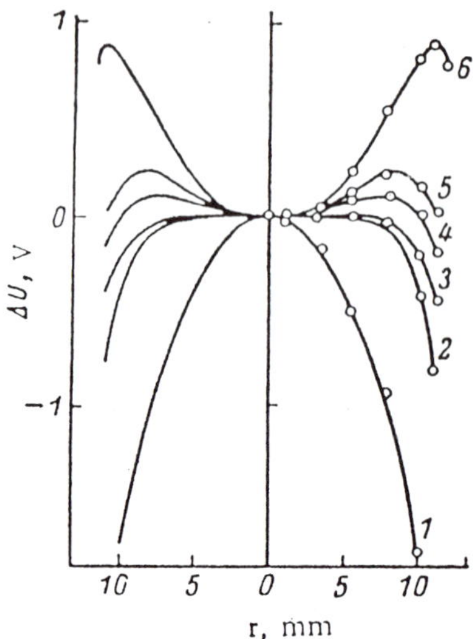

Figure 10.   Radial profile of the plasma potential in a cylindrical hollow cathode (Kirichenko et al., 1976).

PHOTONS AND METASTABLES

The role of neutral energetic particles (photons and metastables) on the efficiency of a hollow cathode discharge was discussed by Little and Engel (1954) and Oskam and Sturges (1967). Since their motion is not affected by the electric field they will move, starting from the position or origin into the full solid angle 4π. The efficiency of utilizing these particles therefore depends on the solid angle under which the cathode surface is seen from different positions inside the negative glow.

UV Radiation: The importance of photoemission depends on the intensity and spectral distribution of the UV radiation emitted from the negative glow and the material of the cathode (Oskam and Sturges, 1967). Due to the electron energy distribution with the high energy tail and the higher electron density in a hollow cathode it is expected that the influence of photoemission is more important in a hollow cathode discharge that in a glow discharge. The strong emission of VUV radiation from hollow cathode discharges has been demonstrated (Gabrielyan et al., 1983; Holmgren et al., 1984; Eichler et al., 1985).

Metastables: Lawler et al. (1983; 1985) demonstrated the importance of metastables for the emission of electrons from the cathode. In experiments in which the density of metastables was depleted in front of the cathode surface by optical excitation into a higher energy state which can decay radiatively the cathode fall voltage was increased significantly.

In hollow cathode discharges with rare gases high metastable densities ($\sim 10^{12} cm^{-3}$) have been measured (de Hoog et al., 1977). Numerous laser systems based on energy transfer collisions or on Penning ionization plus excitation of metal atoms by rare gas metastables, have been operated with hollow cathodes.

MULTI-STEP PROCESSES

Sturges and Oskam (1964;1967) discussed the importance of cumulative ionization processes to the production of electrons. The rate of such processes strongly depends on the electron and excited states density of the negative glow. Electron density measurements have been performed by many authors. Maximum electron densities achieved in continuous operation are $n_e \sim 10^{14} cm^{-3}$ (Gill and Webb, 1978; Gerstenberger, 1980). This is orders of magnitude higher than can be achieved in the negative glow of a glow discharge.

For electronic excitation processes however only the density of fast electrons is important. Handle et al. (1984) measured the density of fast electrons on the discharge axis depending on gas pressure (Figure 5) showing that in the limit of the existence of a hollow cathode discharge the density on axis decreases with pressure. This result is in agreement with the radial distribution of the spontaneous emission of specific spectral lines which can only be excited by fast electrons. Figure 11 shows the radial profiles of some Zn Ii lines which can either be excited through electron collisions or Penning collisions (Gill and Webb, 1978).

SPUTTERING

Already Paschen (1916,1933) realized that hollow cathode discharges are ideal plasma sources to study the spectra of metals at temperatures much lower than required to evaporate these materials. The mechanism found to generate the vapor pressure of these metals (either the cathode material or a sample on the inside wall of the hollow cathode) was found to be sputtering. Sputtering is a mechanism in which the momentum of a bombarding ion or atom is transferred in a wall collision to one or a few surface atoms of the cathode material. These atoms are then reflected in the asymmetric potential at the surface. It is therefore obvious that the sputtering coefficient has some threshold with respect to the momentum of the incoming ion and increases

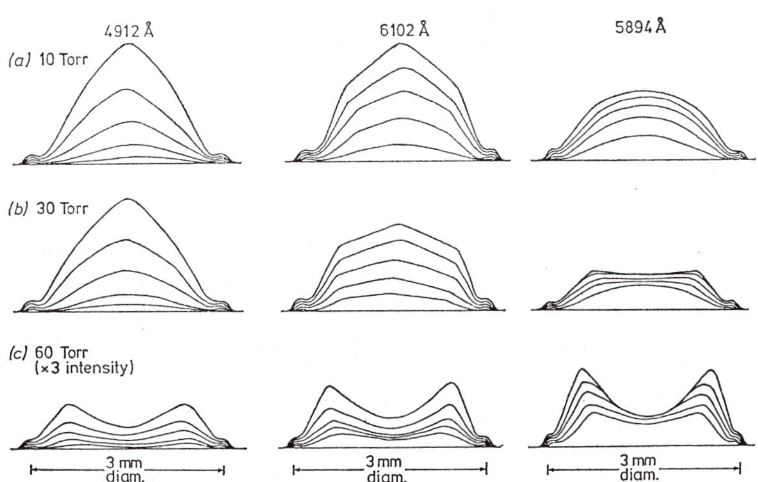

Figure 11.   Radial profile of Zn II spontaneous emission in a cylindrical hollow cathode for discharge currents from 0-1000 mA (Gill and Webb, 1978).

strongly with increasing atomic weight of the ion and ion energy. The sputtering coefficient also varies significantly for different materials.

In a hollow cathode this material either diffuses back to the wall or the metal atoms are ionized and return to the cathode surface as positive ions, thus contributing to further sputtering of the cathode. In cathodes made of materials with high sputtering coefficient this process can change the geometry until the cathode approaches a stable configuration. White (1959) found this configuration to be of nearly spherical shape with an aperture to sphere radius ratio of $\sim$0.25 (see Figure 12). Designers of Cu II lasers using copper hollow cathodes have used such geometries to improve the lifetime of such lasers (van der Hoeven and Jones, 1984).

It is interesting to realize that the sputtering rate is a direct consequence of the current density in the hollow cathode. Areas in which material is removed have current densities larger than the average current density and the final geometry is a configuration with approximately constant current density inside the cathode.

The dominant influence of the sputtered metal vapor on the hollow cathode characteristic stems from the fact that the metal atoms have much lower ionization energies than the atoms or molecules of most fill gases. If operated with rare gases Penning ionization can become a major ionization mechanism (White, 1959; Musha, 1962).

Especially if operated with one of the lighter rare gases (helium or neon) and a cathode material with high sputtering coefficient (copper or molybdenum) the hollow cathode can have a negative V-I characteristic in an intermediate current regime caused by sputtering (Musha, 1962). At low current densities sputtering through the rare gas atoms provides a low density of sputtered atoms (and ions) in the plasma. With increasing current the density of sputtered atoms and ions increases until the dominant sputtering rate is caused by sputtering through the metal ion which are much heavier than the rare gas ions. This transition causes the negative V-I characteristic.

Sputtering also causes a hysteresis in the current voltage characteristic. Figure 13 shows a characteristic for a system with four parallel cylindrical copper cathode measured for increasing and decreasing current (Salk and Schaefer, 1977).

Metal vapor densities have been measured for several cathode materials and fill gases. In copper hollow cathodes copper vapors densities up to 10^{14}cm^{-3} have been reported (deHoog et al., 1977). The radial dependence of the vapor

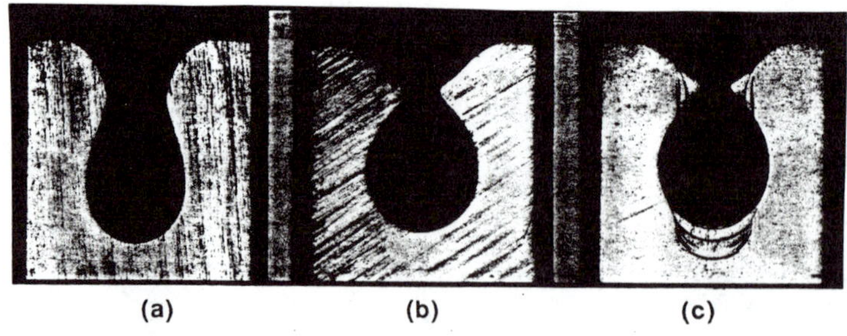

Figure 12.   Formation of the hollow cathode cavity by sputtering. The original hole was cylindrical; a) 100 hr; b) 500 hr; c) is b) after etching (White, 1959)

density depends on the pressure of the fill gas. At low pressures the metal vapor density is uniform within the negative glow while at higher pressures a maximum exist close to the boundary to the cathode fall (Vaessen et al., 1977; Warner et al., 1979).

HOLLOW CATHODE GEOMETRY

A large variety of hollow cathode geometries have been used for a variety of investigations and applications as shown in Figure 1. First investigations comparing the discharge characteristics for a variety of geometries have already been performed by Guentherschulze (1923). The two most common geometries used to investigate properties of hollow cathode discharges are the parallel plate cathode and the cylindrical hollow cathode. The parallel plate cathode has the advantage that the potential of each plate can be adjusted independently and that the plate distance can be adjusted. Therefore it is the ideal setup to investigate the transition from a plane to a hollow cathode, to measure the potential distribution, and to investigate the influence of the different mechanisms contributing to the hollow cathode effect.

Higher plasma densities are achieved with cylindrical hollow cathodes. This property in combination with the cylindrical symmetry makes the optimum geometry for many applications. We will therefore only discuss the cylindrical hollow cathode. Although the most important parameter of a cylindrical hollow cathode is its diameter D there are a number of additional geometric parameters which influence the discharge characteristic such as the cathode length, both sides open or one side closed, the aperture, the shape and distance of the anode, and shapes of insulating walls.

The optimization of the cathode length requires the knowledge of the current distribution along the axis of the cathode (Weizel and Mueller, 1956; Howoraka and Pahl, 1972). Figure 14 shows the experimental setup using a

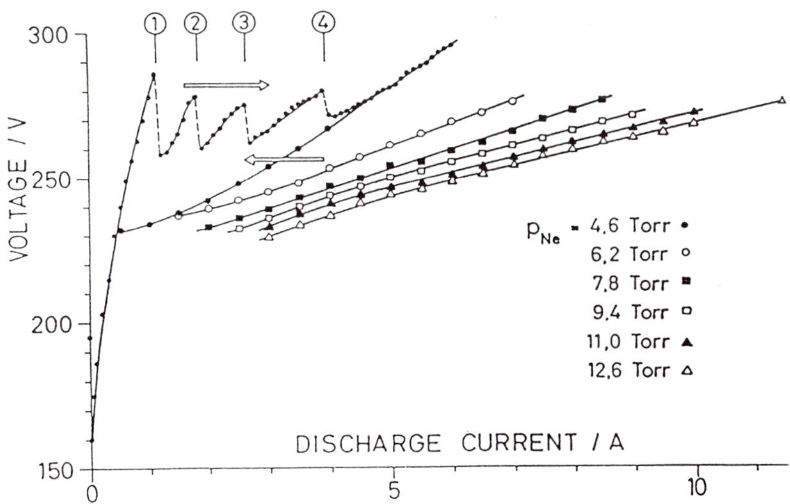

Figure 13.   Voltage current characterization for four parallel operated cylindrical copper hollow cathodes measured for increasing and decreasing current. The numbers indicate the initiation of the hollow cathode discharge in the individual hollow cathode (Schaefer and Salk, 1977).

segmented cathode. The axial current distribution with the total current and the pressure as the variable parameter is shown in Figure 15. It demonstrates that the maximum of the current density moves further into the cathode if the current is increased or the pressure is decreased.

The influence of the cathode aperture was investigated by Geranyi and Metel (1984). Sturges and Oskam (1966) investigated the influence of the anode fall on the hollow cathode discharge characteristic.

THE INFLUENCE OF A MAGNETIC FIELD

The parallel plate double cathode allows to investigate the influence of a magnetic field applied in both, parallel and transverse directions to the electric field in the cathode fall region (Popovici and Somesan, 1966). In both cases an amplification of the discharge current is observed; however, the mechanisms are quite different.

Parallel Field: In this field configuration the acceleration process of the electrons in the cathode fall is not influenced by the magnetic field, and also not their motion with constant velocity in the negative glow until their first collision. In any collision, elastic or inelastic, the electrons acquire a velocity component transverse to the electric field. This component contributes to losses of fast electrons to the outside of the hollow cathode. These losses are more severe in the outer parts of the cathode. This is one of the reasons why the cathode fall has its shortest thickness in the center of the cathode as shown in Figure 8. The main influence of the magnetic field parallel to the electric field is to reduce the losses of fast electrons to the outside of the cathode. Its effect is therefore more pronounced if the ratio of cathode diameter to plate separation decreases.

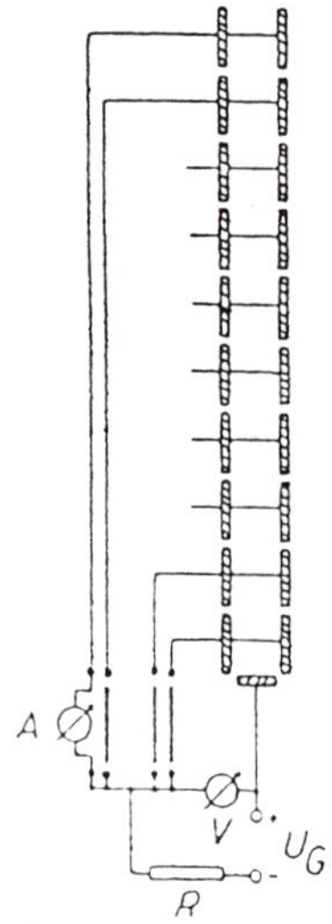

Figure 14. Schematic experimental setup with a segmented zylindrical hollow cathode (Weizel and Mueller, 1956).

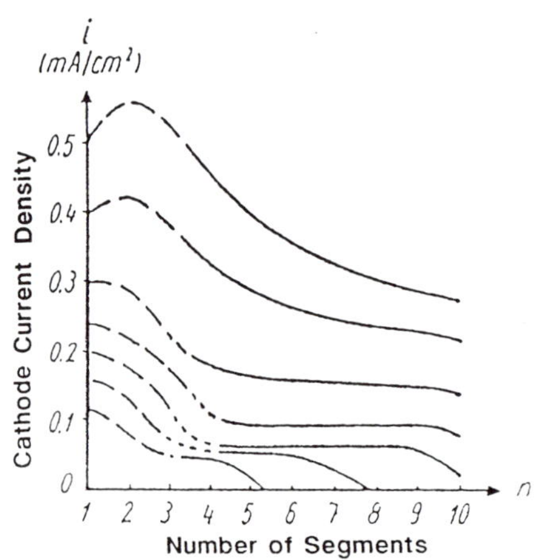

Figure 15. Current density distribution in the segmented hollow cathode shown in figure 14 with the discharge current as the variable parameter (Weizel and Mueller, 1956).

Transverse Field: A transverse magnetic field forces the electrons accelerated in the cathode fall into cycloide type trajectories in which they are directed back to the cathode surface. Abramovich et al. (1966) pointed out that the effect of the magnetic field by itself is similar to the effect of a hollow cathode generating pendular electrons. In both cases fast electrons moving back towards the cathode surface reduce the positive space charge in front of the cathode. The combination of a hollow cathode with a transverse magnetic field is therefore expected to enhance the HCD effect on the discharge characteristic. In a parallel plate hollow cathode the ExB drift will, at one boundary of the plates, move the electrons out of the hollow cathode allowing some losses of fast electrons. This effect can for specific operating conditions even reduce the hollow cathode effect (Badareu et al., 1967). An axial magnetic field in cylindrical hollow cathode however generates a tangential ExB drift and does not generate losses of fast electrons. A cylindrical hollow cathode is, therefore, the better device to investigate the influence of a transverse magnetic field on a hollow cathode discharge.

Experimental investigations have been performed on the magnetic field dependence of discharge parameters such as current, voltage or discharge power and on discharge characteristics and pressure dependencies with the magnetic field as the variable parameter. Figure 16 shows as an example the magnetic field dependence of the discharge voltage with the pressure as the variable parameter (Trachenko and Tyutyunnik, 1972). A decrease of the voltage by a factor of ~0.6 is observed which is achieved with lower magnetic fields if the discharge is operated at lower pressures.

The transverse magnetic field also restricts the trajectories of the beam electrons to a volume closer to the surface of the cathode. This can be seen from its influence on the space dependence of excitation rates for specific

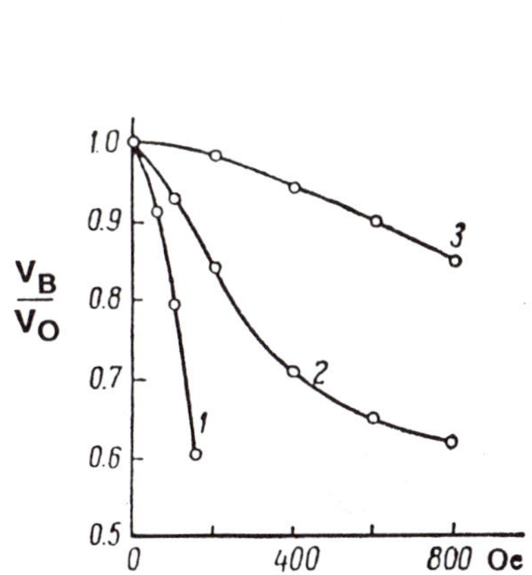

Figure 16. Relative decrease of cathode fall voltage versus magnetic field in a discharge in helium; 1) 0.1 torr; 2) 0.5 torr; 3) 3.0 torr (Tkachenko and Tyutyunnik, 1972).

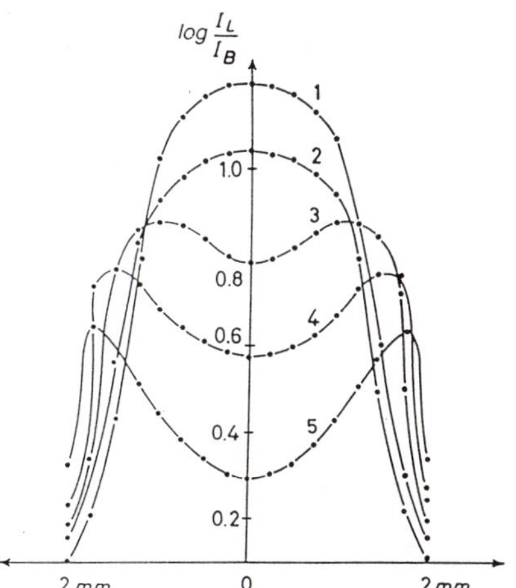

Figure 17. Relative intensity distribution of the He I line 318.8 nm versus radius with the magnetic field as the variable parameter; 1) 0 Oe; 2) 600 Oe; 3) 680 Oe; 4) 850 Oe; 5) 1150 Oe (Rudnevsky and Maksimov, 1985).

spectral lines as shown in Figure 17 (Rudnevski and Maksimov, 1985). This also demonstrates that modest transverse magnetic fields can be used to homogenize the plasma inside the hollow cathode.

PULSED HOLLOW CATHODE DISCHARGE INITIATION

The common problem of pulsed discharge initiation in all hollow cathode devices is that a plasma has to be generated inside the hollow cathode before the cathode fall can develop. Unless very high voltages are available during discharge initiation there is essentially no field inside the hollow cathode before this plasma has developed (see Fig. 18). All methods to achieve fast and low jitter discharge initiation are therefore based on some additional mechanism to generate carriers inside the hollow cathode. Some of these mechanisms are presented in Table I.

Table I. Pulsed HCD Initiation

Predischarge	
DC + Pulsed	Aksenov et al., 1970.
DC + Pulse Charged	Schaefer et al., 1984.
Pulsed	Mechtersheimer et al., 1986.
Plasma Injection	Koltypin et al., 1971.
Magnetic Field	Koltypin et al., 1971.
Surface Discharge	Bloess et al., 1983.
Laser, Flashlamp	Kirkman and Gundersen, 1986.
Electron Beam	Billault et al., 1987.
	Gundel et al., 1989.

Since most of these methods have been developed to operate some type of hollow cathode switch, they will be discussed in other papers of this workshop.

A few experiments have been performed on the space-time dependent development of the plasma inside the hollow cathode. First investigations were performed using segmented cylindrical cathodes in which the time dependent current was measured for each segment (Stoeri et al., 1975). A pulsed voltage was applied without using an additional mechanism to generate carriers inside the cathode. After a statistical delay time the discharge started at the entrance

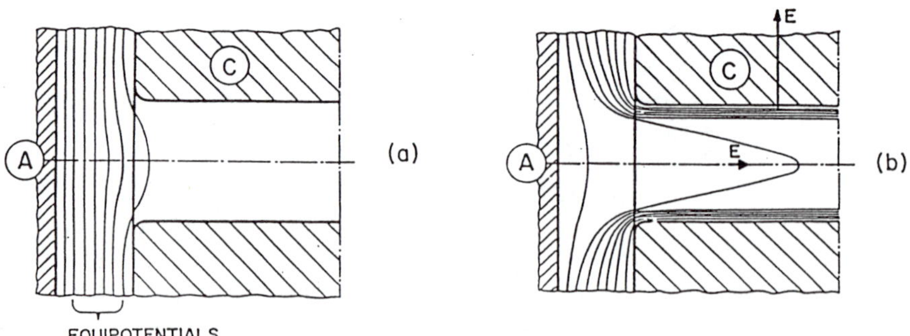

Figure 18.   Schematic potential distribution in a cylindrical hollow cathode, a) before breakdown and b) steady state discharge.

of the cathode. The propagating velocity of the discharge pulse from the entrance into the cathode was in the range from 0.5 to 2.5 times 10^6 ms^{-1}. The mechanism responsible for the propagation was found to be photoemission.

Other experiments on the space-time dependent development of the plasma in a pulsed HCD were based on Streak photographs of the light emitted from the negative glow through side windows in a cylindrical hollow cathode (Schaefer and Schoenbach, 1987; Schaefer and Wages, 1988). In these investigations a high voltage pulse ($\sim$1–2 ns risetime) was applied to a hollow cathode which was already operated at low currents (few mA). The current increased by at least four orders of magnitude as a consequence of the high voltage pulse limited by the transmission line pulser. Figure 19 shows the propagation of the plasma inside the hollow cathode. At low preionization currents ($<$0.5mA) the discharge starts at the cathode entrance while at higher preionization currents ($>$5mA) the discharge starts inside the cathode and spreads in both directions with a velocity of up to 5 10^6m s^{-1}. Under these conditions the delay time between the application of the high voltage pulse and the high current discharge was found to be 3 ns and the current risetime (0.1 to 0.9 I_{max}) was found to be 2 ns (Schaefer et al., 1984). The application of an axial magnetic field did not improve this performance if a sufficient preionization current was used (Schaefer et al., 1989).

The dynamics of the initiation of a hollow cathode discharge is at this time not well understood. No measurements of the time dependent cathode fall potential distributions exist.

CONCLUSION

The definition of a steady state hollow cathode discharge is well established through distinct external properties of the discharge. One is the current voltage characteristic which is shifted to lower voltages (down to 0.5) and significantly higher currents (up to 10^3) compared to discharges with a planar cold cathode. The other important feature is the confinement of the negative glow inside the

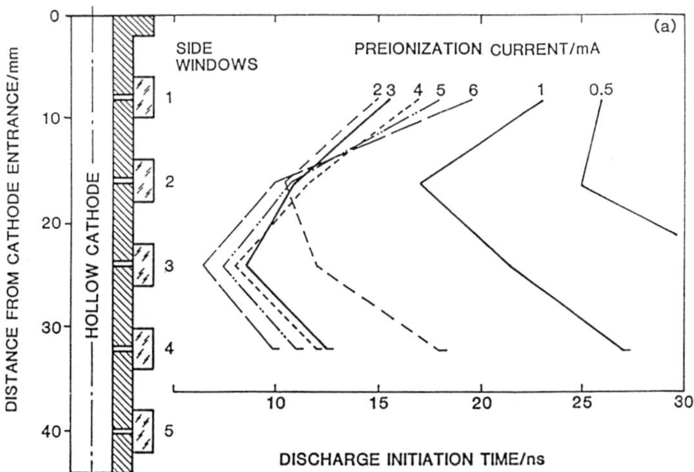

Figure 19. Discharge initiation time versus distance from the cathode entrance. The variable parameter is the preionization current (Schaefer et al., 1988).

hollow structure including the merging of the glows from opposite surfaces with significantly higher emission intensities.

Several mechanisms have been considered to contribute to this hollow cathode effect. A summary was given in the introduction. At this time it is agreed that the pendular electrons play a dominant role in the hollow cathode geometry while all other effects can enhance this effect depending on specific operating conditions.

One indication for this conclusion is that the effect of pendular electrons is also found in discharges with planar electrodes and a transverse magnetic field, a discharge condition in which the other effects do not apply. This also makes the hollow cathode operation extremely sensitive to small geometric changes since the electron trajectories strongly depend on the electric field distribution.

A direct consequence of the pendular electrons is the reduction of the cathode fall thickness and an increase of the electric field at the cathode surface and as a consequence higher ion energies at the cathode surface.

The influence of the other mechanisms such as electron emission through collisions of metastables with the cathode surface, photoemission, and sputtering can be influenced through variations of operating parameters such as fill gas and pressure, cathode material and cathode geometry. The evaluation of the importance of multistep processes is very difficult since most cross section for processes starting from excited states are missing.

Especially the dynamics of the hollow cathode discharge operation is at this time not well understood since some essential experimental investigations are missing.

Due to the large number of hollow cathode parameters it is at this time very difficult to compare experimental results of different authors or to select experimental investigations which are suitable for checking the validity of theoretical and numerical investigations.

Overall one can see that a hollow cathode discharge is a physical system of extreme complexity. The individual processes contributing to the hollow cathode effect are quite well understood but the importance of any of these processes is not clear for a specific situation.

ACKNOWLEDGEMENT

This work was supported by AFOSR, ARO and NSF.

REFERENCES

Abramovich, L. Yu., Kylarfeld, B.N. and Nastich, Yu. N., "Superdense Hollow Cathode Glow Discharge," Soviet Phys.-Tech Phys., vol. 25, pp. 528-532, 1966.

Aksenov, I.I., Belous, V.A., and Smirnov, S.A., "Controlled Gas Discharge Instrument with a Cold Cathode," translated from Pribory i Technica Experimenta, vol. 4, pp. 184-185, 1970.

Badareu, E. and Popescu, I. "Research on the Double Cathode Effect," J. Electr. Contr., Vol. 4, pp. 503-514, 1958.

Badareu, E., Popescu, I. and Iova, T. "Beitraege zur Klaerung des Mechanismus des Doppelkathodeneffektes," Ann. Physik, vol. 7-5, pp. 308-326, 1960.

Badareu, E., Popovici, C., and Somesan, M., "Contributions to the Study of Hollow Cathode Effect," Rev. Roum. Phys., vol. 12, pp. 1-11, 1967.

Badareu, E. and Waechter, F. "The Hollow Cathode Discharge in Mercury Vapour," J. Electr. Contr., vol. 4, pp. 539-543, 1958.

Bespalov, O.G., Knyazyatov, A.S., Nastyukhova, A.I., Smirnov, P.A. and Udovenko, A.N., "High-Current Switching Devices with a Hollow Cold Cathode III," Sov. Phys.-Tech. Phys., vol. 26, pp. 1051-1053, Sept. 1981.

Bespalov, O.G., et al., "Six-Anode High-Power Switching Device" translated from Pribory Teknika Experimenta, vol. 1, pp. 149-151, 1979.

Billault, P. et al., "Pseudospark Switches," CERN Report, CERN 87-13.

Bloess, D. et al., "The Triggered Pseudo-Spark Chamber as a Fast Switch and as a High Intensity Beam Source," Nucl. Instr. Meth., vol. 205, pp. 173, 1983.

Borodin V.S., and Kagan, Yu M., "Investigation of Hollow-Cathode Discharge. I. Comparison of the Electrical Characteristics of a Hollow Cathode and a Positive Column," Sov. Phys.-Tech. Phys., vol. 11, pp. 131-134, 1966.

Burkes, T.R., "Switches for Directed Energy Weapons - An Assessment of Switching Technology in the USSR - Supplement II," Final Report, Contract No. DAAK 21-82-C-0111, Feb. 1985.

Caroli, S., ed., "Improved Hollow Cathode Lamps for Atomic Spectroscopy," John Wiley and Sons, New York, 1985.

Crane, J.K. and Verdeyen, J.T. "The Hollow Cathode Helium-Fluorine Laser," J. Appl. Phys., vol. 51, pp. 123-129, Jan. 1980.

Chebotayev, V.P., "Operating Condition of an Optical Maser Containing a Helium Neon Mixture," Radio Eng. Electron Phys., vol. 10, pp. 314-316 and pp. 316-318, (1963) 1965.

Christiansen, J. and Schultheiss, C., "Production of High Current Particle Beams by Low Pressure Spark Discharges," Z. Physik, vol. A290, p. 36, 1979.

Ciobotaru, D., "Der Doppelkathodeneffect und die Randbedingungen" Rev. Roum. Phys., vol. 17, pp. 1191-1198, 1972.

Circel, H.J.,"High Power Excimer Lasers," Proc. Conf. LASERS 1987, Lake Tahoe, 1987.

Csillag, L., Janossy, M., Rosza, K. and Salamon, T., "Near Infrared CW Laser Oscillation in Cu II," Phys. Lett., vol. 50 A, pp. 13-14, 1974.

Den Hartog, E.E., O'Briam, T.R. and Lawler, "Electron Temperature and Density Diagnostics in a Helium Glow Discharge," Phys. Rev. Lett., vol. 62, pp. 1500-1503, 1989.

Den Hartog, E.E., Doughty, D.A. and Lawler, J.E., "Laser Optogalvanic and Fluorescence Studies of the Cathode Region of a Glow Discharge," Phys. Rev. A, vol. A38, pp. 2471, 1988.

de Hoog, F.J., McNeil, J.R., Collins, G.J. and Persson, K.B., "Discharge Studies of the Ne-Cu Laser," J. Appl. Phys., vol. 48, pp. 3701-3704, 1977.

Deininger, W.D., Aston, G., and Pless, L., "Hollow Cathode Plasma Source for Active Spacecraft Charge Control," Rev. Sci. Instr., vol. 58, pp. 1053-1062, 1987.

Doughtry, E.H., Donohue, D.L., Slevin, P.J. and Harrison, W.W., "Surface Sputter Effects in a Hollow Cathode Discharge," Analytical Chem., vol. 47 pp. 683-688, 1975.

Falcone, R.W. and Pedrotti, K.D., "Pulsed Hollow-Cathode Discharge for

Extreme-Ultraviolet Lasers and Radiation Sources," Opt. Lett., vol. 7, pp. 74-76, 1982.

Gerstenberger, D., Solanki, R. and Collins, G.J., "Hollow Cathode Metal Ion Lasers," IEEE J. Quantum Electron., vol. QE-16, pp. 820-834, 1980.

Gewartowski, J.W. and Watson, H.A., "Principles of Electron Tubes." Princeton, N.J., Van Nostrand, 1965, p. 56.

Gill, P. and Webb, C.E., "Electron Energy Distribution in the Negative Glow and their Relevance to Hollow Cathode Lasers," J. Phys. D: Appl. Phys., vol. 10, pp. 299-311, 1977.

Gill, P. and Webb, C.E., "Radial Profiles of Excited Ions and Electron Density in the Hollow Cathode He/Zn Laser," J. Phys. D: Appl. Phys., Vol. 11 , pp. 245-254, 1978.

Grechanyi, V.G. and Metel, A.S., "Influence of a Magnetic Field on the Cathode Current Distribution in a Glow Discharge with Oscillating Electrons," Translated from Teploflzika Vysokikh Temperatur, vol. 21, no. 6, pp. 1071-1075, Nov. - Dec. 1983.

Grechanyi, V.G., and Metel, A.S., "Hollow Cathode Glow Discharge with Vacuum Operation of the Cathode Holow," Translated from Teploflzika Vysokikh Temperatur, vol. 22, pp. 444-448, 1984.

Green, J.M., and Webb, C.E., "The Production of Excited Metal Ions in Thermal Energy Charge Transfer and Penning Reaction," J. Phys. B: Atom. Molec. Phys., vol. 7, pp. 1698-1711, 1974.

Guentherschulze, A., "Die Stromdichte des normalen Kathodenfalles," Z. Physik, vol. 19, pp. 313-332, 1923.

Gundel, H., Riege, H. Handerek, J. and Zioutas, K., "Low Pressure Hollow Cathode Switch Triggered by a Pulsed Electron Beam Emitted from Ferroelectrics," Appl. Phys. Lett., vol. 54, pp. 2071-2073, 1989.

Handle, F., Lindinger, W., Howorka, F., and Pahl, M., "Density of Fast Electrons on the Axis of a Zylindrical Hollow-Cathode Discharge," Beitr. Plasmaphys., vol. 24, pp. 407-416, 1984.

Helm, H., "Experimental Evidence of the Existence of the Pendel Effect in a Low a Pressure Hollow Cathode Discharge in Argon," Z. Naturforschung, vol. 27a, pp. 1812-1819, 1972.

Holmgren, D.E., Falcone, R.W., Walker, D.J, and Harris, S.E., "Measurement of Lithium and Sodium Metastable Quartet Atoms in a Hollow Cathode Discharge," Opt. Lett., vol. 9, pp. 85-87, 1984.

Howorka, F. and Pahl, M., "Experimentelle Bestimmung innerer und aeusserer Parameter des negativen Glimmlichts einer zylindrischen Hohlkathodeneutladung in Argon," Z. Naturforsch., vol. 27, pp. 1425-1433, 1972.

Kagan, Yu. M., Cohen, C., and Avivi, P., J. Appl. Phys., vol. 63, pp. 60, 1988.

Kielkopf, J.F., "Pulsed Hollow Cathode Light Sources," Spectrochimica Acta, vol. 26B, pp. 371 -390, 1971.

Kirichenko, V.I., Tkachenko, V.M. and Tyutyunnik, V.B., "Influence of Dimensions, Cathode Material, and Gas on the Optimum Pressure for a Cylindrical Hollow Cathode Glow Discharge." Sov. Phys.-Tech. Phys., vol. 21, pp. 1080-1086, 1976.

Kirkman, G.F. and Gundersen, M.A., "Low Pressure Light Initiated Glow Discharge Switch," Appl. Phys. Lett., vol. 49, pp. 494-495, 1986.

Koltypin, E.A., Nastyukha, A.I., and Smirnov, P.A., "Rectification in a Low-Pressure Arc with a Hollow Cold Cathode. I. and II.," Sov. Phys.-Tech. Phys., vol. 15, pp. 1703-1709, and pp. 1710-1715, April 1971.

Kostin, V.N. and Tkachenko, V.M., "Effect of a Constant Magnetic Field on a Stationary Corona Discharge," Soviet, Phys.-Tech. Phys., vol. 9, pp. 968-972, 1965.

Kuzmichev, A.I. and Shendakov, A.I., "Restoration of the Electric Strength of Solid Cold Cathode Gas Discharge Switching Instrument," translated from Radiotekhnika i Electronica, vol. 23, pp. 2247-2250, 1978.

Lawler, J.E. and Doughty, D.K., "Amplification of Optogalvanic Effects in the Cathode Fall," Proc. 4th IEEE Pulsed Power Conf., pp. 33-36, 1983.

Lawler, J.E. and Guenther, A.H., "Application of Optogalvanic Effects in Opening Switches," Proc. 3rd IEEE Int. Pulsed Power Conf., Albuquerque, New Mexico, pp 147-150, 1984.

Little, P.F. and von Engle, A., "The Hollow Cathode Effect and the Theory of Glow Discharges," Proc. Roy. Soc., vol. A224, pp. 209-227, 1954.

Lompe, R., Seeliger, R. and Wolter, E., "Untersuchungen an Hohlkathoden," Ann. Phsik, vol. 36, pp. 9-37, 1939.

Mavrodineanu, R., "Hollow Cathode Discharges - Analytical Applications," J. Research NBS, vol. 89, pp. 143-185, 1984.

Mechtersheimer, G., Kohler, R., Lasser, T., and Meyer, R., "High Repetition Rate, Fast Current Rise Pseudospark Switch," J. Phys. E, vol. 19, p. 466, 1986.

Menown, H. and Newton, B., "A Multigap Double Ended Hydrogen Thyratron," Conf. Record, 11th IEEE Power Modulator Symp., IEEE, New York, NY, 1973.

Metel, A.S., "Effect of Ionization in the Cathode Layer on the Characteristics of a Penning Discharge: I. Hollow Cathode Discharge," Sov. Phys.-Tech. Phys., vol. 30, pp. 1133-1136, 1985.

Metel, A.S., and Nastyukha, A.I., "Estimation of Energy Losses of Fast Electrons in the Hollow Cathode Discharge Plasma," Proc. 15th ICPIG, vol. I, p. 99, Lenningrad 1981.

Metel, A.S. and Nastyukha, A.I., "Hollow Cathode Discharge in the Transverse Magnetic Field with a Foreign Body in the Cavity," Proc. 15th ICPIG, Vol. I, p. 100, Leningrad 1981.

Mizeraczyk, J., "Electron Energy Distribution in Helium in a High Voltage Discharge in a Hollow Cathode Used for Lasers," J. Phys. D: Appl. Phys., vol. 17, pp. 1647-1656, 1984.

Musha, T., "Cathode Sputtering in Hollow Cathode Discharges," J. Phys. Soc. Japan, vol. 17, pp. 1440-1446, Sept. 1962.

Novikov, A.A., "Nonself-Sustaining Hollow Cathode Cold Discharge," Soviet Phys.-Tech. Phys., vol. 16, pp. 1216-1217, 1972.

Okhmatovskii, G.V., "Fast Electron Current in a Hollow- Cathode Glow Discharge," Sov. Phys.-Tech. Phys., vol. 23, pp. 552-554, 1978.

Pahl, M., Lindinger, W., and Howorka, F., "Massenspectrometrie am negativen Glimmlicht einer zylindrischen Hohlkathode," Z. Naturforsch., vol. 27, pp. 678-692, 1972.

Popovici, C., Krejci, V., and Stirand, "Influence of Hollow Cathode with Magnetic Field on Ionization Waves in a Neon Discharge Plasma," Rev. Roum. Phys., vol. 13, pp. 423-426, 1968.

Popovici, C., and Somesan, M., "On the Emission Spectrum of the Negative Glow Plasma of a Hollow Cathode Discharge in Magnetic Field." Appl. Phys. Lett., vol. 8, pp. 103-105, 1966.

Rudnevsky, N.K. and Maksimov, D.E., "Hollow Cathode Discharge in a Magnetic Field," in "Improved Hollow Cathode Lamps for Atomic

Spectroscopy," S. Caroli, Ed., John Wiley and Sons, New York 1985.

Salk, J. and Schaefer, G., "Copper Ion Laser Action in Cylindrical Copper Hollow Cathodes," Topical Conf. Quantum Opt., German Phys. Soc., 1977.

Schaefer, G., " Fast Plasma Mixing - A New Excitation Method for CW Gas Lasers," IEEE J. Quantum Electron., vol. QE-22, pp. 2022-2025, 1986.

Schaefer, G., Husoy, P.O., Schoenbach, K.H.,and Krompholz, H., "Pulsed Hollow Cathode Discharge with Nanosecond Rise Time," IEEE Trans. Plasma Sci., vol. PS-12, pp. 271-274, Dec. 1984.

Schaefer, G. and Schoenbach, K.H, "Fast Pulsed Hollow Cathode Discharges - Time and Space Dependence and Influence of an Axial Magnetic Field," Proc. AIAA Aerospace Science Meeting Reno, Nevada, Jan. 1987.

Schaefer, G. and Wages, M., "Space-Time Dependent Development of the Plasma in Pulsed Hollow Cathode Discharges," IEEE Trans. Plasma Sci., vol. 16, pp. 54-56, 1988.

Schuler, H., "Uberine neue Lichtquelle und ihre Anwendungsmoglichkeiten," Zeitschr. für Physik, vol. 35, pp. 323-337, 1926.

Sena, L., J. Phyf., USSR, Vol. 10. pp. 179-182, 1946.

Stoeri, H., Maerk, T.D., Allis, W.P., and Pahl, M., "Investigation of the Breakdown in an Argon Low Pressure Hollow Cathode," Proc. I, p. 96, 15th ICPIG, Leningrad 1981.

Sturges, D.J. and Oskam, H.J., "Studies of the Properties of Hollow Cathode Glow Discharges in Helium and Neon," J. Appl. Phys., vol. 35, pp. 2887-2894, 1964.

Sturges, D.J. and Oskam, H.J., "Hollow-Cathode Glow Discharge in Hydrogen and Noble Gases," J. Appl. Phys., vol. 37, pp. 2405-2412, 1966.

Sturges, D.J. and Oskam, H.J., "A Qualitative Theory of the Medium Pressure Hollow Cathode Effect," Physica, vol. 37, pp. 457-466, 1967.

Tkachenko, V.M., and Tyutyunik, V.B., "Hollow-Cathode Discharge in a Magnetic Field," Soviet Phys.-Tech. Phys., vol. 17, pp. 49-52, 1972.

Vaessen, P.H.M., de Hoog, F.J., and Mc Neil, J.R., "Temperature of Neutral Copper and Neon Atoms in a Hollow Cathode Laser Discharge," Phys. Lett., vol. 68A, pp. 204-206, 1978,

Van der Hoeven, C.J. and Jones, E.G., "Short Active Length CuII Laser Excited by a Spherical Hollow Cathode Discharge," Opt. Commun. vol. 52, pp. 292-294, 1984.

Warner, B.E., Persson, K.B., and G.J. Collins, "Metal- Vapor Production by Sputtering in a Hollow Cathode Discharge: Theory and Experiment," J. Appl. Phys., vol. 50, pp. 5694-5703, 1979.

White, A.D., "New Hollow Cathode Glow Discharge," J. Appl. Phys., Vol. 30, pp. 711-719, 1959.

Weizel, W., and Mueller, G., "Die Verteilung der Stromdichte ueber die Kathode einer Glimmentladung," Ann. Physik, vol. 6-18, pp. 417-440, 1956.

Willet, C.S., "Introduction to Gas Lasers: Population Inversion Mechanisms." Pergamon Press, 1974.

H. L. Witting, "Hollow Cathode Discharge with Thermionic Cathodes," J. Appl. Phys., vol. 42, pp. 5478-5482, Dec. 1971.

Znamenskii, V.B., "An Investigation of the Concentration Dependence of Metastable Helium and Neon Atoms on the Parameters of a Hollow Cathode Discharge," Opt. Spectrosc., vol. 25, pp. 7-10, 1969.

CATHODE-RELATED PROCESSES IN HIGH-CURRENT DENSITY, LOW PRESSURE

GLOW DISCHARGES

Werner Hartmann

M. A. Gundersen

Physikalisches Institut
Universität Erlangen-Nürnberg
8520 Erlangen, FRG

Dept. of Electrical Engineering
University of Southern California
Los Angeles, CA 90089, U.S.A.

INTRODUCTION

In transient high current, low pressure gas discharges like the pseudospark /1/ and other hollow cathode switching devices /2/, a diffuse glow discharge mode has been reported with current densities of the order of 10^4 A/cm^2 at cathode surface - plasma interaction cross-sectional areas of the order of 1 cm^2 /3-7/. This extraordinary high current density from an initially cold metal cathode surface cannot be explained in terms of secondary electron emission yield due to particle impact or the photo effect. To find an explanation for this high cathode emission, a much more detailed investigation of the plasma - cathode interaction is necessary, taking into account non-ideal cathode surface properties which are usually neglected in gas discharge modelling /8/. In this contribution, an attempt is made to explain the experimentally observed diffuse, non-arcing glow mode of very high current density in terms of field-enhanced thermionic electron emission from the initially cold cathode surface. Heating of a thin layer at the cathode surface is thought to be due to ion bombardment by ions which are accelerated in the cathode fall voltage drop. Although a strong ion pumping effect is expected to occur at the plasma boundary facing the cathode, long pulse duration current pulses can be transfered due to the thermal desorption of adsorbed gas layers from the cathode surface; this leads to an effective ion replenishment through electron impact ionization of the resulting dense gas layer.

The discharge model presented hereafter requires a very high degree of symmetry and discharge uniformity during the initial phase of the discharge, and involves a build-up of the discharge in several successive steps, which can only be accomplished in a certain range of discharge parameters.

From the comparison of predictions which can be made from the model with experimental data, conclusions can be drawn on the nature of the dominant electrode erosion mechanism, which in turn leads to a set of preferred material parameters for the cathode material in order to achieve a long electrode lifetime in high-power plasma devices. This makes it possible to design high-power plasma devices for switching, charged particle (particularly electron) beam production, microwave generation, etc. Peak power levels may be achievable exceeding those of existing devices by orders of magnitude, while maintaining a reasonable electrode lifetime and eliminating the need for externally heated thermionic oxide cathodes.   The proposed discharge mechanism is general enough to be also applied, within certain limitations, to high-pressure gas

Physics and Applications of Pseudosparks
Edited by M. A. Gundersen and G. Schaefer
Plenum Press, New York, 1990

discharge devices like spark gaps, and to explain the erosion rate of electrodes in those cases as well as providing a 'tool' for the design of cathode materials and cathode - working gas combinations that give optimum performance and minimal electrode wear.

EXPERIMENTAL WORK

Experimental investigations were done in setups similar to that described in ref. /3/ and shown in fig. 1. The discharge geometry consists of 2 metal cups with the plane bottoms facing each other at a distance of 3-5 mm. Both electrode faces are usually made of metals like molybdenum, and have a circular center hole of 3-5 mm in diameter each. The electrodes are held in place, insulated, and enclosed in a vacuum environment with a glass or alumina cylinder sealed with O-rings. The discharge circuit comprises of a pulse-discharge capacitor connected to the main electrodes through low-inductance leads, a high-voltage power supply, and eventually a low-inductance series load resistor.

The gas discharge can be initiated by either releasing photoelectrons inside the hollow cathode through the injection of an intense UV-light pulse (compare fig. 1 and ref. /3/), or by the injection of electrons into the hollow cathode. In the latter case, the electrons are produced in a pulsed glow discharge /5,9/.Typical operating parameters are:
-- cold cathode (room temperature)
-- low pressure hydrogen (30 - 60 Pa)
-- electrode distance, electrode hole diameter: 3-5 mm
-- hold-off voltage: 5 -30 kV
-- reduced electric field strength: $> 10^{-11}$ V*cm^2 ($> 10^6$ Td)
-- discharge current: 1-30 kA
-- rate of current rise: $> 10^{11}$ A/sec
-- discharge duration: 10^{-7} - 10^{-5} sec .

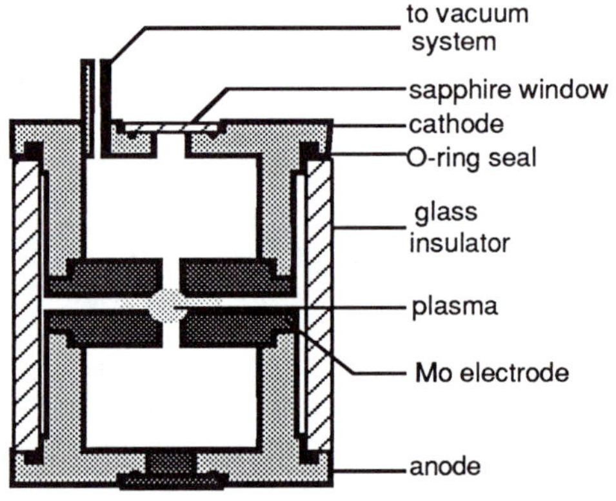

Figure 1.   Schematic drawing of the experimental discharge geometry. The setup consists of two metal cups bearing the discharge electrodes which are made from molybdenum. The diameter of the central holes is 5 mm, and the electrode distance is usually between 3.5 and 5 mm. Two Viton O-rings form a vacuum seal to the glass or alumina insulating cylinder. Triggering of the discharge can be accomplished by photoelectrons released by a UV-light pulse inside the hollow cathode (hc) /3/, or by electrons which are injected into the hc from a pulsed glow discharge behind the hc (not shown) /5/,/9/   .

The temporal development of the discharge occurs in several different steps, which are described only qualitatively in the following. With the application of the high voltage, a very low current (typically $<10^{-6}$ A) Townsend discharge develops on axis (fig. 2a) due to the focusing effect of the electric field around the symmetry axis of the system. A positive space charge builds up inside the hollow cathode as a result of this discharge due to the low mobility of the ions produced by ionizing electron - neutral collisions.

The release of a sufficient number of starting electrons inside the hollow cathode initiates a transient hollow cathode discharge (hcd), shown schematically in fig. 2b. Due to the extremely high reduced electric field, an axial electron beam is formed during this phase of the discharge, consisting mainly of runaway electrons. Typical total discharge currents are of the order of several tens to several hundreds of amperes. The main electron emission processes at the (hollow) cathode surface are photoemission, and secondary electron emission due to the impact of fast ions, neutrals, and metastables onto the cold hollow cathode surface. The duration of this phase is of the order of several 10^{-8} sec. As the hcd current increases, the plasma density on axis increases correspondingly, and the cylindrical plasma column starts to expand in the radial direction. As soon as the plasma boundary ´contacts´ the wall of the cathode hole due to this radial expansion, the plasma prevents the electric field from penetrating into the hollow cathode any longer, and begins to electrically shield the hollow cathode. Therefore, the hcd mechanism breaks down, and the cold cathode surface facing the anode has to take over the discharge current. This is possible above a certain power level and within a certain range of discharge parameters without transition into a metal vapor arc. The resulting diffuse, homogeneous glow discharge is what is called hereafter a ´superdense´ glow; the discussion of the physical mechanisms leading to this kind of high-current density, low pressure glow discharge is the purpose of this contribution.

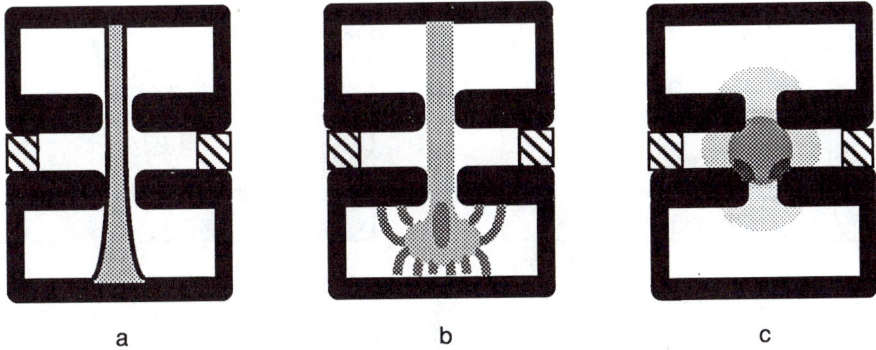

Figure 2.   Schematic drawing of the different phases of the discharge during its development to a superdense glow. 2a) Electrostatically focused high-voltage glow discharge (Townsend discharge); 2b) hollow cathode discharge with run-away electron beam formation; 2c) A superdense glow discharge has developed in the central region around the cathode hole.

In fig. 3, a 3D-plot of the radial distribution of the light intensity in the middle of the anode-cathode gap is shown as a function of time for the hydrogen Balmer alpha transition (λ=486.1 nm). It is evident that the light emission starts on the symmetry axis of the discharge geometry before switching to the edge of the cathode hole; from there, it starts to spread radially, and later homogeneously fills the central part of the electrodes up to a radius of the order of 5 mm, as can be seen in fig. 4. The electron density and temperature shown in fig. 4 indicate a homogeneous, ´cool´ (1-2 eV) discharge plasma of an electron density of $(1-3)*10^{15}$ cm^{-3}. The cross-sectional area of the dense, hot core of the discharge plasma is of the order of 0.5 cm^2, from which an average discharge current density of the order of 10^4 A/cm^2 can be concluded. For an explanation of such a high current density, several electron emission processes have to be taken into consideration.

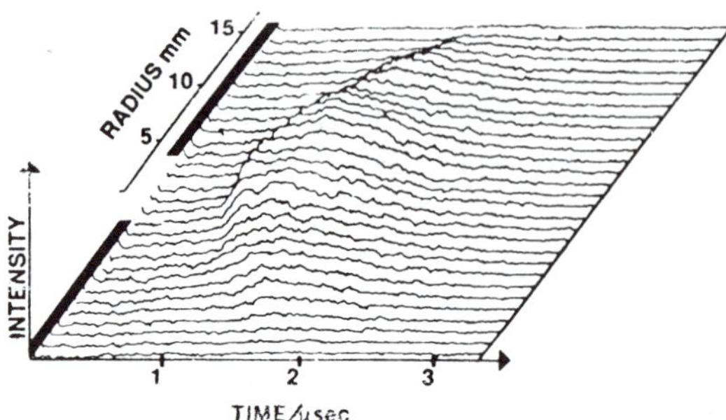

Figure 3. Three-dimensional plot of the radial distribution of the light intensity in a superdense glow as a function of time, for emission at λ=486.1 nm (hydrogen Balmer beta transition). The data are taken from a 450 nsec duration (FWHM) discharge of ~8 kA peak current in a 27 Pa hydrogen atmosphere, in a discharge geometry as of fig. 1.

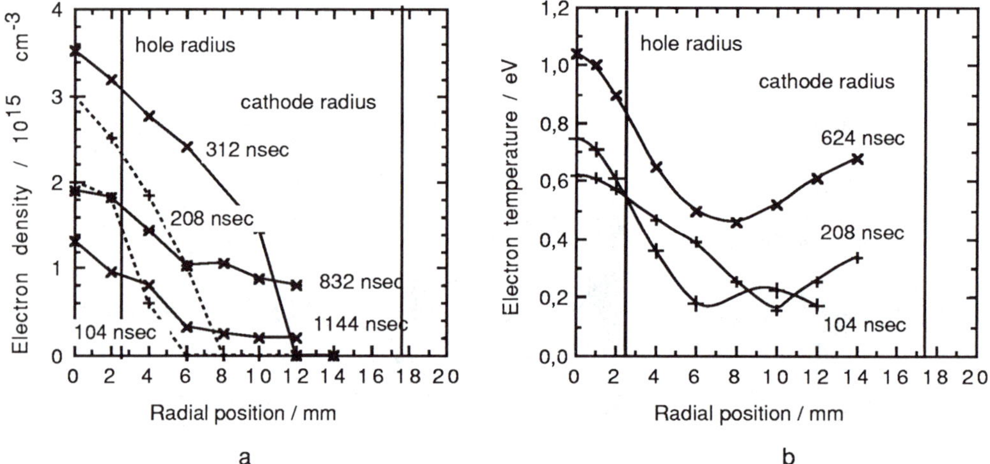

Figure 4. Radial electron density (a) and temperature distribution (b) for a discharge as of fig. 2) and 3), for different times (in nsec). The measurement was made in the mid-plane between anode and cathode.

ELECTRON EMISSION PROCESSES AT THE CATHODE

As mentioned above, a detailed investigation of the plasma-cathode interaction must be based on a careful investigation of the various electron emission processes relevant at a metal cathode surface which is in contact with a plasma, and has to take into account the ion current which is impinging on the cathode surface. The emission processes analyzed in the following are:

      secondary electron emission due to particle impact,
      field emission of electrons,
      photoemission, and
      field-enhanced thermionic electron emission from a heated metal surface.

If other secondary electron emission processes can be neglected, the sum of the current densities of the above processes plus the ion current density at the cathode surface must equal the local total current density:

$$J_{total} = J_{photo} + J_{sec.(impact)} + J_{field\ emission} + J_{thermionic} + J_{ion}$$

Photo effect

For most metals, the photoelectric work function F is of the order of 4-5 eV, i.e. 4.26 eV for molybdenum. Only UV- and VUV-radiation from the plasma can thus contribute significantly to the photoemission of electrons from the cathode surface. The bremsstrahlung continuum can completely be neglected due to its low intensity in the photon energy range of 5-10 eV. Only UV line radiation from atomic hydrogen (in particular, the Lyman series) and molecular hydrogen can significantly produce photo electrons with high efficiency. As the quantum efficiency for 10 eV photons, however, is at best 10^{-2} electrons per incident photon, a power flux of over 5 MW/cm^2 of UV- and VUV-radiation would be necessary to account for 50% of the 10^4 A/cm^2 of current density observed in the experiment. This extremely high VUV light flux cannot be provided from a plasma of electron temperature of 2-5 eV /8/. Therefore, photoelectric emission of electrons obviously does not account for the observed high discharge current density.

Secondary electron emission from particle impact

A much more relevant source of electrons from the cathode surface is the impact of energetic particles on the surface, and the subsequent release of secondary electrons. In principal, two different sources of particles can contribute to this kind of emission:
   -- particles with sufficient **potential** energy to release electrons from the
      metal, in particular metastables;
   -- particles with sufficient **kinetic** energy: fast ions and neutrals
Although the secondary electron emission coefficient for metastables impact can be as high as unity, a metastable production rate in excess of 10^{22} sec^{-1}cm^{-3} would be necessary in order to account for only 10% of the discharge current density; this can obviously not be provided from a plasma of 10^{15} cm^{-3} neutral density. Therefore, metastable impact does not play a role in the production of secondary electrons at these current densities.

Fast ions, however, are produced in the cathode fall due to the high electric field strength of the order of 10^5 to 10^6 V/cm. The cathode fall voltage drop can be estimated to be of the order of 300 - 500 V. The secondary electron emission coefficient for protons in this energy range is about 0.1 /10/, from which a secondary electron current density of ~10% of the ion current density can be concluded. Therefore, the secondary electron emission yield due to proton impact will be neglected in the following, as it gives only a minor correction to the total discharge current density.

A similar treatment is applicable for electrons released by the impact of fast neutrals on the cathode surface. due to the fact that fast neutrals can only be produced to a significant amount from fast ions by charge exchange collisions with slow neutrals, the flux of fast neutrals in the cathode fall is rather small as can be shown by comparing the cathode fall width and the mean free path of ions in regard of charge exchange collisions. We thus conclude that the electron flux from the cathode surface which is due to the impact of fast neutrals is negligible, and that the total secondary electron current density from the cathode surface produced by impact of particles does not significantly contribute to the observed total current density.

<u>Field emission of electrons</u>

Field emission of electrons from a metal surface can lead to very high current densities, if the electric field strength at the surface exceeds $\sim 4 \ast 10^7$ V/cm /10/. In order to estimate whether field emission plays a role in the discharge under consideration or not, the electric field at the cathode surface must be known. An upper limit for the surface electric field strength can be given if the cathode fall width and voltage drop are known. The cathode fall width d_c can be estimated to be several times the Debye length λ_D of the plasma:

$$\lambda_D = \sqrt{(kT_e/4\pi n_e e^2)}$$

In our case, with $T_e = 1$ eV and $n_e = 10^{15}$ cm^{-3}, we get

$$\lambda_D = 2 \ast 10^{-5} \text{ cm} \quad \text{and} \quad d_c = 10^{-4} \text{ cm} .$$

The cathode fall voltage drop has been measured to be in the range of 300 to 500 V /8/ /11//12/, from which follows a surface electric field strength of

$$E_{surface} \leq (3\text{-}5) \ast 10^6 \text{ V/cm,}$$

which is an order of magnitude less than that required for the onset of strong field emission /13/. A field enhancement factor of ~ 3 can be estimated /14/ for the surface irregularities of radius of curvature around ~ 1 μm as observed experimentally /4/, which is still not enough to explain an average current density of the order of 10^4 A/cm^2 over an area of ~ 0.5 cm^2. Although field enhancement of electron emission at the cathode surface seems to be significant, pure field emission alone cannot explain the high average cathode emission alone.

<u>Field-enhanced thermionic electron emission</u>

If we consider a surface electric field of the order of $\sim 3 \ast 10^6$ V/cm at the plasma edge which is facing the cathode surface, then we can calculate the ion (proton) current density which is impinging on the cathode surface due to the extraction and acceleration of ions from the plasma sheath. The ion current density is described by Langmuir's formula /15/ under the assumption of a plasma acting as an unlimited source of ions, and a simultaneous flow of electrons from the cathode:

$$J_{ion} = 1.86 \ast \sqrt{(2e/m_i)} \ast U^{3/2}/d^2 \quad \text{(in esu).}$$

Here, U is the cathode fall voltage drop in volts, d is the cathode fall width, and m_i is the ion mass. With the above assumptions for the cathode fall width and voltage drop (10^{-4} cm and >300 V, respectively), we get an upper limit for the ion current density that can be drawn from the plasma edge of $J_{ion} \geq 10^5$ A/cm^2. (The corresponding upper limit for the electron current density that can be drawn from the cathode is $J_e \geq 2 \ast 10^6$ A/cm^2, assuming an unlimited cathode emissivity.)

During the early part of the discharge, i.e. as long as the cathode surface remains at essentially room temperature, the electron emission from the cathode is mainly due to secondary emission from particle impact, and to some extend to photo emission, as pointed out before. Therefore, the electron current density at the cathode surface is comparable to, or even only a fraction of, the ion current density at the cathode surface; thus, the ion current density must be a large fraction of the total discharge current density. The power load at the cathode surface caused by this ion "beam" is then of the order of several 10^6 W/cm^2, or even over 10^7 W/cm^2, if we take into account that the cathode fall voltage drop during this phase can be as high as several kV for time interval of several 10^{-8} sec (which is a typical time constant for the anode fall time).

The temporal evolution of the cathode surface temperature is shown in fig. 5 as an example for a constant power load of 13 MW/cm^2 on a molybdenum electrode.

A surface temperature of ~3,000 K is reached after 100 nsec, after which the power density is allowed to decrease hyperbolically while still maintaining a ±constant surface temperature. (The cathode surface temperature is calculated from the formula

$$T(x,t) = \sqrt{(c/\pi^* \kappa^2)} * \int_0^t P(t-\tau)/\sqrt{t} \; * \; \exp(-x^2/4^*\kappa^*t)d\tau$$

where $\kappa = k/\rho^*c$ k = heat conductivity
         ρ = mass density   c = specific heat
         x = depth (distance from cathode surface,
which describes the heat conduction of a power flux density P(t) into a solid of thermophysical material constants k, c, and ρ /16/.)

Such a high surface temperature is sufficient to provide a thermionic electron emission current density of the order of 10^3 to 10^4 A/cm^2, particularly in connection with a surface electric field of over 10^6 V/cm, as can be seen from the formula describing the Schottky-emission or field-enhanced thermionic electron emission /17/:

$$J_{thi} = 4\pi m_e e/h^3 (kT)^2 \exp\{-[\Phi - \sqrt{(e^3 E/\varepsilon_0)}]/kT\}$$

Here, the following definitions are used:
         h = Planck's constant m_e = electron rest mass
         k = Boltzmann's constant   E = surface electric field strength
         ε_0 = permittivity of space Φ = thermionic work function

Above ~2,500 K, J_{thi} is a violent function of the temperature, which means that a relatively small variation in temperature of some 100 K can adjust J_{thi} over a large range (i.e., more than one order of magnitude). For most metals, the surface temperature will thus be found in the range between 3,000K and 4,000K, if the principal cathode electron emission process is field-enhanced thermionic electron emission, and if the discharge current density is of the order of 10^4 A/cm^2 or more. The surface heating is believed to be caused by the energy deposition from the 'ion beam' produced in the cathode fall, and is limited by heat conduction to a thin (~1μm thickness) surface layer at the immediate cathode surface.

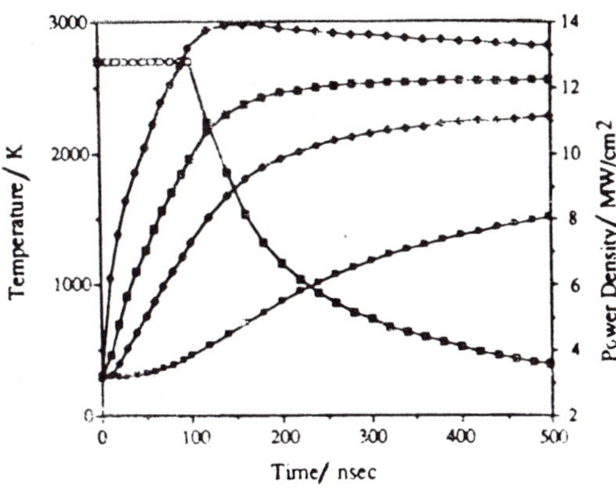

Figure 5.   Temporal development (calculated) of the surface temperature of a Mo electrode exposed to a rectangular power load of 13 MW/cm^2, for different depths x (in μm from the cathode surface at x=0 μm). After 100 nsec, the surface power density is allowed to decrease exponentially while a ± constant surface temperature is still maintained.

<u>Limitation of ion current density</u>
Up to now, we assumed that the plasma is able to supply an ion current density of over 10^3 A/cm^2; an estimation of the space-charge limitation of the ion current density has shown that the space-charge limit lies well above the observed total discharge current density.

A severe limitation of ion current density, however, is found in the limited ion drift velocity. As the ions extracted from the plasma edge must be replenished by an equal amount of ions per unit time drifting into the plasma edge region, the thermal flux of ions from the plasma which is described by

$$J_{thermal} = 0.5en_{ion}\sqrt{(2kT_i/m_i)}$$

limits the maximum ion current density that can be drawn from the plasma edge to values of the order of several 100 A/cm^2. The mean ion drift velocity inside the plasma which is directed towards the plasma edge, however, can be appreciably higher than the thermal drift velocity in the case of a plasma pre-sheath, which is formed as a consequence of the high current density and the high electric field at the plasma boundary. Therefore, the maximum ion current density that can be supplied by the plasma might be well in excess of 10^3 A/cm^2.

The resulting ion pumping effect, however, leads to a rapid decrease of plasma density near the plasma edge, if there is no additional source of ions (or neutrals, which would be ionized by electron impact). An analysis of the ion-induced desorption yield at the cathode surface shows, that the number of desorbed atoms per incident ion is of the order of or even less than unity. Due to the low kinetic energy of the desorbed atoms, a severe ion deficiency can develop within a short time during the initial, transient phase of the discharge. This could lead to either a transient or a long-lasting limitation of the discharge current, depending on a large number of discharge parameters. Even an almost complete current interruption (current quenching) should be possible, if the secondary processes are insufficient to sustain a high current discharge. Both of these cases ('weak' and 'strong' current limitation) have been found experimentally, which supports these assumptions /19/ (compare also fig. 6).

With a sufficiently high power flux onto the cathode surface, however, the surface temperature would quickly exceed 1,600K-1,800K, a temperature above which the surface adsorbate (which is present on any surface exposed to a gaseous environment) is desorbed almost completely by thermal desorption /18/. Even the desorption of only one single monolayer (which represents a particle density of the order of $3*10^{15}$ cm^{-2}) would lead to a very high density gas layer immediately above the heated cathode surface, and could act as an effective ion source, balancing the limitations imposed by the ion pumping effect.

The mechanism described implies that the occurrence of current quenching should be limited to cases where the gas density is low (with a more rapidly increasing depletion of ions at the plasma edge), and for the case of medium or low current density (where surface heating due to ion bombardment is insufficient for the onset of thermal desorption and field-enhanced thermionic emission of electrons). These findings fit well with experimental results (compare fig. 6) showing that quenching occurs for a range of

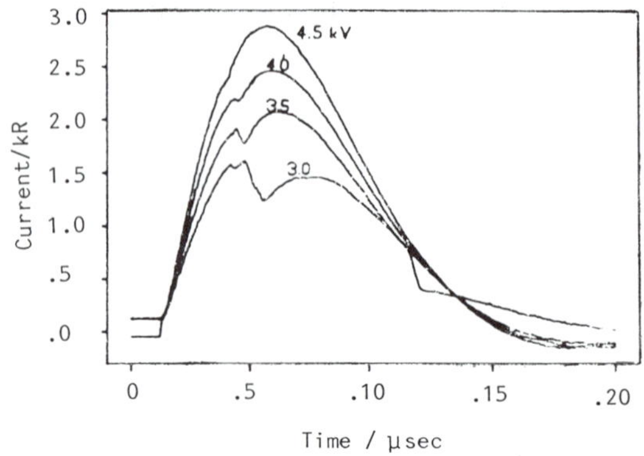

Figure 6. Discharge current traces showing current quenching in a certain range of discharge parameters. Parameter: capacitor charging voltage (3.0, 3.5, 4.0, 4.5 kV); 21 Pa hydrogen; The discharge circuit is critically damped by a matched series load resistor.

discharge currentsbetween ~0.5 kA and 3 kA, at a hydrogen pressure below ~15 Pa, for a setup similar to that shown in fig. 1 /19/.

We thus conclude that most likely field-enhanced thermionic electron emission, caused by the heating of a thin cathode surface layer due to ion bombardment, is the principal electron emission process at the cathode surface, thereby explaining the observed high current density of the superdense glow discharge.

Up to now, no final decision can be made as to whether the superdense glow is homogeneously distributed over the whole cathode surface from a microscopic point of view, as it appears in the macroscopic electron temperature and density distributions, or if electron and ion current density (and, consequently, the cathode surface heating) are concentrated at the convex microscopic surface irregularities where the surface electric is maximal. In the latter case, however, the resulting miniature glows from the multiply, closely spaced emission sites would quickly merge into one single, homogeneous glow within a few mean free paths, which would thus be indistinguishable from a 'real' homogeneous glow.

CATHODE EROSION

The erosion of the cathode surface, as observed experimentally, appears to be rather homogeneous over an electrode area of ~0.5 cm^2 around the cathode hole; this result complies well with the observed cross-sectional area of the dense core of the superdense glow plasma /4/,/8/. Even at an average current density well above 10^4 A/cm^2 at the cathode surface, and after several million discharges, the total erosion appears symmetric around the cathode hole (compare fig. 7). This even erosion pattern suggests an erosion mechanism which is homogeneously distributed around the edge of the cathode hole in every single discharge, and it seems to be different from the erosion observed in an arc mode /20/.

Three different erosion mechanisms should be considered to be responsible for the observed electrode wear: <u>physical sputtering</u>, <u>chemical sputtering</u>, and <u>evaporation</u>. Out of these erosion mechanisms, the second - chemical sputtering - is the least well known, and probably the most difficult one to be characterized. In the experiments considered, however, which are done in an environment apt for UHV-conditions, with high-purity hydrogen (or deuterium) as the working gas and high-purity metals as electrode materials, chemical sputtering can only be caused by gaseous impurities like nitrogen and oxygen. A comparison of the electrode lifetime for a hydrogen gas fill and a nitrogen gas fill shows, that an increase in cathode erosion rate of the order of a factor of two occurs for the case of nitrogen /11/; this increase is of the order of that expected for the ion-induced physical sputter rate, and much less than that if chemical sputtering would be a considerable effect. Physical sputtering from fast ions, however, is supported by these results as well as by crude estimations of the total sputtering rate in the experiments. Assuming a cathode electron emission process which is based mainly on field-enhanced thermionic emission, we should also consider simple evaporation as an effective erosion mechanism because of the high surface temperature necessary for thermionic emission. Both erosion processes - physical sputtering, and evaporation - will therefore be discussed in the following in more detail.

From the total mass eroded in the experiment of fig. 7, an erosion rate of the order of ~10^{-5} grams per coulomb of charge transported over the entire cathode surface can be estimated, which refers to approximately 10^{-2} Mo-atoms per elementary charge. This erosion rate is compared with the sputtering yields R_S for ion-induced sputtering of the following species:

We can easily exclude hydrogen and deuterium as the species responsible for the observed cathode erosion due to ion-induced sputtering (even if the ion 'beam' current would equal the total discharge current at the cathode surface), because the low

Projectile	R_s(in atoms/ion, at 500 eV)	
H⁰, H⁺, D⁰, D⁺	< $(3-6)*10^{-3}$	(working gas ions)
N⁺, O⁺, C⁺, ...	< $8*10^{-2}$	(impurities)
Mo⁺	< 0.4	(self-sputtering)

sputtering yield can, by no means, explain the observed erosion rate. Impurity sputtering, however, could explain the erosion rate, but only under the assumption that **at least 10% of the total discharge current** is carried by impurity ions in the cathode fall. In a hydrogen discharge which is run under UHV conditions, this would be comparable to the total ion current expected in the cathode fall, the principal species being H^+ (D^+) ions! (The ratio between ´working gas ion´ current density J_{wg} and impurity ion current density J_{imp} is given by

$$J_{wg}/J_{imp} = \sqrt{(m_{imp}/m_{wg})} \qquad /15/,$$

where m_{imp} and m_{wg} are the impurity ion and working gas ion mass, respectively; in our case, this ratio is of the order of 2.5-3, assuming unlimited ion sources.) Thus, most of the ion current should be carried by protons (deuterons), which makes impurity-ion sputtering unlikely to be the main cathode erosion mechanism.

A similar estimation can be made for the case of self-sputtering. Although the self-sputtering yield is ~0.4 Mo-atoms per incident Mo-ion of ~500 eV kinetic energy, the ratio of deuteron ion current and Mo-ion current density is ~7 even in the case of unlimited Mo^+ supply! Therefore, a maximal amount of only ~10% of the total cathode erosion observed can be contributed to self-sputtering.

In order to estimate the contribution of the third erosion mechanism envolved (evaporation), the cathode surface temperature should be known to within ±10%. A first order approximation of the surface temperature can be made from the Schottky-emission formula and an estimated surface electron current density of ~10^4 A/cm². (A change of the temperature of ~±15%.leads to an order of magnitude variation of the thermionic electron current density. Therefore, an average electron current density around 10^4 A/cm² is justified to span a wide range of operating parameters.) For a surface electric field below $5*10^5$ V/cm, the corresponding surface temperature was ~4,000 K. This can be compared to the surface temperature which is necessary to achieve an evaporation rate of ~10^{-5} g/C in an experiment as of fig. 7. The total ´on-time´ of the cathode when acting as a thermionic electron emitter in that case was ~2.7 sec, as a result of ~5 million short-duration discharges.

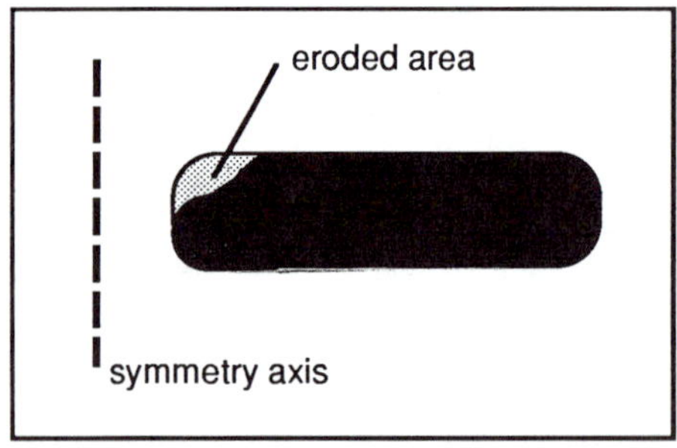

Figure 7. Shape (cross-section, schematic) of a molybdenum cathode after exposure to 5 million discharges of 90 nsec duration, ~20 kA peak current at >90% current reversal (slightly damped ringing discharge), and a deuterium gas fill. The initial shape of the hole was straight with slightly rounded edges at both faces of the electrode.

From these numbers, an evaporation rate of $2.2*10^{21}$ atoms/cm²*sec is calculated for the total erosion of the molybdenum cathode. (The other erosion processes, like sputtering are neglected, as they do not contribute to more than ~10% of the total erosion losses in this order-of-magnitude estimation of the erosion rate.) From this evaporation rate, a cathode surface temperature of 4,050 K is derived (see, for example, ref. 22), which is very close to that expected for a thermionic emission of the order of 10^4 A/cm^2. This coincidence makes us believe that the principal electrode erosion mechanism in a superdense glow discharge in hydrogen and deuterium is evaporation.

CONCLUSIONS

We have shown that in rotational symmetric low-pressure gas discharges a so-called superdense glow discharge mode exists which is characterized by current densities of the order of 10^4 A/cm^2 and more, and by a diffuse, homogeneous current distribution across the cathode surface, in contrast to a constricted arc or spark. The high current density at the cathode surface is explained in terms of field-enhanced thermionic emission of electrons from a thin, hot surface layer. Heating of the cathode surface is accomplished by ion bombardment, mainly during the initial (transient)phase of the discharge, with ions extracted from the bulk plasma due to the large electric field strength at the plasma edge. These ions are accelerated in the cathode fall to energies of 300-500 eV, according to the cathode fall voltage drop. The resulting thermal desorption of the surface adsorbate seems to be necessary to explain the ability of the discharge to transport several mC (or even C!) of electric charge without current quenching and without evolving into a vacuum spark or an arc. The experimental observation of current quenching in an intermediate range of power density supports this interpretation.

A comparison of expected erosion rates for ion-induced sputtering, chemical sputtering, and evaporation, with experimental results confirms the assumption that evaporation of cathode material due to the ion-bombardment heating of the surface is the principal cathode erosion mechanism. This result implies that high melting-point refractory metals like molybdenum, niobium, tantalum, and tungsten (in about this order) should be the best choice of high-durability cathode materials for devices operating with a thermionic cathode emission. A more detailed investigation, both experimental and theoretical /23/, will give a more precise insight into the interaction between plasma and cathode in high current density discharges, and offer the possibility to choose the optimal cathode material for a specific application.

ACKNOWLEDGEMENTS

We appreciate valuable discussions with J. Christiansen, K. Frank, M. Kushner, J. Lawler, and G. Schäfer. Part of the computing and experimental work were done by V. Dominic and G. F. Kirkman. The test series resulting in the erosion rate for molybdenum was performed by A. Görtler, R. Kowalewicz, and C. Kozlik. This work was supported by the AFOSR, U.S.ARO, and the Deutsche Forschungsgemeinschaft.

REFERENCES

/ 1 /   J. Christiansen, C. Schultheiß, Z. Physik **A 290**, 35 (1979)

/ 2 /   L. Y. Abramovich, B. N. Klyarfel´d, Y. N. Nastich, Sov. Phys. Techn. Phys. **11**, 528 (1966), and
G. I. Nosov, S. A. Smirnov, Instr. Exp. Techn. **20**, 1147 (1977)

/ 3 /   G. F. Kirkman, W. Hartmann, M. A. Gundersen, Appl. Phys. Lett. **52**(6), 613 (1988)

/ 4 /   W. Hartmann, V. Dominic, G. F. Kirkman, M. A. Gundersen, Appl. Phys. Lett. **53** (18), 1699 (1988)

/ 5 /   K. Frank, J. Christiansen, O. Almen, E. Boggasch, A. Görtler, W. Hartmann, C. Kozlik. A. Tinschmann. G. F. Kirkman. Proceed. SPIE vol. **871**. 173 (1988)

/ 6 / W. Hartmann, M. A. Gundersen, Phys. Rev. Lett. **60**, no. 23, 2371 (1988)

/ 7 / W. Hartmann, G. F. Kirkman, V. Dominic, and M. A. Gundersen, IEEE Transact. Plasma Sci. **36**, no.4, 825 (1989)

/ 8 / W. Hartmann, V. Dominic, G.F. Kirkman, M. A. Gundersen, J. Appl. Phys. **65**(11), 4388 (1989)

/ 9 / G. Mechtersheimer, R. Kohler, T.Lasser, R. Meyer, J. Phys. E: Sci. Instr. **19**, 466 (1986)

/ 1 0 / S. Flügge, "Handbuch der Physik vol. XXI"(Springer, Berlin, 1956)

/ 1 1 / W. Hartmann, M. A. Gundersen, unpublished data

/ 1 2 / J. Bernardes, NSWC, private communication

/ 1 3 / E. L. Murphy, R. H. Good, Phys. Rev. **102**, 1464 (1956)

/ 1 4 / F. Rohrbach, CERN Report 71-5/TC-L (1971), CERN, Geneva, Switzerland

/ 1 5 / I. Langmuir, Phys. Rev. **2**, 450 (1913), and Phys. Rev. **33**, 954 (1929)

/ 1 6 / H. W. Carslaw, J. C. Jaeger, "Conduction of heat in solids" (Clarendon, Oxford, 1959)

/ 1 7 / W. Schottky, Z. Physik **14**, 63 (1923)

/ 1 8 / J. P. Cowin, D. J. Auerbach, C. Becker, L. Wharton, Surf. Sci. **78**, 545 (1978); see also
G. Wedler, H. Ruhmann, Surf. Sci. **121**, 464 (1982), and
R. B. Hall, A. M. Desantolo, Surf. Sci. **137**, 421 (1984)

/ 1 9 / M. A. Gundersen et al., to be published

/ 2 0 / J. Christiansen, K. Frank, W. Hartmann, C. Kozlik, W. Krauss-Vogt, R. Michal, "Comparison of electrode effects in high-pressure and low-pressure gas discharges like spark gaps and pseudospark switch", this vol.

/ 2 1 / W. Hartmann et al., to be published

/ 2 2 / R. Behrisch, J. Nucl. Mat. **93&94**, 498 (1980)

/ 2 3 / W. Hartmann et al., to be published

COMPARISON OF ELECTRODE EFFECTS IN HIGH-PRESSURE AND LOW-PRESSURE GAS DISCHARGES LIKE SPARK-GAP AND PSEUDO-SPARK SWITCH

J. Christiansen, K. Frank, W. Hartmann, C. Kozlik
Univ. of Erlangen, FRG

W. Krauss-Vogt, R. Michal
DODUCO GMBH + CO, Pforzheim, FRG

ABSTRACT

The mechanisms of the interaction between plasma and electrode material in a pseudo-spark-discharge seem to be of fundamental interest. Therefore, molybdenum electrodes, stressed by long term operation in a pseudo-spark switch (PSS) as well as in a spark gap switch, were studied with a scanning electron microscope. In addition, electrodes with a 10μm molybdenum coating deposited by a PVD-process, were also applied, both in the PSS and in a spark-gap configuration. In this contribution observations about the influence of the pseudo-spark discharge plasma on the electrode surface are reported. By comparing the results with the effects caused by the high pressure spark discharge, conclusions concerning the discharge mechanism can be drawn.

1. INTRODUCTION

Previous research about the temporal and spatial development of the pseudo spark discharge has led to a certain understanding of the plasma behaviour (1). Up to now, the interaction between plasma and electrode material has only been considered in connection with the extremely high current density, which was observed at the initially cold cathode surface (2-4). More generally, the influence of the cathode material in any gas discharge switching device is mainly given by two different time dependent aspects. During repetitive switching, the rate of current rise as well as the plasma decay time are -among others- important parameters, which are influenced by the electrode material. On the other hand, electrode erosion is the limiting factor, which mainly determines the life of a gas discharge tube.

As a starting point for a further investigation of these topics bulk molybdenum electrodes were stressed by repetitive switching in both a low-pressure discharge (pseudo-spark) and in a high-pressure spark-gap discharge. In order to get also information about the interaction between plasma and electrode material in a layer very close to the surface, electrodes with a 10μm-molybdenum coating were tested both in the pseudo-spark (PSS) and in a spark-gap test device.

2. EXPERIMENTAL SETUP

Three experiments were performed using different test-devices and pulse circuits in order to obtain test conditions, which are relevant for typical switch applications. The respective data are summarized in table 1.

Physics and Applications of Pseudosparks
Edited by M. A. Gundersen and G. Schaefer
Plenum Press, New York, 1990

Table 1. Experiment data

	low-pressure discharge		high-pressure discharge
	experiment 1 metal ceramic PSS	experiment 2 modular PSS	experiment 3 modular 2-electrode spark-gap
U (kV)	32	10	10
I (kA)	25	10	1.8
dI/dt (A/sec)	$2.5*10^{11}$	$4*10^{10}$	$8.6*10^9$
rep. rate (1/sec)	19	1	2
pulse length (μsec)	4.5(*)	0.8(*)	1
charge/shot (mCoul)	25	0.65	1
energy/shot (Joule)	25.6	3.25	3.75
number of pulses	$8*10^6$	$1*10^4$	$5*10^4$
waveform	weakly dam- ped sine	reverse current 40%	asypmtot. damped
gas/pressure (pascal)	D_2; 40	H_2; 25	He/H_2, $6*10^5$
gap distance (mm)	3	5	5

(*) in case of oscillating current (l/e decay), the pulse length is taken until current-zero.

2.1 Experiment 1 (pseudo-spark switch)

The geometry of the pseudo-spark switch used in experiment 1 is shown in fig 1. This version is made of OFHC-copper electrodes brazed directly to insulating Al_2O_3-rings. The anode and cathode parts, which directly interact with the discharge-plasma, are made of bulk molybdenum. The switch was connected to a H_2-gas buffer in order to prevent changes in the gas atmosphere during repetitive switching.

The electric test circuit for this device (see fig. 2) consists of a 1Ω–waterline, which can be charged up to 35kV. The switch under test (PSS 2) represents an unmatched load in the pulsforming network producing a weakly damped oscillating current pulse (up to 20 current-oscillations). Thus the stored energy of the circuit is almost completely transferred into the switch resulting in a strong stress for the electrode material. Because of the alternating current it could be expected, that cathode and anode will show nearly the same structure on the electrode surface. Typical current and voltage waveforms are shown in fig. 3.

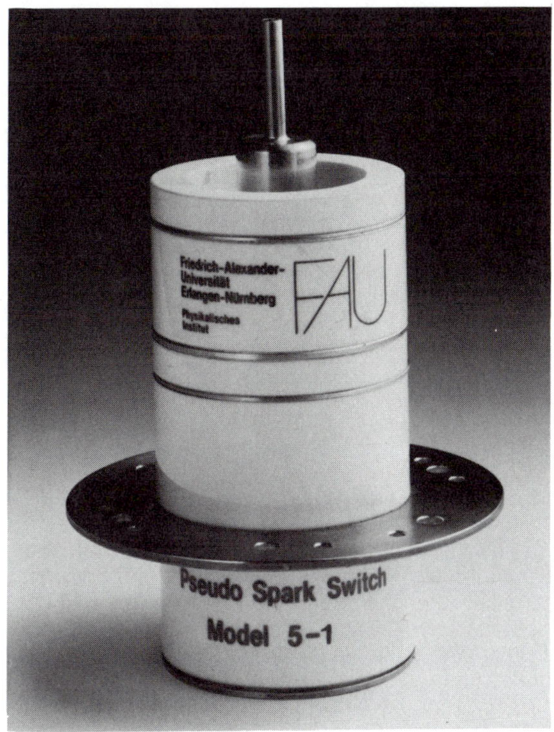

Figure 1. Metal-ceramic version of Pseudo-spark switch. The metal rings at the top
are related to the trigger unit; large electrode ring: cathode connector;
bottom: anode flange.

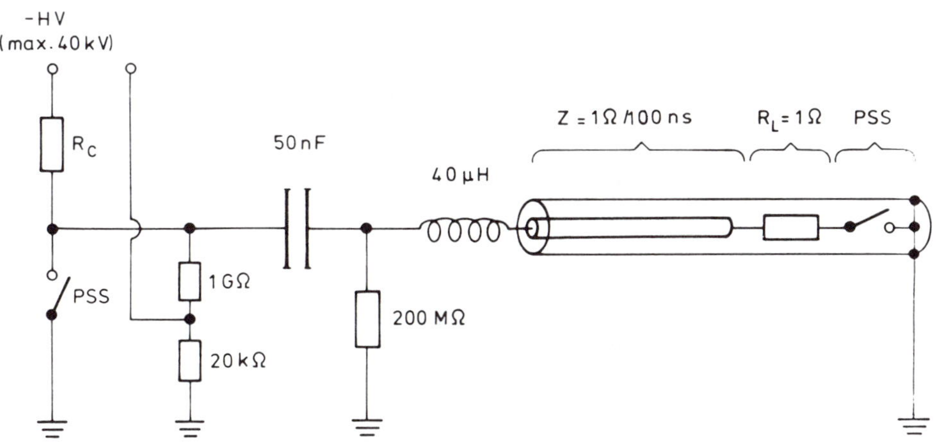

Figure 2. Schematic drawing of the experimental setup of experiment 1.

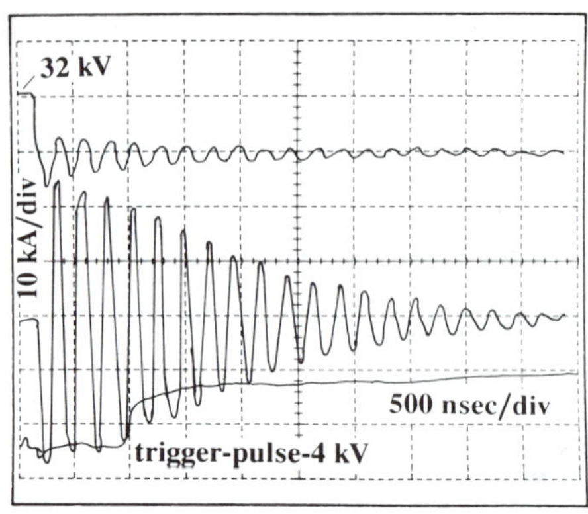

Figure 3. Typical current and voltage waveform of experiment l; ringing frequency: 4MHz. There are about 20 sine waves until the current has come to zero.

2.2 Experiment 2 (pseudo-spark switch)

For the investigation of electrode effects and for the optimization of gap geometry, a demountable modular metal-ceramic version of the pseudo-spark switch was used (fig. 4). This flexible test device enabled an uncomplicated variation of the electrode materials.

The modular system was operated in a separate test-circuit illustrated in fig. 5. In this case, several bunches of high voltage cables are used as energy storage device and pulsforming network. The resulting current pulse is shown in fig. 6. The first half cycle transfers about 60% of the total energy.

Figure 4. Photograph of modular Pseudo-spark switch based on CF-vacuum flange system. Upper part: trigger unit; lower part: main gap insulator.

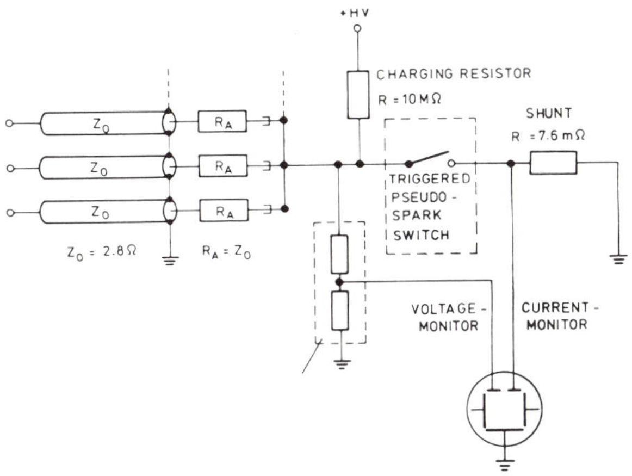

Figure 5. Schematic drawing of the experimental setup of experiment 2, which is based on the arrangement of HV-cable bunches. There are 7 bunches in parallel Z(0)= 2.8Ω; each bunch consists of 14 cables => Z(total)= 0.4Ω.

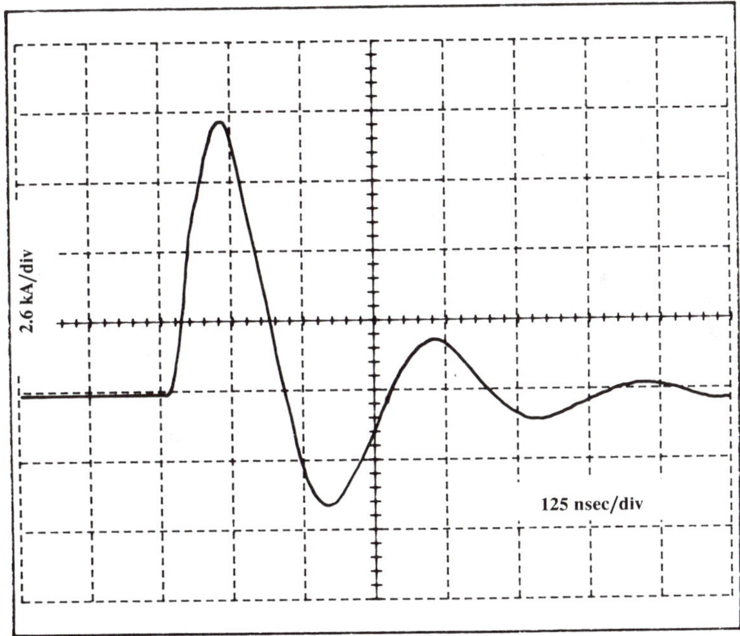

Figure 6. Typical current waveform of experiment 2. The first half-cycle transfers about 60% of the stored energy.

2.3 Experiment 3 (overvolted spark-gap)

Fig. 7 shows the geometry the 2-electrode spark-gap used for these test series. The features of this device are uncomplicated exchange of electrodes and adjustment of electrode distance and gas pressure. Corresponding to a sealed-off system, the gas volume in the spark-gap was closed by a UHV-valve after evacuation and subsequent filling with the working gas of a proper type (He/H2 mixture).

93

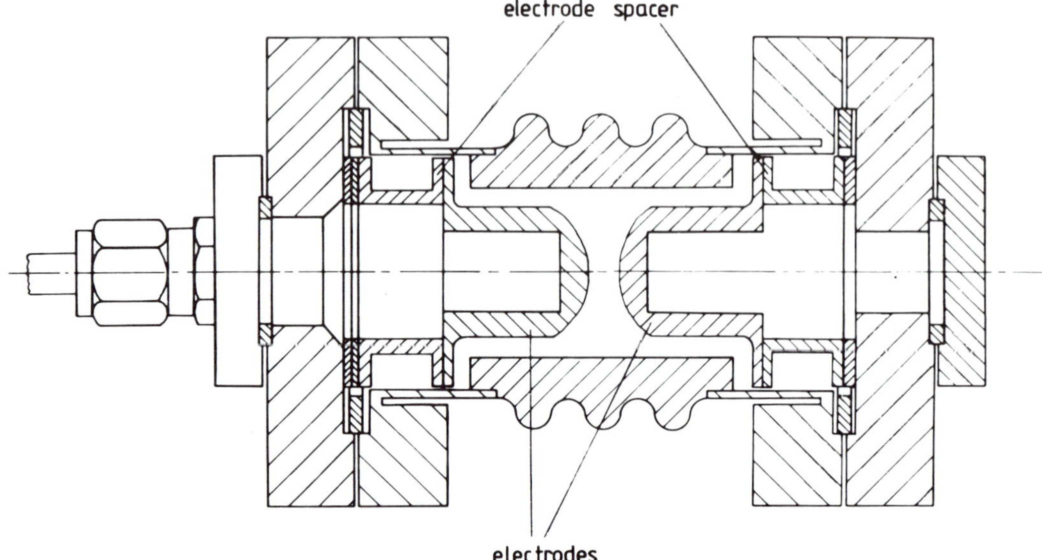

Figure 7. Cross sectional drawing of modular 2-electrode spark-gap based on CF-flange system.

In experiment 3 the spark-gap was applied in a coaxial low-inductance circuit, which consists of energy storage capacitor and resistive load (fig. 8). The circuit generates an asymptotically damped current pulse without any current reversal (fig. 9) producing different surface modifications on both anode and cathode. The parameters of this experiment, which are important for plasma-material interaction (charge/shot, energy/pulse), are comparable to experiment 2. The spark-gap fires when the self-breakdown voltage of the gap is exceeded.

2.4 Materials

For the tests reported in this paper only electrode materials based on molybdenum were used. Such materials are commonly used in many pulse-plasma applications. In one case the electrodes were made of bulk molybdenum with a purity of 99.95%. For the second material, the molybdenum was deposited by a PVD-process onto a copper insert.

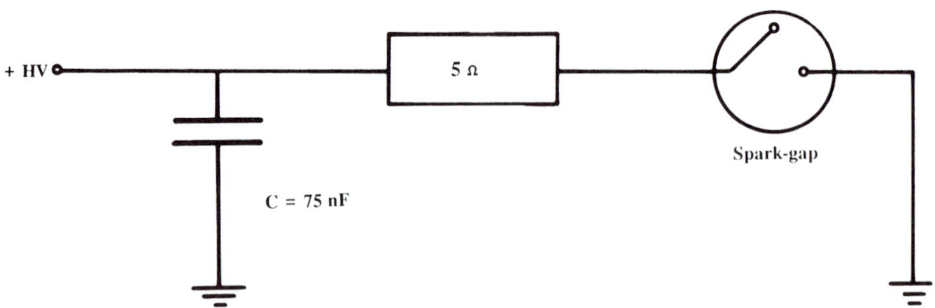

Figure 8. Test-circuit of experiment 3.

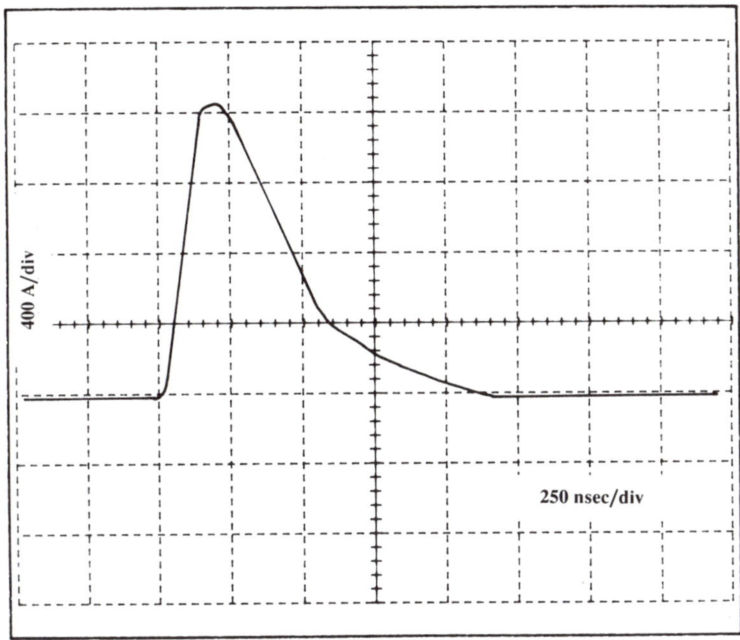

Figure 9. Typical current pulse in experiment 3; U = 10kV; I = 1.8kA; rep-rate 2Hz; dI/dt = 8.6*10^9 A/sec; charge/shot = 1mCoul; energy/shot = 3.75mCoul.

3. PHYSICAL ASPECTS IN THE ELECTRODE REGIONS

Electrode surfaces of gas-filled switch devices in general, and the cathode surface in particular, are subject to intense power loads during operation, leading to melting and evaporation loss of electrode material in the active region of the electrodes. At the cathode surface, electrons have to be released by some physical mechanism in order to supply free charge carriers necessary for conduction at the interface between plasma and solid material. At low current densities (i.e., lower than ≈ 100 A/cm^2), secondary electron emission due to the photoeffect, due to impact of fast ions and neutrals, and due to the impact of metastable ions, atoms or molecules is sufficient to maintain a self-sustained gas discharge. Much higher current densities, however, require electron emission mechanisms different from these. Up to several 100 A/cm^2, pure thermionic emission can provide a sufficient amount of electrons; at even higher current densities (several 10^3 - 10^5 A/cm^2), field enhanced thermionic emission are mechanisms for the production of a sufficiently high electron curre density at the cathode surface.

3.1 Pseudo-Spark

In low pressure gas discharges, the comparably large mean free path of charged particles causes a larger cathode fall width, which results in a lower electric field strength at the cathode surface. Although the cathode fall voltage drop might be as high as ≈ 500V, the resulting electric field at the cathode surface (which is in the order of 10^6 V/cm) is too low to explain current densities of more than 10^4 A/cm^2 mainly by field emission. In this case, field enhanced thermionic emission of electrons seems to be more appropriate to describe the cathode emission mechanism. It requires, however, an intense heating of the cathode surface to temperatures above 3000°K, which can only be explained in terms of ion bombardment heating of the cathode surface by ions, which are accelerated within the cathode fall. This mechanism implies ion current densities of the order of several 10^3 A/cm^2 at the "active" part of the cathode surface in order to overcome the heat conduction losses into the bulk cathode material; even then, only a thin ($\approx \mu$m) layer at the cathode surface is heated to the temperature required for field enhanced thermionic emission of electrons of the order of 10^4 A/cm^2. As a consequence of this cathode emission mechanism, several conclusions concerning the physical properties of the electrode material can be drawn:

a) **Work function**: With a low work function of the cathode material, an onset of thermionic emission at lower temperatures is achieved. Therefore, a low work function is desirable in order to achieve less cathode heating and less power losses in the gas discharge.

b) **Melting temperature**: For most of the pure metals and alloys, the cathode surface working temperature will be well above 3000°K, up to almost 4000°K for some metals. This temperature range is well above the melting point of most of the commonly used electrode materials, and even higher than the boiling point of some of them. In order to avoid excessive material loss due to evaporation, high-melting materials like molybdenum, tantalum and tungsten seem to be applicable as electrode material for the low-pressure discharge like the pseudo-spark.

c) **Physical sputtering**: Due to the high ion current density of $>10^3$ A/cm^2, ion-induced sputtering of cathode material can significantly contribute to electrode erosion. Therefore, sputter resistant metals like Mo, Ta and W are preferable to other materials.

d) **Thermal shock resistance**: Because of the high cathode surface temperature necessary, only transient discharges of the kind discussed can be sustained without macroscopic loss of cathode material. The resulting extremely high temperature gradients of several 10^6K/cm within the first 10-20 μm of the cathode surface consequently exposes this cathode surface to a strong thermal and mechanical stress. If sandwich layered cathode designs are to be used, care has to be taken for a close matching of the expansion coefficients of the involved materials in order to avoid the destruction of the layer structure by repeatedly stressing the layer boundaries.

3.2   Spark-Gap

In high-pressure sparks, thermally enhanced field emission is more likely to be the predominant electron emission process due to the high electric field in the cathode fall, which is a consequence of the very short physical length of the cathode fall. Thermal enhancement of the field emission current is accomplished by Joule-heating, and by heating due to bombardment of the cathode fall with ions which are accelerated in the cathode fall voltage drop.

In a separate experiment, which was performed with a copper-tungsten composite material, it has been observed that the arc is formed by a number of individual filaments starting at the cathode. In the region between the electrodes these filaments form a narrow spark-channel leading to a bright spot in front of the anode. From the size and the number of the filament pattern on the cathode a microscopic current density of about 10^{12} A/m^2 can be deduced in this case. On the other hand the macroscopic current density of $5*10^9$ A/m^2 is determined by the area of the entire cathode spot. This current density is sufficient to initiate the transfer of electrode material by an evaporation process. The material transfer predominantly goes from the cathode to the anode which was shown in another spark-gap test. In this experiment cathode and anode were made from different electrode materials. The experience from long term tests with sealed-off metal-ceramic spark-gaps is that the life of these devices is indeed limited by the observed material transfer, which leads to a successive metalisation of the insulator wall.

4. RESULTS

The main purpose of the work reported here is to investigate the correlation between the alteration of electrode surfaces and the electrical behaviour of the discharge device. Therefore an optical microscope, a scanning electron microscope as well as Auger-electron-spectrometry (AES) combined with electron- spetroscopic-chemical-analysis (ESCA) are available. The applied microanalytical techniques can contribute to a more detailed characterisation of electrode materials.

4.1 Behaviour of bulk molybdenum electrodes

4.1.1. Pseudo-spark switch

The following description of electrode surface effects refers to experiment 1. The bulk molybdenum electrodes of the metal-ceramic PSS-device were aged according to the data given in table 1. After about 8 million pulses cathode and anode were investigated with a SEM. Although the experiment reported here was performed at higher voltage and current level the surface structures are similar to those observed in ref 2. The micrograph in fig. 10 gives a survey of the edge of the cathode hole, where the strongest effect of the plasma could be expected.

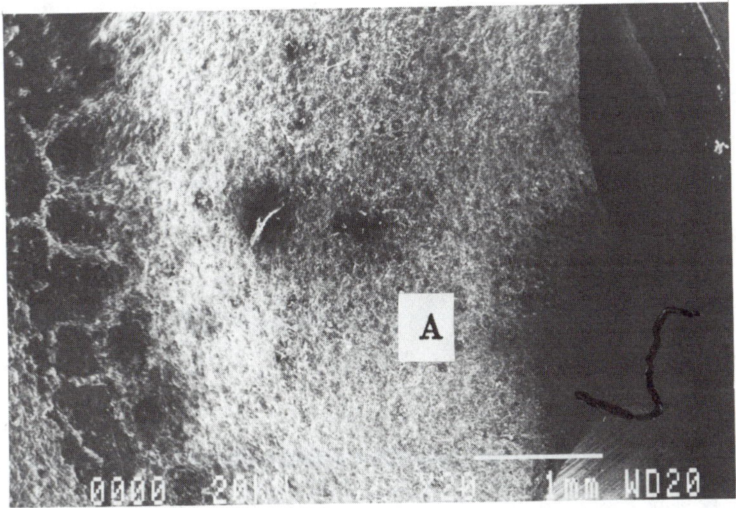

Figure 10. Micrograph of the electrode surface at the edge of the cathode hole; electrode material: bulk molybdenum; pulse data according to experiment 1.

The magnification of the indicated region (A) is shown in fig. 11. The surface is covered with relatively large sheets. Their rounded edges indicate, that the material has been at least partly melted to a thin layer. Another feature of the plasma-electrode interaction is the occurrence of several cracks in the surface. These cracks might be due to rapid melting within a short time. This is followed by a fast resolidification process because of the heat conduction into the cold bulk material of the electrode. The cracks run through all surface structures for a length of about 100µm, which again implies, that relatively large parts of the electrode had reached the melting temperature of molybdenum (2600°C) at the same time. The few small splashes which can be seen in the center of fig. 11 indicate that some liquid material has been transferred across the main gap of the switch. Fig. 12 shows the corresponding sector on the anode surface. The slightly different appearance compared to fig. 11 results from a smaller angle of view. There are no significant differences of electrode modifications between cathode and anode. This was expected because the ringing current will smooth out different cathode and anode effects.

97

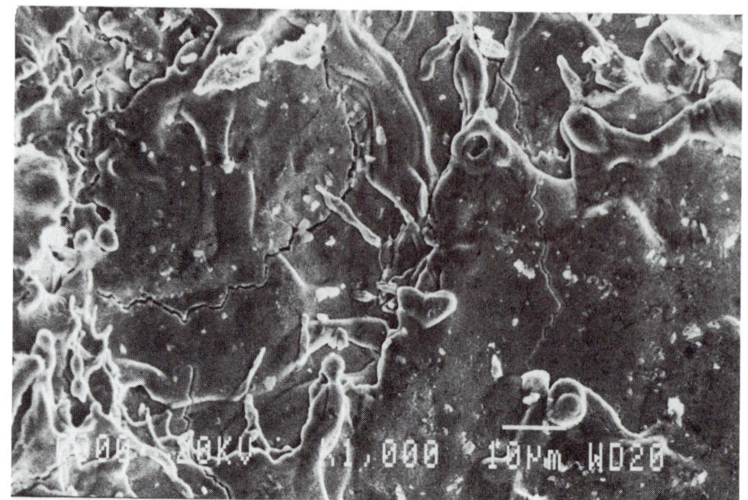

Figure 11. SEM magnification (1:1000) at indicated position (A) in fig. 10.

4.1.2. Overvolted spark-gap

4.1.2.1. Cathode

The bulk molybdenum electrodes were analyzed after 50.000 pulses according to the data of experiment 3.

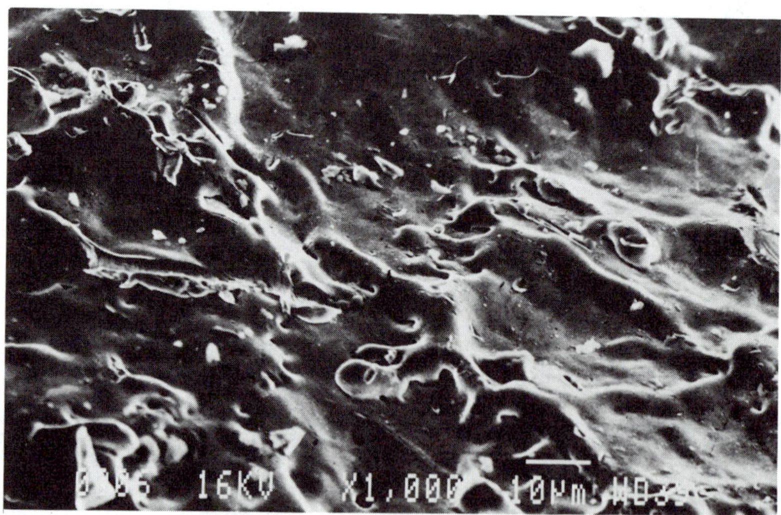

Figure 12. SEM magnification (1:1000) of Pseudo-spark anode at a position opposite to (A).

The center of the cathode surface (fig. 13) is covered with many randomly distributed droplets, which have been in the liquid state followed by a rapid solidification on the cold base material. However, the most striking effect on the cathode is the appearance of numerous holes. They are 1 - 2 µm in diameter and are scattered all over the surface. The chemical analysis showed, that the level of low melting elements is in the ppm-region. Therefore, these bubble-like structures could not be explained by the boil-off of impurities in the bulk molybdenum. It rather seems to be possible, that some part of the working gas is dissolved in the liquid state of the electrode material and is then again evaporated during solidification of the surface. In contrast to the pseudo-spark electrodes an increased number of tree-like branched cracks of about 100µm length can be found on the surface of the

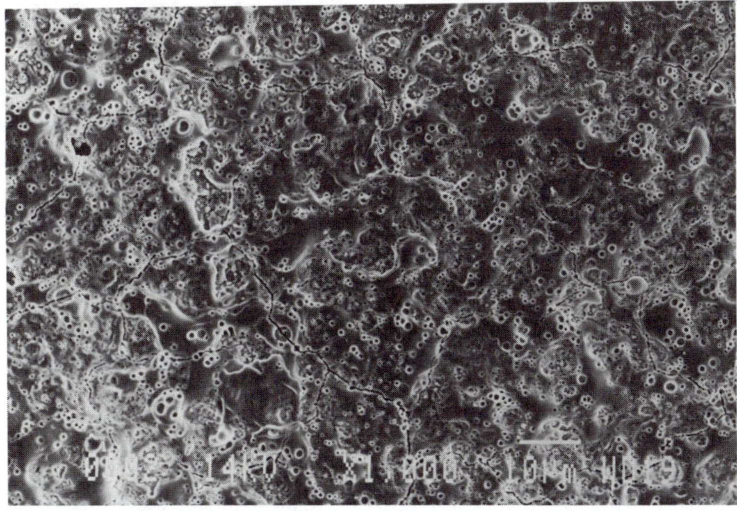

Figure 13. SEM-photograph (1:1000) of the center of the cathode in the modular 2-
electrode spark-gap after operation according to the data of experiment 3;
material: bulk molybdenum.

spark-gap cathode. This implies that a higher temperature of the surface has led to the observed stress of the material.

4.1.2.2. Anode

The anode of the high-pressure spark-gap (fig. 14) is covered by round-topped hill-like structures which are nearly regularly spaced over the entire center region of the electrode. The annular rings on these "hills" lead to the assumption that this structure has been grown by successive discharges. In another experiment it had been shown, that the hill-like structures partly consist of condensed metal vapour which was transferred from the cathode.

4.2 Behaviour of electrodes with a PVD-coating of molybdenum

4.2.1. Pseudo-spark switch

For the application in gas discharges, electrodes with a coating of highly arc-resistive materials deposited by a PVD-process have been developed (5). Typically, the base material is OFHC-copper and the thickness of the coating is about 10μm. These PVD-electrodes have already shown a superiour performance compared to commonly used bulk electrodes in the application of a TEA-C02-laser with a diffuse gas discharge. It is possible that inspite of the PVD-surface layer, the good heat conduction of the base material is maintained. Therefore it was obvious that these new kind of electrodes could be concerned for the application in other types of gas discharges.

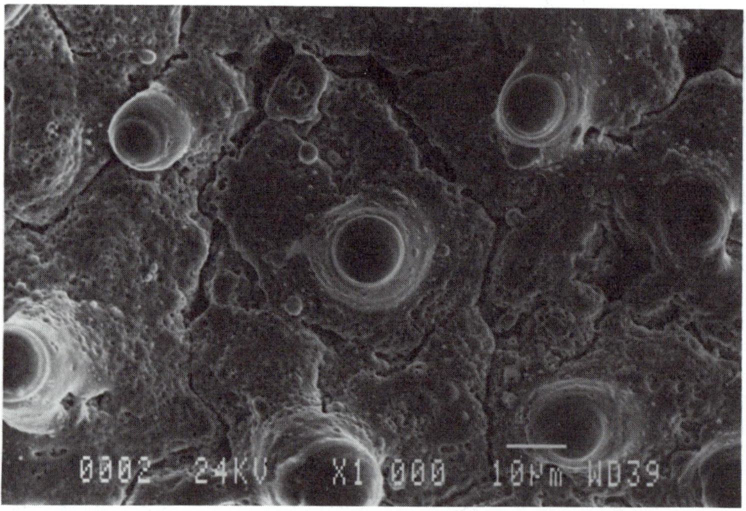

Figure 14. SEM-photograph (1:1000) of spark-gap anode made of bulk molybdenum.

The molybdenum PVD-coated electrodes were fitted into the modular pseudo-spark test device, which was part of the experimental setup shown in fig. 5. The electrical data of the corresponding experiment 2 are given in table 1. After a series of 10.000 pulses the electrodes were dismounted and investigated by the microanalytical techniques mentioned above. The microsections of cathode and anode are given in fig. 15 and in fig. 16 respectively. The first survey shows that there is no position, where the plasma interaction has completely removed the molybdenum coating. It even comes out that the original crystalline structure of the PVD-coating has been maintained. From the comparison with a sample of untreated material it was deduced, that some material transfer occurred from the edge of the upper electrode surface facing the cathode-backspace in the direction of the trigger unit (fig. 15a). The thickness of the removed layer is estimated to be in the order of 1-2μm. On the other hand a sheath of molybdenum which has grown up onto the rounded edge of the opposite anode hole can be recognized (Fig. 16a, b). The structure of this sheath clearly differs from the structure of the original PVD-coating. The microsections also show that during switch operation the PVD-coating has partly come off from the copper base material in the region of the edges and in the center of the electrode holes (Fig. 15b, c and fig. 16c, d). Obviously the adhesive strength of the PVD-coating was not sufficient.

4.2.1.1 Cathode

Fig. 17a shows a photograph of the entire cathode inset of the modular pseudo-spark switch, looking onto the surface opposite to the anode. The center region around the cathode hole is covered with distinct patterns of arc impact within a diameter of 20mm. As it can be seen in fig. 18a there is a zone on the anode, which also indicates the appearance of arcing.

The SEM-photographs in fig. 17b show a magnification of an arc spot in the vicinity of the cathode hole. Inside the spot the molybdenum surface has been molten and formed to small spherical droplets. It seems to be that the melting only occurs in a very upper sheath (1-2µm) of the molybdenum coating. From fig. 17c it even seems to be possible that arc interaction had taken place up to the inner edge of the cathode hole, because there are some indentations in the surface layer.

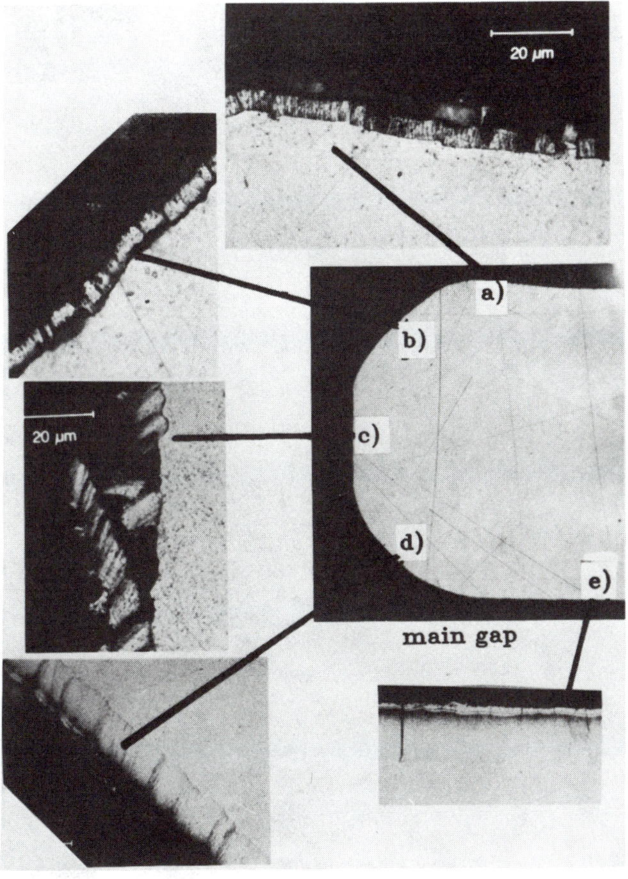

Figure 15. Microsections of modular PSS-cathode with PVD-coating of molybdenum

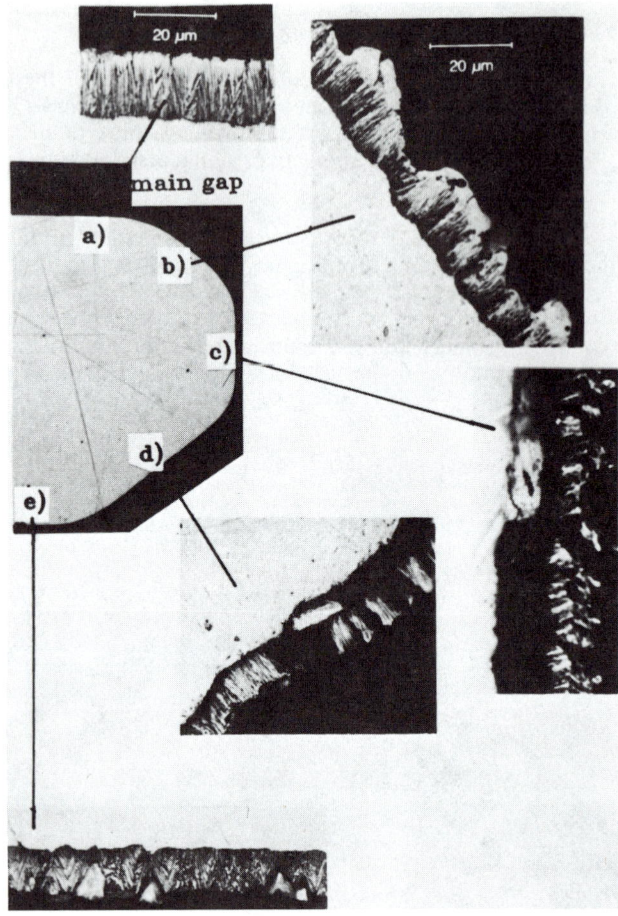

Figure 16. Microsections of modular PSS-anode with PVD- coating of molybdenum.

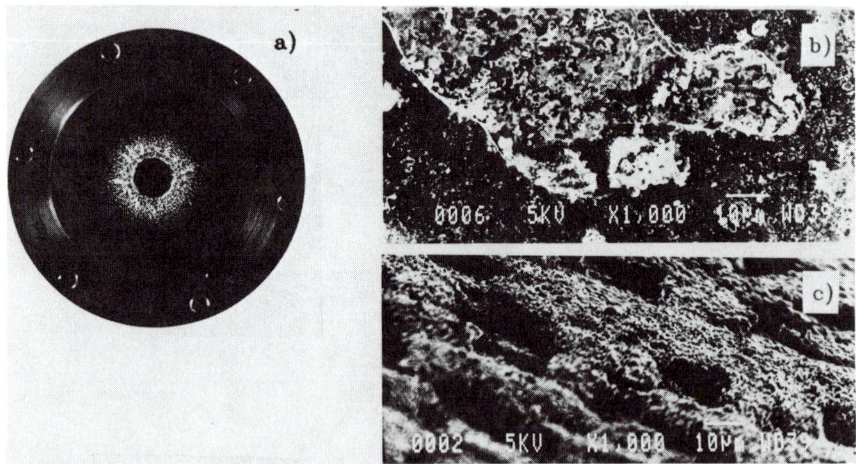

Figure 17. PVD-cathode after operation in experiment 2:
   a) overview
   b) micrograph from the vicinity of the electrode hole
   c) micrograph taken from the inner edge of the cathode hole

Up to now it is not clear, whether the appearance of these arcs simply results from insufficient cleaning of the electrodes. It is also possible, that the diffuse character of the pseudo-spark in the initial state changes into a hollow cathode arc in course of the pulse duration.

4.2.1.2 Anode

Inside the anode hole an overall melting of the surface occurred within an estimated depth of 1-2 µm (fig. 18b). The molybdenum is formed to small droplets. No indication for arcing in this region could be found. In the vicinity of the anode hole (fig. 18c) the material has been in the liquid state forming relatively large, scaled structures, which are similar to those observed with the bulk material of the PSS (compare to fig. 12).

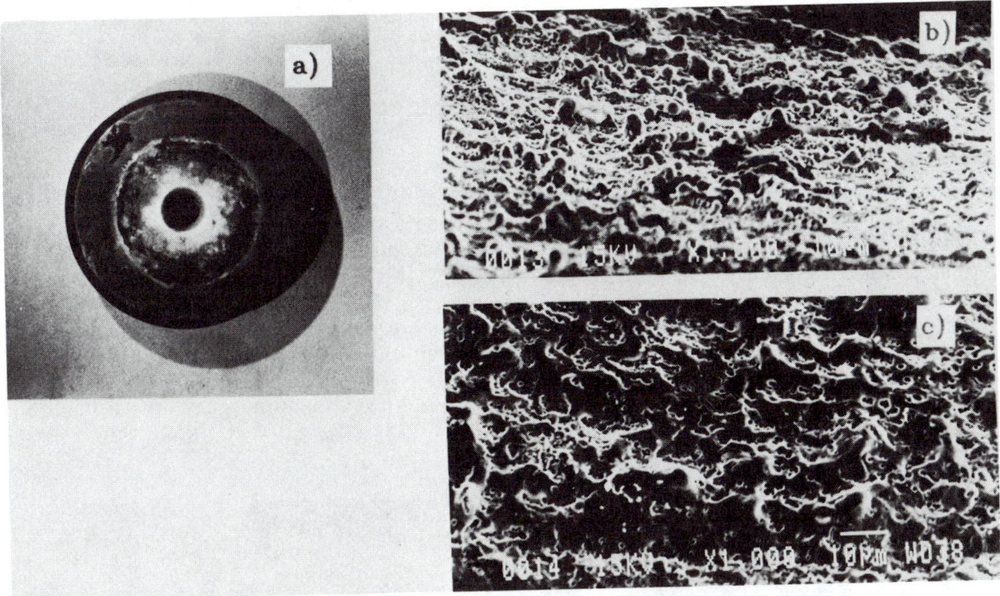

Figure 18. PVD-anode after operation in experiment 2:
a) overview
b) micrograph taken from the inner edge of the hole
c) micrograph taken from the vicinity of the hole

4.2.2 Overvolted spark-gap

The molybdenum coated electrodes were also tested in the modular spark-gap according to the data of experiment 3 (see table 1). The essential result is, that after a series of 50.000 pulses the molybdenum coating was completely removed in the center of the cathode. Within a diameter of about 2.5mm the pure copper was facing the discharge region. Outside the center area the surface is still covered with the molybdenum layer which is shown in fig. 19. Again it can be seen that the molybdenum layer has been molten and formed to flat droplets. They seem to have been rapidly frozen on the cold copper ground. From the fact that almost no bubbles appear it can be concluded, that the temperature was relatively low compared to the bulk material.

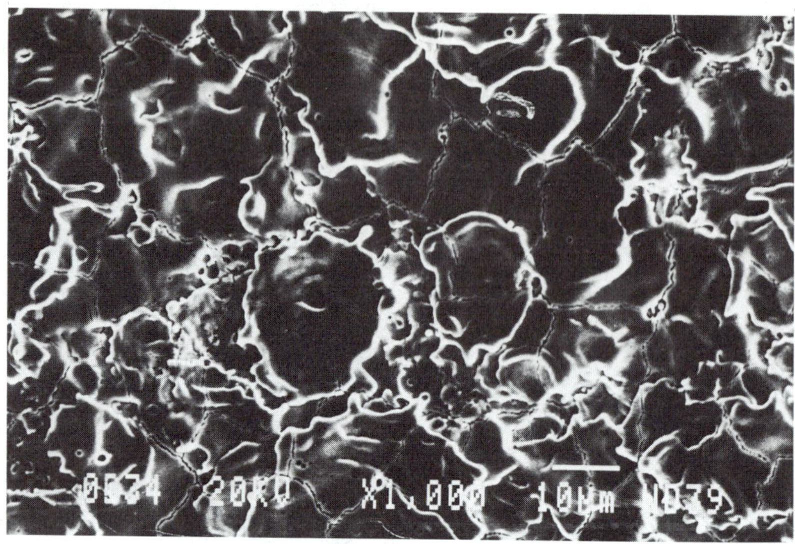

Figure 19. SEM micrograph 1:1000. Cathode of 2-electrode spark-gap after
application in experiment 3. Material: bulk copper coated with a
10 µm-layer of molybdenum

The comparison between bulk material and coated material shows, how the penetration of the discharge plasma into the electrode surface is influenced by the different heat conductivity of the base materials.

The anode of the PVD-coated spark-gap electrodes (fig. 20) shows the same hill-like structures as in the case of the bulk molybdenum, but their number is significantly increased. This might be due to the surface topology of the untreated material, which is quite different in both cases. The hills consist of pure molybdenum as a result of material transfer from the cathode. The smooth background is covered with an extremely thin copper-layer.

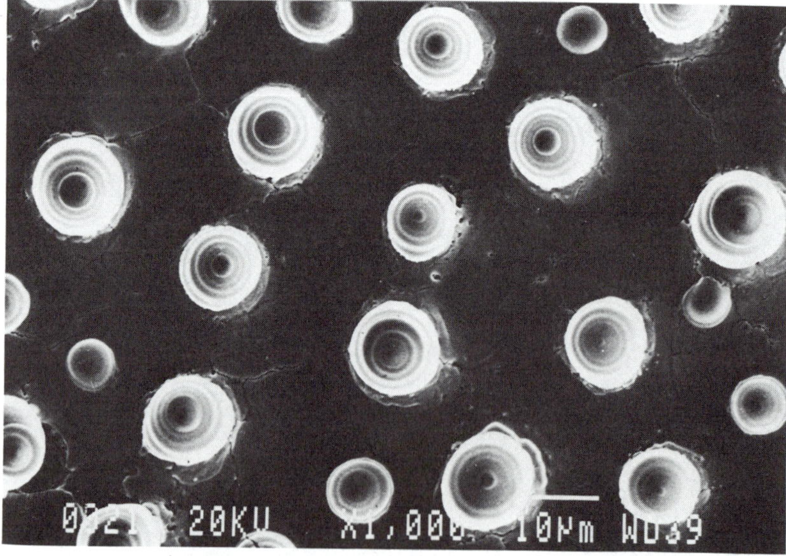

Figure 20. SEM micrograph 1:1000. Anode of 2-electrode spark-gap. Material:
Mo-coated copper. Note: the "hills" consist of pure Molybdenum. The
bottom is covered with a thin copper-layer.

5. DISCUSSION

By comparing the SEM photographs of bulk metal surfaces, which were aged in two different types of gas discharges, one can summarize, that in the case of the pseudo-spark there is no evidence for arcing inside the holes of the main electrodes. This confirms the results of earlier work about the diffuse character of this discharge at very high current densities. Concerning the electrical data of the experiments 1 and 3 it becomes quite obvious, that the electrodes in the high-pressure spark gap had been exposed to a much lower stress compared to the pseudo-spark. However in the spark-gap much more severe modifications of the electrode surface occurred. It seems to be, that the high current being concentrated in a narrow spark channel is one of the main reasons for this.

When the micrographs of the bulk cathodes in pseudo-spark and spark-gap are compared (fig. 11 and fig. 13) one can recognize that there are significantly more cracks on the surface of the spark-gap electrode. This can be correlated to the fact that these gas discharges produce different microstructures of the cathode surface layer, which is shown in fig. 21 for spark-gap and in fig. 22 for the pseudo-spark respectively. Fig. 21 shows that the interaction of the spark-gap discharge goes into the bulk material up to the order of a few micrometers. In contrast to that, the microsection of the pseudo-spark (fig. 22) clearly shows that inside the cathode hole a surface layer has been grown by the long term operation of the switch. The layer has a thickness of about 20μm. This considerable amount of electrode material was transferred during the operation of the device. It could be established that this material has been removed from the backspace-side of the cathode leading to an increase of the hole diameter, which results in a drift of switch performance in terms of delay and jitter. This in turn can be the reason for a premature end of life of the pseudo-spark switch. It becomes quite clear that this problem gives rise for further optimisation of switch geometry and electrode material.

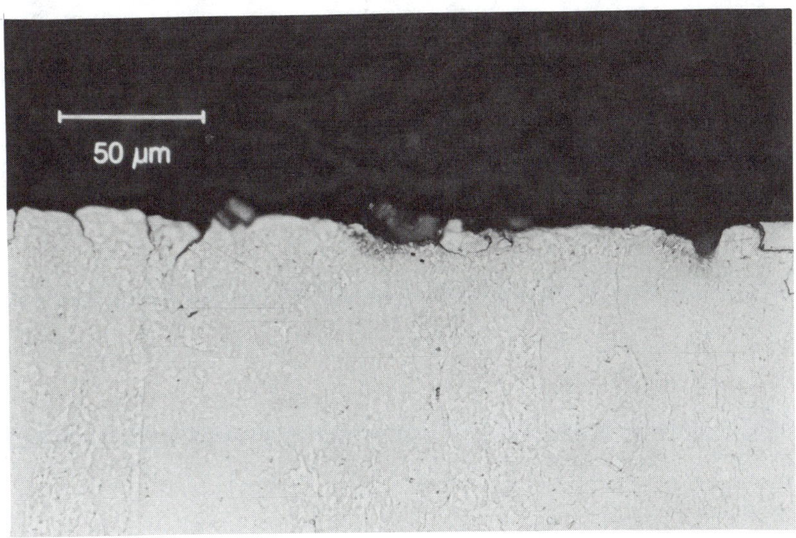

Figure 21. Microsection of spark-gap cathode. The cracks which are shown in fig. 13 run into the bulk material up to a depth of about 10μm.

Assuming that the transferred charge and energy are the dominant parameters for material erosion, the PVD-coated electrodes were stressed under comparable conditions in the pseudo-spark and in the spark gap application (experiment 2 and 3). In the case of the pseudo-spark the electrodes showed relatively low degradiation effects on the surface layer. The comparison of the micrographs taken from the pseudo spark (fig. 17 and fig. 18) and from the spark-gap (fig. 19 and fig. 20) give the impression that in the case of the pseudo spark the interaction of the plasma had only influenced the upper sheath of the PVD-coating. From the study of the micrographs it becomes quite obvious that only a relatively small amount of the upper material concentrated in small spots has reached the melting temperature, whereas in the case of the spark-gap the discrete surfaces structures are distributed over a larger area.

Further investigations are necessary in order to decide whether PVD-electrodes could be applied for long term operation at low current levels in a PSS-gas discharge device.

In the case of the spark gap the PVD-coating was completely destroyed in a center region. The thickness of the sheath was obviously not sufficient in order to withstand the strong arc interaction during long term repetitive switching, even at a low peak current level.

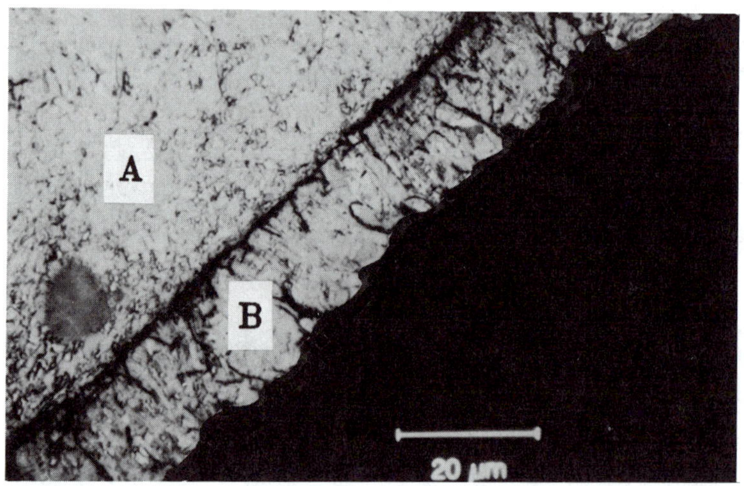

Figure 22. Microsection of Pseudo-spark cathode at the edge of the electrode hole; (A) bulk material (Mo); (B) molybdenum sheath deposited during successive discharges.

6. CONCLUSIONS

It has been shown, that the influence of the plasma on the electrodes leads to significant surface modifications in different kinds of gas discharges like low-pressure pseudo spark and high-pressure spark gap. By comparing the SEM-photographs it can be stated, that in the case of the pseudo-spark the electrode surfaces give no rise for the appearance of arc interaction inside the hole of the main gap. On the other hand the micro-sections showed, that transfer of electrode materials occurred. The resulting variation of the gap geometry leads to a drift of switch parameters in terms of delay and jitter. In case of the spark-gap the result of long term tests is that the life of these devices is mainly limited by removal of electrode material and subsequent metalisation of the insulator wall.

From these facts we conclude that, in order to optimize the reliability of gas discharge devices, material research will become more and more important in the near future. A detailed understanding of the processes correlated to plasma-electrode interaction is necessary in order to improve the switching capabilities. For the systematic material development modular test devices of spark-gap and pseudo-spark switches are available, which can be operated under sealed-off conditions. These systems can be used in order to establish the specifications for the electrode material.

REFERENCES

(1)   K. Frank, J. Christiansen, "The fundamentals of the pseudospark and its applications", Proc. of XIIIth Int. Symp. on Discharges and Electrical Insulation in Vacuum, Paris 1988.

(2)   W. Hartmann, V. Dominic, G. F. Kirkman, M. A. Gundersen, "Evidence for large-area superemission into high-current glow discharge", Appl. Phys. Lett. **53** (18), 1988.

(3)   W. Hartmann, M. A. Gundersen, "Origin of anomalous emission in superdense glow discharge", Phys Rev Lett. **60** (23), 1988.

(4)   W. Hartmann, G. Kirkman, V. Dominic, M. A. Gundersen, "A super-emissive self-heated cathode for high-power applications", IEEE Trans. on Electron Devices **36** (4), 1989.

(5)   DODUCO GMBH + CO, D7130 Pforzheim, FRG, patents pending.

MAPPING AND MODELING OF THE CATHODE FALL
AND NEGATIVE GLOW REGIONS

J. E. Lawler*, E. A. Den Hartog*, W. N. G. Hitchon[†], T. R. O'Brian*, and T. J. Sommerer*

*Department of Physics
[†]Department of Electrical and Computer Engineering
University of Wisconsin, Madison, WI 53706, U.S.A.

ABSTRACT

Experimental and theoretical work on the cathode region of a cold-cathode helium glow discharge is described. A current balance, or ratio of ion to electron current, at the cathode is deduced from precise electric field maps and gas density measurements for five discharge current densities. Monte Carlo simulations are compared to the empirical current balance for the five cases. New discharge diagnostics based on LIF are described and are used to determine the density and temperature of the low energy electrons trapped in the negative glow. The Convective Scheme is described and used for a self-consistent (electric field) kinetic model of the cathode fall region. Prospects are good for self-consistent kinetic models from electrode-to-electrode based on the Convective Scheme.

I. INTRODUCTION

Many important pulsed power devices are based on diffuse discharge plasmas. The performance and stability of these devices is often limited by the cathode region in the discharge plasma. The cathode region is the least understood part of a typical discharge plasma. Experimental work on the cathode region is difficult for many reasons, such as the order-of-magnitude variation of key parameters in short distances. Theoretical work is difficult because essentially none of the standard approximations of plasma physics or gaseous electronics can be used from the surface of the cathode continuously through the negative glow.[1,2]

The research described in this paper was performed at the University of Wisconsin in recent years. The goal of the continuing work is to develop a detailed quantitative understanding of the cathode region. We chose to study a clean, reproducible, nearly one dimensional glow discharge plasma in helium. We developed and applied new laser-induced fluorescence[3] and optogalvanic[4,5] diagnostics in this work. These diagnostics were used to produce spatial maps of key parameters such as the electric field, charged particle densities, and charged particle fluxes in the cathode region. The maps provide important insight into the cathode region and provide a means for testing various models.

The cathode region is conveniently divided into a cathode fall and negative glow region. Large electric fields and a low density of high energy "beam" electrons are found in the cathode fall region. The negative glow region is characterized by very small electric fields and a high density of low energy electrons. We define the boundary between these two regions to be the position where the cathode fall electric field extrapolates to zero, since the field magnitude decreases almost linearly with distance from the cathode.

Physics and Applications of Pseudosparks
Edited by M. A. Gundersen and G. Schaefer
Plenum Press, New York, 1990

A realistic model of electrons in the cathode region must avoid the hydrodynamic equilibrium approximation, which is a central approximation in much of gaseous electronics.[1,2] In this approximation the velocity dependence of the electron distribution function is determined by the local E/N (ratio of electric field to gas density) and the electron flux density is modeled using a density gradient expansion. Coefficients in the density gradient expansion are the familiar transport coefficients such as: Townsend alpha coefficients, drift velocities, and diffusion coefficients.[6] The failure of this hydrodynamic equilibrium approximation (or local field approximation) in the cathode fall region is due to the large, rapidly changing field, and due to the proximity of the cathode. The negative glow is also a nonhydrodynamic region because it has a low density of high energy beam electrons injected from the cathode fall region. Space charge neutrality is maintained in the negative glow by a high density of low energy electrons which are trapped in a weak potential energy well. A realistic model of the cathode region should start from the Boltzmann equation (kinetic theory).

We have used Monte Carlo simulations of electron avalanches in the cathode region to interpret experimental results and to test theoretical models. The simulations are equivalent to exact solutions to the Boltzmann equation. Monte Carlo simulations are inexpensive and relatively fast when performed using a null collision algorithm.[7] They are unfortunately not yet fast enough for fully self consistent (electric field) calculations. They are also not suited to studying the low energy electrons which are trapped in the negative glow. The Convective Scheme, which we recently introduced to gaseous electronics, preserves the attractive features of Monte Carlo simulations, especially the realistic detail.[8] The Convective Scheme was recently used in a self-consistent kinetic model of the cathode fall. There are good prospects for using the Convective Scheme for self-consistent electrode-to-electrode calculations at the kinetic theory level.

The structure of this paper is as follows. Section II is a brief description of the experiment. Section III and IV address two old and difficult questions on the cathode region: "What fraction of the discharge current is carried by ions at the cathode surface?" "Where are those ions created?" Section V describes research on measuring the density and temperature of the low energy electrons trapped in the negative glow. Finally, Section VI is an introduction to the Convective Scheme.

II. EXPERIMENTAL

The experiments described in this paper were performed in a clean, reproducible, nearly one dimensional He discharge. Helium was chosen because it is ineffective at sputtering the cathode, and because the cross section data base is better for He than for almost any other gas.

Figure 1 is a schematic of the experimental apparatus. The discharge used in these studies is produced between flat circular Al electrodes 3.2 cm in diameter and separated by 0.62 cm. The electrodes are water-cooled to minimize gas heating. The discharge tube is made primarily of glass and stainless steel. Most of the large seals are made with knife edge flanges on Cu gaskets. The only exceptions are the high vacuum epoxy seals around the fused silica Brewster windows. A liquid N_2 trapped diffusion pump evacuates the tube to 2×10^{-8} Torr. When no liquid N_2 is in the trap an ion pump maintains the vacuum to prevent oil from back-diffusing into the system. The leak rate into the sealed discharge tube is approximately 3×10^{-4} Torr/day. For discharge operation ultra high purity (0.999999) He is slowly flowed through the system. A capacitive manometer monitors the pressure which is maintained at 3.5 Torr. The He first passes through a cataphoresis discharge to remove any residual contaminants before entering the main discharge tube. Emission spectra reveal only weak Al and H impurity lines. Although He is much less effective at sputtering than heavy inert gases, some sputtering of the Al cathode does occur. We suspect that the slow erosion of the cathode gives rise to both the weak Al and H impurity lines. Many metals absorb hydrogen. The surface of the Al cathode is cleaned in-situ by running an Ar discharge. Argon is very effective at sputtering most metals including Al. A carefully prepared cathode provides discharge V-I characteristics which are stable to a few percent over a period of a month.

Experiments are carried out over a range of discharge current densities from 0.190 mA/cm^2 to 1.5 mA/cm^2. The low current density corresponds to a near normal cathode fall voltage of 173 V and the high current density corresponds to a highly abnormal cathode fall voltage of 600 V. The discharge current is spread uniformly across the surface of the electrodes. The absence of significant fringing was verified by segmenting the 3.2 cm diameter cathode into a 1.6 cm diameter disc and a close fitting annulus with an outside diameter of 3.2 cm. Each part of the cathode was maintained at the same potential during operation and the average current density on each was measured. The current density was found to be uniform across the cathode. This measurement is one of the justifications for using one dimensional models of the cathode fall region. The segmented cathode was replaced by a solid cathode after the current density measurements.

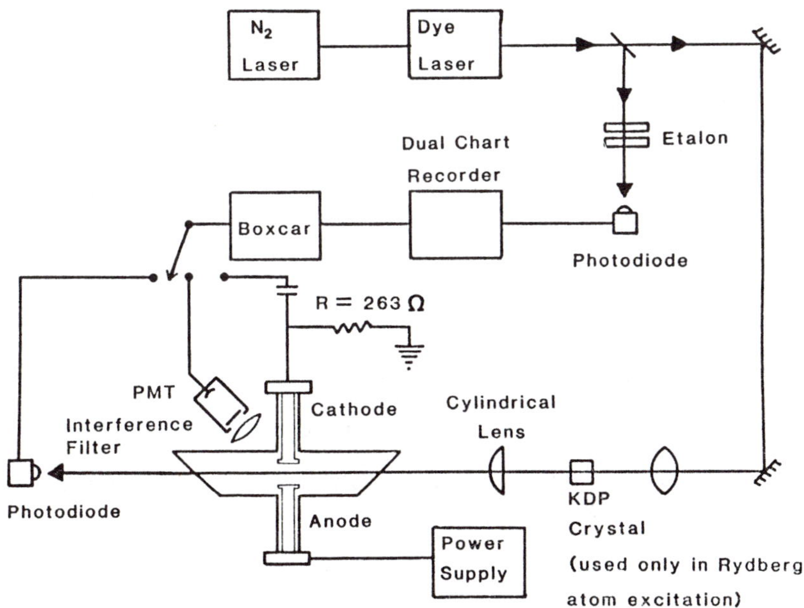

Fig. 1. Schematic of the experimental apparatus showing three detection methods: optogalvanic, fluorescence, and absorption detection.

The laser used is a N$_2$ laser pumped dye laser. Several different configurations for the dye laser are used, depending on the specific experiment. The dye laser bandwidth is 0.3 cm^{-1} without an etalon. An etalon is used to reduce the bandwidth to 0.01 cm^{-1} (300 MHz) for some measurements. The dye laser with frequency doubling is tunable over a wavelength range 200-700 nm.

Figure 1 shows all three detection schemes used in these experiments. Optogalvanic detection, fluorescence detection, and absorption detection each have unique advantages which are discussed in Ref. 10. In order to perform spatially resolved measurements without disturbing laser alignment the discharge is mounted on a precision translation stage. Good spatial resolution is achieved, when necessary, by passing the laser beam through a narrow slit and imaging the slit into the discharge region.

III. EMPIRICAL CURRENT BALANCE AT THE CATHODE SURFACE

What fraction of the discharge current is carried by ions at the cathode surface? This is an old question, central to any discussion of the cathode region, but it has not been an easy question to answer. Vacuum measurements of γ°, the electron emission coefficient for ion bombardment, do not directly answer the question. Electron emission due to metastable atom and UV photon bombardment is not included in a vacuum measurement of γ. Furthermore, there is evidence that cathode surface conditions change when a discharge starts.

Doughty, Den Hartog, and Lawler answered the question by an analysis of precise electric field maps and gas density measurements.[9,10] They determined the current balance or ratio of ion to electron current at the cathode surface in a clean He discharge. A spatial gradient of the electric field determines the ion charge density near the cathode by Poisson's equation. The electron density is negligible in the high field cathode fall region. This analysis depends on a rigorous proof that the ions are in hydrodynamic equilibrium throughout most of the cathode fall region. The E/N near the cathode surface determines the ion drift velocity because the ions are in hydrodynamic equilibrium. An ion current density determined as the product of the ion charge density and ion drift velocity was subtracted from the total discharge current density to determine the electron current density.

Optogalvanic detection of Rydberg atoms was used to map the electric field in the cathode fall region. This technique was developed in 1984 by Doughty, Salih, and Lawler.[4,5] Fragile Rydberg atoms are an ideal discharge probe. They exhibit large, linear Stark effects which can be accurately (~1%) measured, and easily interpreted using nearly hydrogenic single-electron calculations. Optogalvanic methods make it possible to detect the Rydberg atoms which do not survive long enough to fluoresce in a typical discharge. Laser techniques have a number of important advantages over the traditional electron beam deflection technique.[11] Laser techniques are useful at higher pressures and discharge current densities. It is easier to use a spectroscopic technique in a high purity discharge system. A laser technique is truly nonperturbing when performed with a nanosecond pulsed laser because the field is measured when the atoms or molecules absorb the laser light, and any perturbation to the discharge fields occurs on a longer time scale. Lasers have the potential for making measurements with a few microns spatial resolution and nanosecond temporal resolution. A technique based on Rydberg atoms has broad applicability because all atoms and molecules have Rydberg levels. It also has a wide dynamic range because one can choose an appropriate principal quantum number to achieve a desired level of sensitivity to the electric field.

A much more detailed discussion of electric field measurements using optogalvanic detection of Rydberg atoms has been published.[10] Figure 2 shows the electric field maps for five different current densities. The fields are accurate on average to ~1%. This accuracy is verified by integrating the fields across the cathode fall to determine a potential difference within ~1% of the discharge voltage measured using a digital voltmeter. The field magnitude decreases almost linearly with distance from the cathode. A slight negative curvature is visible in Fig. 2. The spatial gradient of the field gives the ion charge density according to Poisson's equation.

The average ion velocity can be determined from the field maps if the ions are in hydrodynamic equilibrium and if the gas density is known. Ion motion is limited by the symmetric charge exchange reaction,

$$He^+(fast) + He(slow) \rightarrow He(fast) + He^+(slow)$$

This reaction has a large cross section which is only very weakly dependent on energy. Charge exchange effectively stops the ions every collision and wipes out all memory of their history. It seems intuitive that ions, unlike the electrons, should be in hydrodynamic equilibrium throughout most of the cathode fall region, because this region has a thickness equivalent to $50 \rightarrow 100$ mean-free-paths for symmetric-charge-exchange. The large field gradient, and the distributed source of new ions from electron impact ionization throughout the cathode fall, makes one reluctant to rely on an intuitive "proof" that ions are equilibrated. Lawler has published a rigorous proof based on analytic solutions to the Boltzmann equation for ions.[12] In this work exact Green's functions for the Boltzmann equation were used to compute ion equilibration distances for four idealized situations. Equilibration distance is here defined as the distance required for the average velocity of the ions to approach within 10% of the equilibrium drift velocity. The four idealized situations and corresponding equilibration distances (expressed in mean-free-paths) are: A) a planar source of ionization in a uniform field (0.65), B) a uniform source of ionization in a

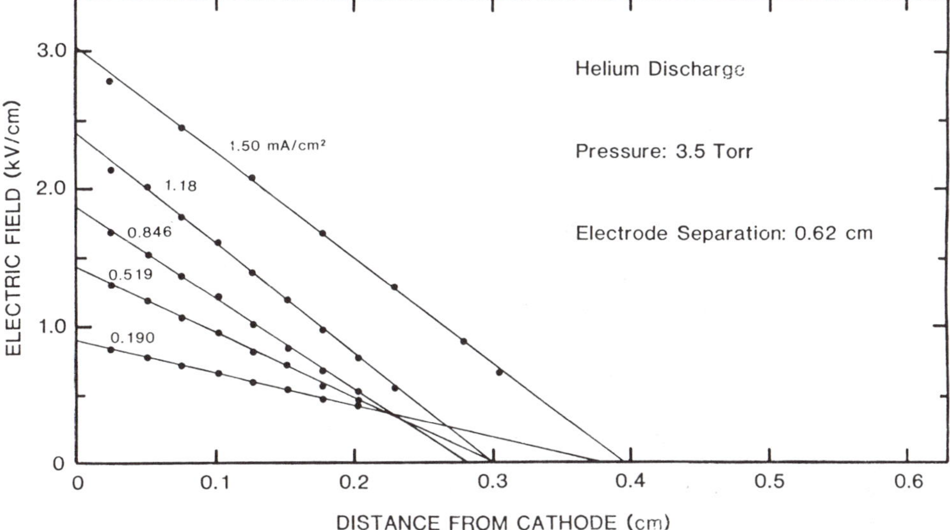

Fig. 2.　Electric field as a function of distance from the cathode for five current densities, all at 3.50 Torr. The lines are linear least-square fits to the data. The anode corresponds to the right-hand side of the figure.

uniform field (4.5), C) a planar source of ionization near the origin of a linearly increasing field (1.7), and D) a uniform source of ionization in a linearly increasing field (5.7). Numerous Monte Carlo simulations of electron avalanches have shown that the ionization rate per unit volume peaks near the cathode fall-negative glow boundary where the electric field extrapolates to zero [1,9,10]. Thus case C) and D) provide a rigorous lower and upper bound for the ion equilibration distance as measured from the cathode fall-negative glow boundary. The ions are in hydrodynamic equilibrium throughout the cathode fall, except for a narrow region within 6 mean-free-paths of the cathode fall-negative glow boundary.

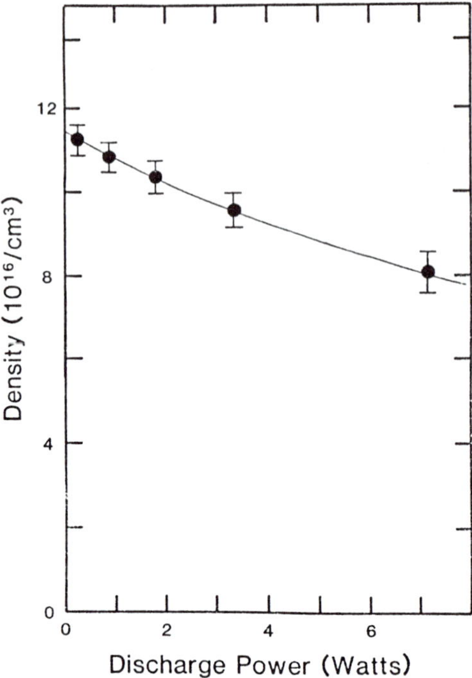

Fig. 3.   Experimentally derived gas density as a function of discharge power measured at a constant pressure of 3.50 Torr.

The final requirement for determining the current balance is a measurement of the gas density in the cathode fall. The discharge tube pressure in these experiments was fixed at 3.5 Torr. In the cathode fall region of an abnormal discharge gas heating is often significant. The symmetric charge exchange reaction which produces a short ion equilibration distance also serves to convert most of the electrical power dissipated in the fall directly into heavy particle (atomic) motion and into heat. Doughty et al. used the Doppler width of a carefully selected He transition to determine the gas temperature and gas density versus discharge power shown in Fig. 3. The linewidth of the 501.6 nm transition from the 2^1S to 3^1P level was measured using laser optogalvanic detection. Various contributions to the observed linewidth including a laser bandwidth of 300 MHz, a natural width of 92 MHz, pressure broadening at 3.5 Torr of 146 MHz, and Stark broadening on the order of 100 MHz, are much smaller than the Doppler width at 293K of 3.66 GHz.[10] It was somewhat surprising that the water cooled electrodes were not more effective in limiting the gas temperature rise, but the electrode assemblies have large thermal impedances. The gas temperature rise is consistent with the expansion of the cathode fall region at high current densities as shown in Fig. 2.

The field of maps of Fig. 2 and the density measurements of Fig. 3 were used to determine the average ion velocity at the cathode. Helm's precise He^+ ion mobilities were used for the lower currents where the E/N is less than 1500 Td.[13] The equilibrium drift velocity $\sqrt{2eE/(m_+\pi\sigma N)}$, where m_+ is the ion mass and e the unit charge was used for higher currents. The symmetric charge exchange cross section, σ, was taken from Sinha, Lin and Bardsley.[14] These theoretical cross sections agreed with Helm's experimentally derived cross sections.

The product of the ion charge density and average ion velocity at the cathode is the ion current density J_+^o. Figure 4 is a plot of the current balance $J_+^o/(J_D\text{-}J_+^o)$ versus the

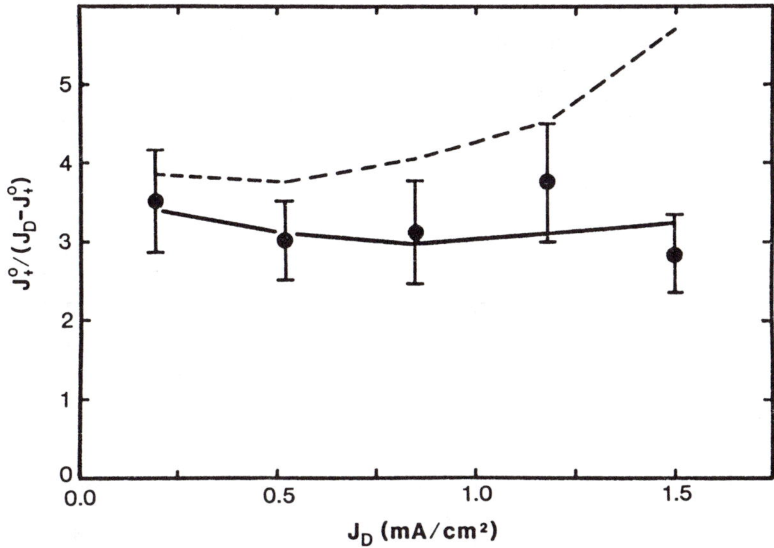

Fig. 4. Region of ion current to electron current at the cathode surface $J_+^O/(J_D-J_+^O)$ as a function of discharge current density. The points are the empirically derived values; the dashed line is the Monte Carlo results assuming the isotropic scattering for ionizing collisions; the solid line is the Monte Carlo results incorporating the anisotropic scattering for ionizing collisions.

discharge current density J_D. It is interesting to note that the current balance, or ratio of ion to electron current, at the cathode is ~3.3 and is essentially independent of total discharge current.

IV. COMPARISON TO MONTE CARLO SIMULATIONS

Where are the ions which hit the cathode produced? It is obvious that all ions formed by electron impact ionization in the cathode fall region must hit the cathode because of the high field driving them to the cathode. The question can be rephrased a number of different ways. Is there enough ionization in the high field cathode fall region to produce all of the ion current at the cathode? Do the electrons carry the discharge current across the cathode fall-negative glow boundary? The question no matter how it is phrased, is also an old, difficult question and is central to any discussion of the cathode region.

Monte Carlo simulations were used to answer the question for He discharges described in the preceeding paragraph. Simulations were performed by launching electrons from the cathode and counting ionization events as an avalanche develops between the cathode surface and cathode fall-negative glow boundary. Empirical field maps and gas densities were used in the Monte Carlo simulations. The powerful null-collision Monte Carlo algorithm developed by Boeuf and Marode[7] was used. Elastic scattering cross sections from LaBahn and Callaway and inelastic scattering cross sections from Alkhazov were used.[15,16] The exact anisotropic angular distribution was used for the elastic scattering events. Two different approximations were used for inelastic events.[10] The first set of simulations used an isotropic scattering approximation for all inelastic collisions. The second set of simulations used a forward peaked anisotropic distribution for ionizing collisions. The number of ionization events in the cathode fall region per electron released from the cathode must be less than or equal to the empirical current balance. The number can be equal to the current balance if the electrons are carrying all of the discharge current across the cathode fall-negative glow boundary.

Figure 5 is a Monte Carlo histogram showing the number of ionization events per electron launched from the cathode as a function of distance from the cathode. This histogram includes both direct electron impact ionization of ground state atoms and associative ionization of highly excited He atoms. The Monte Carlo simulations included excitation to 22 levels, and assumed that 25% of the atoms excited to 1s3p or higher levels are associatively ionized. Direct ionization dominates over associative ionization which was included, and over multistep ionization of metastable atoms which was not included in the simulations.

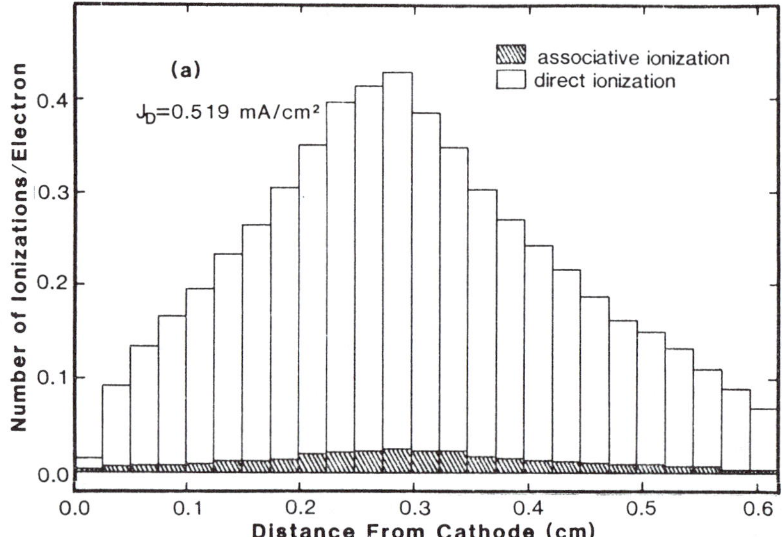

Fig. 5.   Monte Carlo histogram showing the number of ionization events per (net) electron emitted from the cathode as a function of distance from the cathode. The histogram is divided into direct and associative ionization. The Monte Carlo simulation included an anisotropic angular distribution for ionizing collisions.

Figure 5 gives the spatial dependence of the ionization rate per unit volume. It is described by a roughly symmetric function which peaks near the cathode fall-negative glow boundary. This boundary is 0.301 cm from the cathode where the electric field extrapolates to zero. The simulations show that the beam electrons penetrate into the negative glow a distance comparable to the thickness of the high field cathode fall region. The phrase "beam electrons" is used to include all nonhydrodynamic high energy electrons, even if they have suffered inelastic collisions. The histogram of Fig. 5 is from a simulation using the anisotropic angular distribution for direct ionization events.

The cumulative number of ionization events in an avalanche is summed from the cathode surface to the cathode fall-negative glow boundary. These sums are compared to the empirical current balance in Fig. 4. The first set of simulations used the standard isotropic scattering approximation for all inelastic collisions. These first simulations, which are represented by the doted line in Fig. 4, predict a current balance in agreement with experiment for a near normal cathode fall, but predicted too much ionization in the abnormal cathode fall. The second set of simulations use an anisotropic angular distribution for ionizing collisions. These second simulations, which are represented by the solid line in Fig. 4, predict a current balance in agreement with experiment for all discharge currents. Clearly, anisotropic angular distributions for inelastic scattering events must be included in a model of an abnormal cathode fall.

By comparing the empirical current balance and the Monte Carlo simulations we conclude that the vast majority of ions hitting the cathode are created by electron impact ionization between the cathode surface and the cathode fall-negative glow boundary where the electric field extrapolates to zero. This implies that the discharge current is carried almost entirely by high energy beam electrons at the cathode fall-negative glow boundary. The issue of the electron versus ion current at the cathode fall-negative glow boundary was already old and (in)famous when Druyvesteyn and Penning wrote their 1940 review.[17] They did not take a definite position on the issue in their review. Other experimental and theoretical studies addressed this problem in more recent decades. Long, for example, has extensive discussions.[2] He argues convincingly for a field reversal in the negative glow, but does not specify the position of the field reversal. Emeleus denotes an entire section of his 1981 review to the cathode fall-negative glow boundary, or Glimmsaum as he calls it.[18] He also fails to take a definite position on the importance of the ion current at the boundary. The strongest position was taken by Little and von Engel.[19] The arguments used by Doughty, Den Hartog, and Lawler, which involve the empirical current balance and Monte Carlo simulations, are not the same as the arguments used by Little and von Engel. the conclusion of both investigations is the same: the ion current is a negligible fraction of the discharge current at the cathode fall-negative glow boundary.

V. THE NEGATIVE GLOW

The conclusion of the last section leads to more questions. What is the density n_e and temperature T_e of the low energy electron in the negative glow? What happens to the excess ions produced by energetic beam electrons which stream into the negative glow from the cathode fall region? Most of the ions in the negative glow must either recombine or drift to the anode in a process analogous to ambipolar diffusion. The observation of He_2^* emission is unambiguous evidence for recombination in the negative glow.[10] A quantitative assessment of recombination requires knowledge of the density and temperature of the low energy electrons.

During experiments to map the absolute density of He 2^1S and 2^3S metastables we made an unexpected observation that provided critical information on the low energy electrons in the negative glow.[10] We observed a severe suppression of the He 2^1S metastable density in the negative glow caused by the reaction

$$He(2^1S) + e^- \rightarrow He(2^3S) + e^- + 0.79 \text{ eV}. \tag{1}$$

A detailed analysis of the metastables maps enabled us to derive an effective rate for the reaction.[3,10] This rate, which is a product of a temperature dependent rate constant and the electron density, provides a relation between n_e and T_e of the low energy electrons in the negative glow. The relation between n_e and T_e yields a lower limit (2×10^{11} cm^{-3}) on n_e and an upper limit (0.25 eV) on $k_B T_e$.

A second independent relation between n_e and T_e was determined by studying population transfer between low Rydberg levels due to low energy electron collisions.[3] In this Rydberg-atom experiment a low Rydberg level is selectively populated with a laser. The electron collisional transfer rate from the level populated with a laser to a higher-lying Rydberg level is determined by measuring the ratio of the laser-induced fluorescence (LIF) signals from the two levels. Although high Rydberg levels do not survive sufficiently long to fluorescence in a typical discharge, low Rydberg levels do fluoresce. This second rate can also be written as the product of n_e and a rate constant which depends on T_e, thus determining a second relation between n_e and T_e.

The first diagnostic is based on the exothermic metastable spin conversion and yields a relation in which n_e increases with increasing T_e. The second diagnostic, based on the endothermic collisional transfer between Rydberg levels, yields a relation in which n_e decreases with increasing T_e. The intersection of the two relations yields the density n_e and temperature T_e of the low energy electrons in the negative glow.

The following paragraphs contain a more detailed description of the first diagnostic which is based on the metastable spin conversion reaction. The results in this section are for the 0.846 mA/cm^2 discharge described in the previous sections.

The metastable density measurements are made in two steps: 1) LIF is used to make maps of relative metastable densities, and 2) absorption measurements are made to put absolute scales on the density maps. The 2^1S metastables are mapped by driving the 2^1S - 3^1P transition at 501.6 nm with the laser and observing 3^1D - 2^1P fluorescence at 667.8 nm. The 2^3S metastables are mapped by driving 2^3S - 3^3P transition at 388.9 nm with the laser and observing 3^3S - 2^3P fluorescence at 706.5 nm. These laser transitions do not require frequency doubling. The collimated laser is incident on a 0.25 mm slit which is imaged 1:1 into the discharge. A fluorescence collection system composed of a lens, interference filter, and photomultiplier tube is used. Signals are recorded with a Transiac 2001S digitizer with a 4100 averaging memory interfaced with an IBM PC.

The laser absorption measurements, needed for determining an absolute scale for the density maps, are made using the same laser transitions as the LIF measurements. The laser bandwidth is reduced to 500 MHz by introduction of an etalon. Absorption measurements are made at the peak of the spatial profile with 0.50 mm resolution. Saturation of the transition is avoided by reducing the laser power with neutral density filters. The transmitted laser intensity is monitored with a photodiode detector. The photodiode signal is processed with a boxcar averager and plotted on a stripchart recorder as the laser frequency is scanned through the transition. Metastable densities are determined from the integral over frequency of the natural log of the transmittance. Ten absorption measurements are averaged to determine each density. The results of the metastable density measurements are shown as points in Fig. 6.

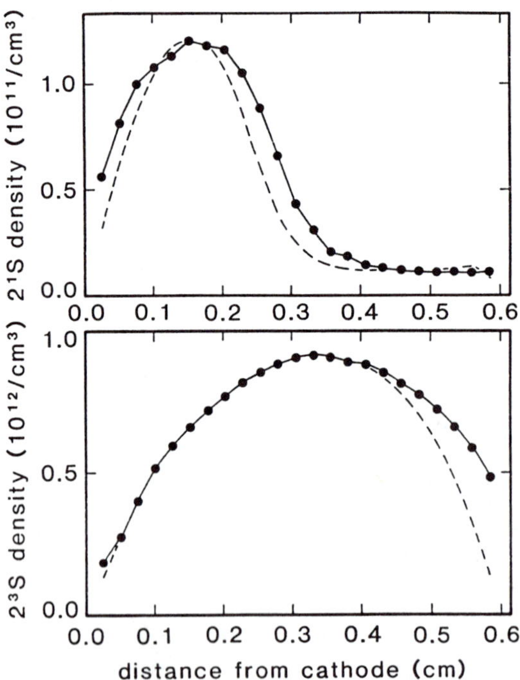

Fig. 6. 2^1S and 2^3S metastable densities vs. distance from the cathode. The points are experimental measurements. The dashed curves are calculated densities using the fitting process described in the text.

The metastable maps are analyzed to determine the net rate of metastable spin conversion. The transport and kinetics of 2^1S and 2^3S metastables in the discharge are modeled using a pair of balance equations which are coupled differential equations,

$$D_s \frac{\partial^2 M_s}{\partial z^2} - \beta M_s^2 - \beta M_s M_t - \gamma M_s N - \kappa n_e M_s + P_s = 0 = \frac{\partial M_s}{\partial t} \qquad (2)$$

$$D_t \frac{\partial^2 M_t}{\partial z^2} - \beta M_t^2 - \beta M_s M_t + \kappa n_e M_s + P_t = 0 = \frac{\partial M_t}{\partial t} \qquad (3)$$

where $M_{s,t}$ is the metastable density, N is the density of ground state atoms, n_e is the density of low energy electrons, D is the diffusion coefficient, P is the production rate per unit volume, γ is the rate constant for singlet metastable destruction due to collisions with ground state atoms, β is the rate constant for the destruction of metastables due to metastable-metastable collisions, κ is the rate constant for net destruction of singlet metastables due to low energy electron collisions, and s and t subscripts indicate singlet and triplet metastables respectively.[10] The first three loss terms in each equation are loss terms arising from diffusion and collisions between metastables. There is an additional loss term in the singlet equation arising from collisions with ground state atoms. The fifth loss term in the singlet equation is due to the metastable spin conversion reaction (1) and a corresponding gain term is included in the triplet equation. Values of DN, γ and β are taken from Phelps.[20] The temperature dependence of D is taken from Buckingham and Dalgarno[21] and that of γ is taken from Allison, Browne and Dalgarno.[22]

The coupled equations are solved numerically for $M_s(z)$ and $M_t(z)$ and compared with the experimental results. In the calculation the spatial dependence of n_e is assumed to be a step function: zero in the cathode fall and uniform in the negative glow. The spatial dependence of the production terms is determined in Monte Carlo simulations to approximate a fundamental diffusion mode. Diffusion modes up to tenth order are included in the solutions to describe the spatial asymmetry of the 2^1S metastables. The values of P_s, P_t and κn_e are varied until a good fit to the experimental metastable maps is obtained. The dashed curves of Fig. 6 are the calculated metastable densities determined in this manner. By this method, κn_e is found to be 7.0×10^4 sec^{-1}.

The rate constant κ_m for metastable spin conversion by room temperature electrons was measured by Phelps.[20] The temperature dependence of the rate constant is $(T_e)^{-1/2}$ because the cross section scales as the inverse of electron energy.[23] The κn_e determined from the metastable analysis must be considered an effective rate for metastable spin conversion since it accounts for both forward and reverse contributions of reaction (1). Setting κn_e equal to the expression for the forward rate minus the reverse rate we obtain

$$\kappa n_e = n_e[\kappa_m \sqrt{0.025 eV/k_B T_e} - (M_t/3M_s)\kappa_m \sqrt{0.025 eV/k_B T_e} \exp(-0.79 eV/k_B T_e)]. \quad (4)$$

This expression yields a relation between n_e and T_e which is shown as the bold line in Fig. 7. This relation has an upper limit for $k_B T_e$ due to thermodynamic considerations. The metastable spin conversion reaction can deplete the singlet population only to the point that the metastables come into thermal equilibrium with the low energy electrons such that exp

$$\left(\frac{-0.79 \ eV}{k_B T_E} \right) \leq 3M_s/M_t.$$

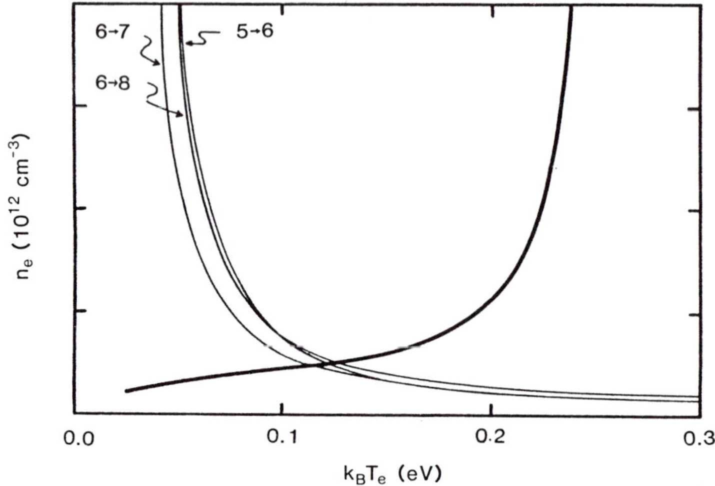

Fig. 7.   Electron density vs. electron temperature for the low energy electrons in the negative glow. The 3 light lines arise from the Rydberg atom diagnostic. The bold line arises from the metastable analysis.

Additional experimental information on the low energy electrons in the negative glow is from the second diagnostic technique based on low Rydberg levels. The experimental set up is much that same as in the preceeding metastable atom experiment except that the laser is frequency doubled and the interference filter is replaced with a compact (0.2 m) monochromator. Higher spectral resolution is necessary for resolving lines in a Rydberg series.

Rydberg levels with the same principal quantum number "n" but different "ℓ" are highly coupled by neutral collisions at 3.5 Torr. Excitation of the np level is effectively excitation of the entire n manifold with the populations of the various orbital angular momentum levels being proportional to their degeneracy. The collisional transfer rate between two Rydberg manifolds is determined by measuring the ratio of their LIF signals when the lower lying manifold is populated with the laser. Three such collisional transfer rates are studied. When the laser is tuned to the 2^3S - 5^3P transition, fluorescence from 5^3D - 2^3P and 6^3D - 2^3P transitions is mapped, yielding the n=5 to n=6 collisional transfer rate. When the laser is tuned to the 2^3S - 6^3P transition, fluorescence from the 6^3D - 2^3P, 7^3D - 2^3P and 8^3D - 2^3P transitions is mapped yielding both the n=6 to n=7 and n=6 to n=8 collisional transfer rates. Figure 8 shows a map of the ratio of LIF from n=7 to that from n=6 when n=6 is populated with the laser. The position at which the cathode fall field extrapolates to zero is indicated on the plot by a vertical dashed line. This is the cathode fall-negative glow (CF-NG) boundary. The enhancement of n=7 fluorescence to the right of this boundary is due to low energy electron collisions in the negative glow.

The short duration (3 nsec) laser pulse produces an initial excess population $N_n(0)$ in the manifold having principal quantum number n. The decay is described by a single exponential because of the weak coupling between adjacent manifolds. The decay rate, τ_n^{-1}, includes both collisional and radiative processes. Under the same conditions the population $N_{n'}(t)$ of a higher lying manifold n' is described by the rate equation

$$\frac{dN_{n'}(t)}{dt} \; = \; -N_{n'}(t)/\tau_{n'} \; + \; R_{nn'}N_n(t) \qquad (5)$$

where $\tau_{n'}^{-1}$ is the depopulation rate of the n' manifold and $R_{nn'}$ is the rate for collisional transfer from manifold n to n', including both neutral and electron collisions. The solution to this equation is

$$N_{n'}(t) \; = \; \frac{R_{nn'}N_n(0)}{\tau_{n'}^{-1} \, - \, \tau_n^{-1}} \; (e^{-t/\tau_n} - e^{-t/\tau_{n'}}) \qquad (6)$$

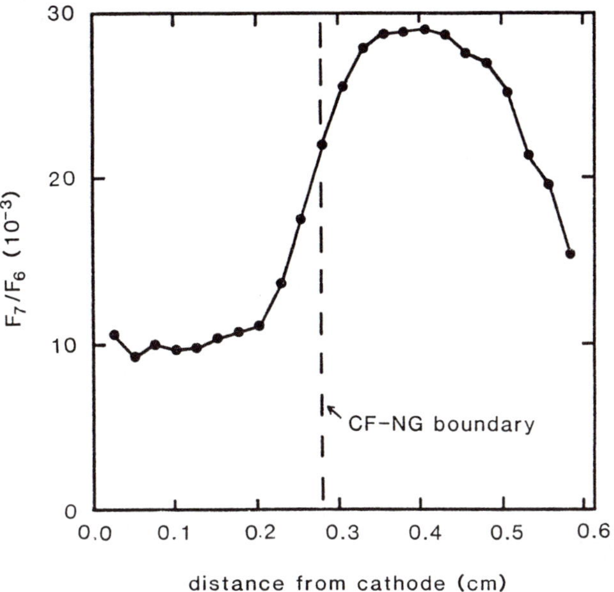

Fig. 8.   Ratio of LIF signal from n=7 to that from n=6 vs. distance from the cathode.

The ratio of time-integrated fluorescence from the n' manifold to that from the n manifold is

$$\frac{F_{n'}}{F_n} = \frac{A_{n'd\text{-}2p}\int N_{n'}(t)dt/(n')^2}{A_{nd\text{-}2p}\int N_n(t)dt/n^2} \qquad (7)$$

where A is a transition rate for a d $\rightarrow$ 2p transition and the observed fluorescence transitions are optically thin. After substituting and carrying out the integrals $F_{n'}/F_n$ is given by

$$\frac{F_{n'}}{F_n} = \frac{n^2 A_{n'd\text{-}2p} R_{nn'} \tau_{n'}}{n'^2 A_{nd\text{-}2p}} \qquad (8)$$

This relation expresses the collisional transfer rate in terms of the measured ratio of fluorescence.

The manifold lifetime $\tau_{n'}$ is measured in situ. This is done by driving the 2^3S - n'^3P transition with the laser and collecting n'^3D - 2^3P fluorescence as described for the fluorescence mapping. The fluorescence decay is time-resolved using a PAR 162/163 scanning boxcar integrator and plotted. Table I shows the results of measurements of $\tau_{n'}$ and the measured $R_{nn'}$ for the three different n$\rightarrow$n' excitation transfers being studied. Because of the spatial asymmetry, the values of $R_{nn'}$ are quoted separately for the cathode fall and for the difference between the peak negative glow signal and cathode fall. The electron density is negligible in the cathode fall so $(R_{nn'})_{CF}$ is interpreted as the collisional transfer rate due to neutral collisions. The neutral transfer rate will be uniform throughout the discharge because the neutral density is uniform. The excitation transfer rate attributable to low energy electrons in the negative glow is the difference between the rate in the negative glow and that in the cathode fall: $(R_{nn'})_{NG}$ - $(R_{nn'})_{CF}$.

121

Table I.   Measured quantities $\tau_{n'}$ and $R_{nn'}$ for the three collisional transfer processes studied. The last column is the excitation transfer rate due to low energy electron collisions in the negative glow.

===

$n \rightarrow n'$	$\tau_{n'}$ $(10^{-9}$ sec$)$	$(R_{nn'})_{CF}$ $(10^5$ sec$^{-1})$	$(R_{nn'})_{NG}$ - $(R_{nn'})_{CF}$ $(10^5$ sec$^{-1})$
$5 \rightarrow 6$	29.1	2.0	3.5
$6 \rightarrow 7$	30.8	6.0	11.1
$6 \rightarrow 8$	33.9	1.2	4.6

===

The measured electron collisional transfer rate can be expressed as the product of a temperature dependent rate coefficient and the electron density n_e. We use the analytical formula given by Vriens and Smeets[24] for the rate coefficient. The resulting relations between electron density and temperature are shown as the thin lines in Fig. 7 for each of the collisional transfer rates measured. The intersection of the two relations yields $n_e = 5 \times 10^{11}$ cm^{-3} and $k_B T_e = 0.12$ eV for the electrons in the negative glow of a 261-V, 0.846 mA/cm^2 discharge in 3.50-Torr helium.

The uncertainty in the metastable atom diagnostic is due mainly to neglect of the 2^1P -2^1S collisional-radiative coupling in the negative glow. Including this rather complex coupling could shift the curve from the metastable experiment to higher electron density, but this effect will be no greater than a factor of two.[10,25] Assuming a factor of two uncertainty in the electron-Rydberg atom rate coefficient also, then n_e will be between 4×10^{11} cm^{-3} and 1×10^{12} cm^{-3}, and $k_B T_e$ will be between 0.07 eV and 0.16 eV. There are good prospects for improved knowledge of such rate coefficients.

These values for n_e and T_e indicate that recombination is not a major loss process for ions in the negative glow because recombination rates are strongly temperature dependent.[26] The uncertainty in the temperature dependence of the recombination rate coefficient makes this conclusion somewhat tentative. If recombination is unimportant then the excess ionization produced in the negative glow by beam electrons must be balanced by an ambipolar-like diffusion to the anode. This mechanism requires a field reversal in the negative glow, and based on the current balance comparison of Section IV we conclude that the field reversal is near the position where the cathode field extrapolates to zero. Field reversals in the negative glow have been observed by R.A. Gottscho, et al., in a DC nitrogen glow discharge.[27]

VI.   THE CONVECTIVE SCHEME

Enormous effort has gone into modeling the cathode region in recent years, and valuable insight has been gained from this work. Although a local field or hydrodynamic equilibrium approximation has recently been used in a two dimensional model of a normal or near normal cathode fall,[28] most modellers avoid this approximation for models in one spatial dimension. A variety of approaches are used to describe the high energy beam electrons including: single beam,[27,29,30,31] multiple beam,[29] nonequilibrium fluid equations,[32] and Monte Carlo simulations.[7] The low energy negative glow electrons are generally modeled using the traditional hydrodynamic equilibrium approximation, where the electron flux is described in terms of drift and diffusion. These very different approximations are then grafted together at the cathode fall-negative glow boundary or in the negative glow. This procedure is not entirely satisfactory and is not likely to provide a very realistic picture of the cathode fall-negative glow boundary. Both beam and plasma approximations fail at the boundary. The boundary region is particularly interesting because it is abrupt; many key parameters have large fractional changes in a mean-free-path, and it is possible that the hydrodynamic approximation fails for ions at this boundary.

The Convective Scheme (CS),[33,34] which was recently introduced to gaseous electronics,[8] is currently the most promising approach for accurately modeling the cathode fall-negative glow boundary. The CS should enable us to produce fully self-consistent electrode-to-electrode models at the kinetic level. The CS is formally a Green's function method applied on a phase space mesh. It involves no assumption about the phase space distribution function because it is a truly kinetic method which solves Boltzmann's equation. It thus preserves all of the rich detail found in Monte Carlo simulations. Anisotropic scattering can easily be incorporated in a CS calculation as in a Monte Carlo simulation. The CS can also provide information on the temporal evolution of a discharge plasma without restarting the calculation as is necessary with Monte Carlo simulations, and the CS can deal with trapped particles. The CS has a significant advantage over finite difference methods in that it is not restricted by the Courant-Friedrichs-Lewy (CFL)[35] criterion on step size. (The CFL criterion in finite difference methods limits a time step so that the fastest particles do not cross more than one mesh cell in a single time step.)

The CS requires the determination of the fraction p of particles in a cell of phase space at $(\vec{r}'',\vec{v}'')$ that move to a new cell of phase space at $(\vec{r}',\vec{v}')$ after a time Δt - the Green's function. (Doubly-primed variables denote values before a movement or collision, and singly-primed variables label the values after the same.) Once this probability is known for all $(\vec{r}'',\vec{v}'')$ and $(\vec{r}',\vec{v}')$, the distribution of particles f at an advanced time $t + \Delta t$ may be determined given the distribution at time t:

$$f(\vec{r}',\vec{v}',t + \Delta t) = \int d\vec{r}''d\vec{v}''p(\vec{r}',\vec{v}';\vec{r}'',\vec{v}'';\Delta t)f(\vec{r}'',\vec{v}'',t).$$

When implemented on a numerical mesh, the integral becomes a sum over all cells of phase space in the mesh.

Only two mechanisms can remove particles from a particular cell of phase space in the kinetic problem:

- Unscattered particles in a cell will move along characteristic trajectories specified by the initial position $\vec{r}''$ and velocity $\vec{v}''$ associated with the cell, the electric field $\vec{E}$ along the trajectory, and the time step Δt.

- Scattered particles will "jump" from the initial cell $(\vec{r}'',\vec{v}'')$ to a new cell at the same spatial location $(\vec{r}'=\vec{r}'')$ but with a new velocity $\vec{v}'$ defined by the nature of the collision.

These two processes are illustrated for a phase space with one spatial and one velocity coordinate in Fig. 9. This figure describes the transport of ions in their parent gas. The only significant ion-atom collisions are symmetric charge exchange collisions which stop the ions in the cold gas approximation.

The two particle-moving mechanisms outlined in the previous section can be implemented independently of each other provided that the distribution function changes on a time scale much longer than the time step and provided that no particle undergoes more than one collision per time step. The latter condition can be approximately enforced by limiting the time step Δt so that the fraction of particles that undergo two collisions in a time step is negligible:

$$[N\sigma_T(v)v\Delta t]^2 \ll 1$$

for every speed v of interest. The left side of this inequality is proportional to the error of the CS, but the full CS error is reduced to the extent that scattering into a cell replaces scattering out of the cell.

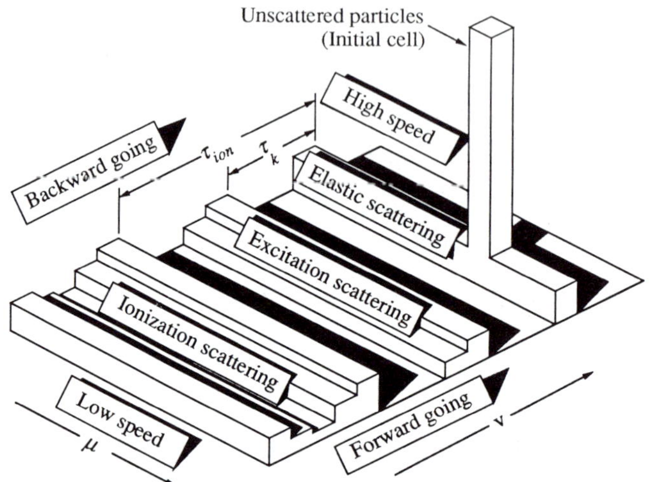

Fig. 9.   An illustration of the ion convective scheme. The problem is one dimensional with two phase space variables, z and v_z. The force (electric field) is assumed constant. The scattering shown is analogous to resonant ion charge exchange; particles completely stop after a collision. Note that the mesh straddles the $v_z = 0$ axis to avoid possible singularities.

A CS model of the electrons in a helium discharge will now be described, building upon the simpler ion model. The physical assumptions here are the same as described earlier in this paper and in Ref. 30. Briefly, the discharge is assumed to be plane-parallel with negligible radial diffusion and an electric field directed along the axis of the discharge: $\vec{E} = E\hat{z}$. The positive z-axis points away from the cathode. The electric field is reversed from that defined for ions in Fig. 9.

An adequate CS model of the electrons demands a realistic description of the angular scattering processes. An extra independent variable is therefore added to the ion model to describe motion transverse to the discharge axis. The independent variables are now the distance from the cathode z, the speed v, and μ, the cosine of the angle θ between the velocity vector and the z axis ($\mu = \cos\theta$). The discharge is azimuthally symmetric, so the azimuthal angle ϕ does not appear.

The ballistic movement of the electrons is more complex than that of the ions described previously. The complications arise out of the nature of the electron mesh, which has a variable Jacobian. The ion CS model is considerably simplified by the fact that the ion phase space is Cartesian, and the size and shape of the cells do not change throughout the mesh. In general, the moved cell overlaps four final cells. The same cannot be said for the electron model; the Jacobian for this mesh is (approximately) $2\pi v^2$. A moved cell may fall completely within one final cell or may overlap many final cells. The

CS accounts for this apparent variation in cell size by independently moving the center of each face of the initial cell to properly calculate the location and size of the moved cell. As before, the particles in the moved cell are distributed to each of the final cells in proportion to the volume of phase space occupied by the moved cell in each of the final cells.

The scattering calculations for the electrons are illustrated in Fig. 10, and described here. All scattering processes are assumed isotropic, with the anisotropic differential elastic cross section being replaced with the isotropic elastic momentum transfer cross section. The assumption of isotropic inelastic scattering for electrons in a normal helium cathode fall is supported by the work of Den Hartog, et al.[10] A detailed investigation of the effects of anisotropic scattering in nitrogen is presented by Phelps and Pitchford.[36] Although the present electron scattering calculations use isotropic scattering, it should be noted that any anisotropic distribution can be included in the CS.

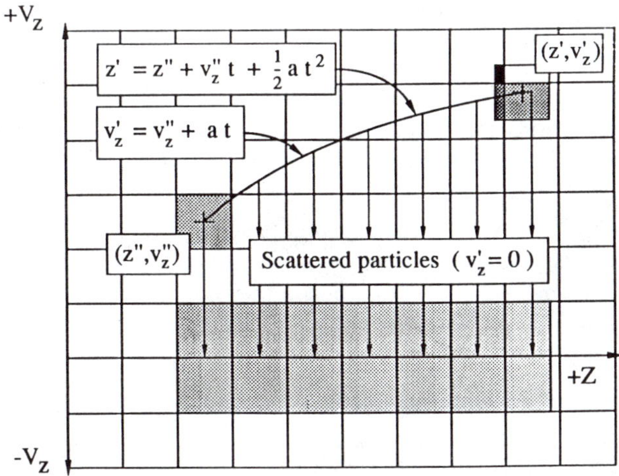

Fig. 10. A schematic of the electron collision operator. Electrons leaving a scattering event are distributed isotropically; anisotropic elastic scattering is described through the elastic momentum transfer cross section. Electrons involved in an ionization event are distributed according to the differential energy scattering cross section. The ionization energy it τ_{ION}. The excitation energy for the single process shown is τ_K. Individual mesh cells have been omitted for clarity.

The isotropic elastic momentum transfer cross section is the simplest to examine. In this process, particles scatter from one polar angle μ'' to another μ' at the same speed ($v' = v''$). Particles scattered out within each cell (z'',v'',μ'') are replaced at the same spatial location z' and speed v' as the unscattered particles, but distributed evenly (isotropically) in the cosine of the polar angle, μ'.

Excitation collision processes are similar, but the scattered particles are replaced isotropically at a final speed v' that is determined by the energy of the kth excitation process, τ_k: $v'^2 = v''^2 = 2\tau_k/m$. The phase space of the electron mesh compresses as one moves toward lower velocities. This is considered again by calculating the positions of the faces of the moved cell when scattering from one speed (v") to another (v'); scattered particles lose a fixed energy τ_k, rather than a fixed speed.

Ionization processes produce scattered particles over a range of speeds $0 < v'^2 <$ v"^{2} - 2τ_{ION}/m, where τ_{ION} is the ionization threshold energy. The number of particles scattered to each speed in this range is found using the differential energy ionization cross section. The scattered particles are then distributed isotropically as before.

As in the ion model, relevant moments of the distribution are calculated and (if self-consistent) the field is adjusted. As a final note, some moments are not found by integrating over the mesh. To improve the accuracy of finding moments like the average electron velocity $<v_z>$ and the electron current density J_e, the electron flux across the z boundaries within the mesh is tracked during the unscattered movement of the electrons used to calculate $<v_z>$ and J_e.

The CS has been used to model electron transport and kinetics in a temporarily fixed spatially uniform reduced field E/N = 282 Td at N = 3.53 x 10^{16} cm^{-3}. It has also been used to model the J_D = 0.190 mA/cm^2 discharge described previously with the temporarily fixed spatially varying field from optogalvanic measurements. These CS calculations are in good agreement with Monte Carlo simulations. Both the CS and Monte Carlo simulations use the same set of elastic cross sections from LaBahn and Callaway[15] and inelastic cross sections for Alkhazov.[16] The CS results in fixed uniform field were also compared to swarm experiments. These results are in the literature[8] and will not be presented here.

The most interesting results from CS calculations are for the cathode fall region with a self-consistent electric field. These first self-consistent CS calculations used the simple mobility parameterization to describe ion transport. The empirical current balance described in preceeding sections ($\gamma = 0.3$) was used as a boundary condition. In this model the ion transport description requires that the ion velocity, and hence the ion current, to vanish any time the electric field nears zero.

A self-consistent field calculation is started by first guessing a field configuration as in   Fig. 11(a) and then running a fixed field calculation for the electrons to near stability. Once the fixed field solution is found, Poisson's equation is combined with the field constraint ($J_+ = 0$ if E = 0) to calculate an electric field. A search is made for the spatial location where the electrons carry the full discharge current; that is, where J_e - J_D. This defines the cathode fall-negative glow boundary. (Such a point may not exist if the guessed field is too weak; the initial guess must then be modified.) The calculated field is set to zero at this point and its behavior in the cathode fall is found by integrating toward the cathode using Poisson's equation, as shown in Fig. 11(b). The field being used in the run (currently the imposed field) is allowed to relax a fixed fraction r (typically r $\approx$ 0.01) of the difference between the field being used and the newly calculated field at each spatial location z (see Fig. 11(c)). A single time iteration of the CS is calculated with the new field configuration, then the field is again relaxed. This scheme of a single CS time iteration followed by a partial field relaxation is continued until a stationary solution is again found, as in Fig. 11(d).

The predicted electric field configuration for the J_D = 0.190 mA cm^{-2} discharge described previously is presented in Fig. 12 along with the field measurements from optogalvanic experiments. The field decreased in a nearly linear fashion in the cathode fall, as expected. The experimental and predicted fields are in good agreement in the high field part of the cathode fall.

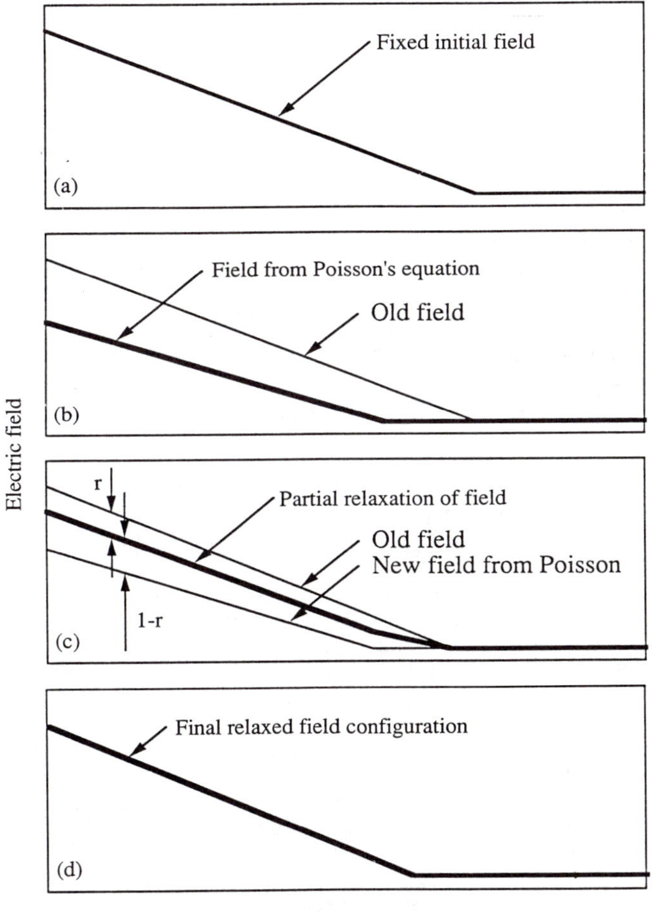

Fig. 11. A schematic of the field relaxation algorithm. An initial fixed field configuration is guessed and the convective scheme run to (near) stability. (a) Poisson's equation is used to calculate a new field configuration. (b) The old field is then allowed to relax a fraction r of the distance toward the new field. (c) The partially relaxed field of (c) becomes the "old" field in (b) during the next iteration and (b) and (c) are repeated until stability is again reached. (d) The separation of the "old" and "new" fields, as well as the amount of relaxation r, have been exaggerated for clarity.

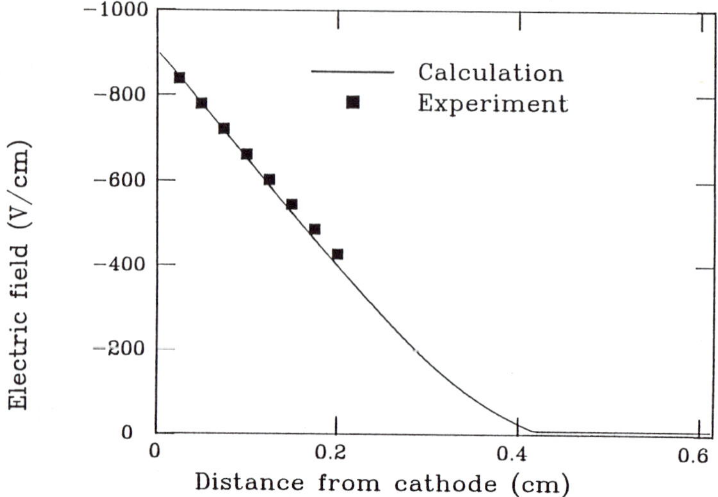

Fig. 12.   The self-consistent electric field configuration as predicted by the convective scheme for the $J_D = 0.190 \, mA \, cm^{-2}$ discharge. The experimental field points are optogalvanic measurements which were described earlier.

To emphasize the fully kinetic nature of this self-consistent calculation, distribution functions at three spatial locations in the discharge are presented in Fig. 13. Electrons that have survived unscattered from the cathode dominate Fig. 13(a), which is taken from near the cathode in the cathode fall. Figure 13(b) is taken from the midst of the cathode fall, showing the expected buildup of inelastically-scattered electrons. Electrons streaming through the negative glow from the cathode fall to the anode are shown in Fig. 13(c); this region is dominated by electrons that have undergone several inelastic collisions. The distribution differs somewhat from a pure Maxwellian - high energy electrons are depleted relative to a Maxwellian due to inelastic collision processes, and close inspection of Fig. 13(c) reveals a slight anisotropy of the negative glow electrons. These electrons are sometimes called "beam" or "ballistic" electrons because they are distinctly nonhydrodynamic electrons in the weak negative glow electric field; they are primarily responsible for excitation and ionization events in the negative glow.

The high density of low energy electrons in the negative glow is not yet included in the CS calculations. Thus these first self-consistent kinetic calculations also do not reliably model the cathode fall-negative glow boundary and the negative glow. The physical assumptions appropriate in the negative glow differ markedly from those of the cathode fall. Several mechanisms have been discounted in the present CS model of the cathode fall that may be important in the negative glow, including energy transfer during elastic collision, a non-zero neutral gas temperature, electron-ion recombination, Coulomb collisions, and the role of metastable atoms. A discussion of these and other mechanisms in a helium afterglow can be found in a comprehensive paper by Deloche, et al.[26] Some of the problems hampering the present model near the cathode fall-negative glow boundary will be avoided with an improved ion transport model.

A more vexing problem is the physical time required to build up the large density of low energy electrons existing in the negative glow. The experiments described in the preceeding section on the $J_D = 0.190 \, mA \, cm^{-2}$ helium discharge being modeled here indicate a large density $(> 10^{11} \, cm^{-3})$ of relatively cold $(< 0.25 \, eV)$ electrons in the negative glow. If this density extends over 0.1 cm (around half) of the negative glow, 10^{10} electrons would exist in the negative glow per square centimeter of electrode area. If the source of these electrons is taken to be the electronic current crossing from the cathode fall into the negative glow ($J_e = J_D = 0.190 \, mA \, cm^{-2}$ at the cathode fall-negative glow boundary), and there are no losses (a best case assumption), it would take a minimum of

10^{-5} seconds for enough electrons to be generated in the discharge to obtain the observed negative glow electron density. Given the CS time step of 0.05 ns (limited by the electron-atom collision rates in helium), this translates into a minimum of 2×10^5 CS iterations. A self-consistent electrode-to-electrode calculation must circumvent this problem.

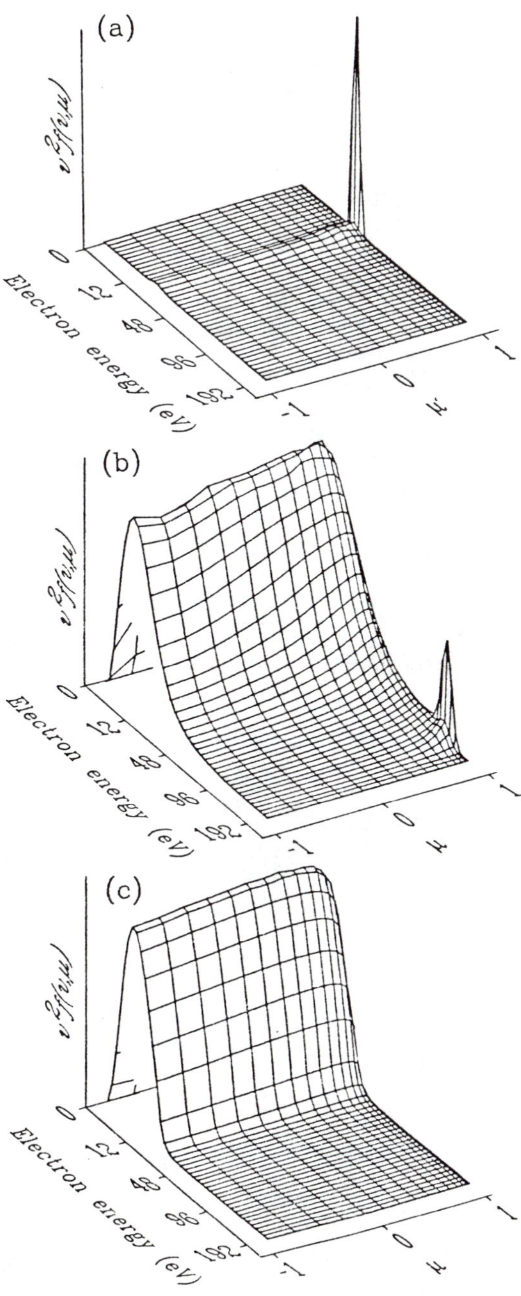

Fig. 13. Distribution functions from various spatial locations within the self-consistent CS calculation shown in Fig. 12. The distances from the cathode are (a) z = 0.028 cm (near the cathode), (b) z = 0.243 cm (midst of the cathode fall) and (c) z = 0.489 cm (midst of the negative glow).

Although this problem and others are formidable, they will be solved. Physical and numerical considerations necessary for a realistic model of the negative glow were outlined. The efficiency of the CS should eventually allow a fully kinetic treatment of both the electrons and ions and a complete electrode-to-electrode calculation, including a detailed description of the true nature of the cathode fall-negative glow boundary.

VII. SUMMARY

In summary, this paper describes experimental work using laser induced fluorescence and laser optogalvanic diagnostics, and theoretical work using advanced techniques, on the cathode region of glow discharges. A current balance or ratio of ion to electron current at the cathode surface is determined empirically and compared to the results of Monte Carlo simulations. The empirical current balance is from precise electric field maps and gas density measurements. Comparisons of Monte Carlo results to the empirical current balance leads to the conclusion that electrons carry most of the discharge current across the cathode fall-negative glow boundary where the cathode fall field extrapolates to zero. New laser induced fluorescence diagnostics are used to determine the density and temperature of the low energy electrons trapped in the negative glow. These results indicate that recombination is not a major loss mechanism for the excess ionization produced in the negative glow by beam electrons from the cathode fall. If recombination is not important, then the current balance implies that field reversal exist near the cathode fall-negative glow boundary where the cathode fall field extrapolates to zero. The Convective Scheme is described and used for a self-consistent kinetic model of the cathode fall region. Work is continuing on electrode-to-electrode, fully self-consistent (electric field) models based on the Convective Scheme.

ACKNOWLEDGEMENTS

This research was supported by the U.S. Air Force Office of Scientific Research under Grant AFOSR 84-0328. E.A. Den Hartog acknowledges the support of the Engineering Research Center for Plasma Aided Manufacturing at the University of Wisconsin-Madison during the preparation of this manuscript.

REFERENCES

1. P. Segur, M. Yousfi, J. P. Boeuf, E. Marode, A. J. Davies, and J. G. Evans, *Electrical Breakdown and Discharges in Gases*, E. E. Kunhardt and L. H. Luessen, eds., NATO ASI Series B **89A**, 331, Plenum, New York (1983).

2. W. H. Long, Plasma Sheath Processes, Technical Report AFAPL-TR-79-2038, Northrup Research and Technology Center (1978).

3. E. A. Den Hartog, T. R. O'Brian, and J. E. Lawler, Phys. Rev. Lett. **62**, 1500 (1989).

4. D. K. Doughty and J. E. Lawler, Appl. Phys. Lett. **45**, 611 (1984).

5. D. K. Doughty, S. Salih, and J. E. Lawler, Phys. Lett. A **103**, 41 (1984).

6. J. Dutton, J. Phys. Chem. Ref. Data **4**, 577 (1975).

7. J. P. Boeuf and E. Marode, J. Phys. D:Appl. Phys. **15**, 2169 (1982).

8. T. J. Sommerer, W. N. G. Hitchon, and J. E. Lawler, Phys. Rev. A **39**, in press (1989)

9. D. A. Doughty, E. A. Den Hartog, and J. E. Lawler, Phys. Rev. Lett. **58**, 2668 (1987).

10. E. A. Den Hartog, D. A. Doughty, and J. E. Lawler, Phys. Rev. A **38**, 2471 (1988).

11.   R. Warren, Phys. Rev. **98**, 1650 (1955).

12.   J. E. Lawler, Phys. Rev. A **32**, 2977 (1985).

13.   H. Helm, J. Phys. B:Atom. Molec. Phys. **10**, 3683 (1977).

14.   S. Sinha, S. L. Lin, and J. N. Bardsley, J. Phys. B: Atom. Molec. Phys. **12**, 1613 (1979).

15.   R. W. LaBahn and J. Callaway, Phys. Rev. A **2**, 366 (1970); Phys. Rev. **180**, 91 (1969); Phys. Rev. **188**, 520 (1969).

16.   G. D. Alkhazov, Zh. Tekh. Fiz. **40**, 97 (1970); [Sov. Phys. Tech. Phys. **15**, 66 (1970)].

17.   M. J. Druyvesteyn and F. M. Penning, Rev. Mod. Phys. **12**, 87 (1940).

18.   K. G. Emeleus, J. Phys. D **14**, 2179 (1981).

19.   P. F. Little and A. von Engel, Proc. R. Soc. London, Ser. A **224**, 209 (1954).

20.   A. V. Phelps, Phys. Rev. **99**, 1307 (1955).

21.   R. A. Buckingham and A. Dalgarno, Proc. Roy. Soc. (London) A **213**, 506 (1952).

22.   D. C. Allison, J. C. Browne, and A. Dalgarno, Proc. Phys. Soc. (London) **89**, 41 (1966).

23.   W. C. Fon, K. A. Berrington, P. G. Burke, and A. E. Kingston, J. Phys. B:Atom. Mol. Phys. **14**, 2921 (1981).

24.   L. Vriens and A. H. M. Smeets, Phys. Rev. A **22**, 940 (1980).

25.   E. A. Den Hartog, Ph.D. thesis, University of Wisconsin-Madison (1989).

26.   R. Deloche, P. Monchicourt, M. Cheret, and F. Lambert, Phys. Rev. A **13**, 1140 (1974)

27.   Richard A. Gottscho, Annette Mitchell, Geoffrey R. Scheller, Yin-Yee Chan, and D. B. Graves, to be published.

28.   J. P. Boeuf, J. Appl. Phys. **63**, 1342 (1988).

29.   A. V. Phelps, B. M. Jelenkovic, and L. C. Pitchford, Phys. Rev. A **36**, 5327 (1987).

30.   T. J. Sommerer, J. E. Lawler, and W. N. G. Hitchon, J. Appl. Phys. **64**, 1775 (1988).

31.   R. A. Gottscho, A. Mitchell, G. R. Scheller, N. L. Schryer, D. B. Graves, and J. P. Boeuf, Proc. 7th Symp. Plasma Processing, Electrochem. Soc. **88-22**, 1 (1988).

32.   P. Bayle, J. Vacquie, and M. Bayle, Phys. Rev. A **34**, 360 (1986).

33.   J. B. Adams and W. N. G. Hitchon, J. Comput. Phys. **76**, 159 (1988).

34.   W. N. G. Hitchon, D. J. Koch, and J. B. Adams, J. Comput. Phys. **82**, in press (1989)

35.   R. Courant, K. Friedrichs, and H. Lewy, Math. Ann. **100**, 32 (1928).

36.   A. V. Phelps and L. C. Pitchford, Phys. Rev. A **31**, 2932 (1985).

EMISSION SPECTROSCOPY

IN OPTICALLY THICK GAS DISCHARGES

Dimitrios Karabourniotis

Physics Department, University of Crete
714 09 Iraklion, Crete, Greece

INTRODUCTION

Information on the processes occurring in a gas discharge can be obtained from the spectral lines emitted by the plasma (see f.e. Refs. 1-5). A line may be optically thin (optical depth much less than unity) or optically thick (optical depth much higher than unity). In the presence of temperature and electron gradients in the plasma, the relative distribution of emitting and absorbing atoms does not remain constant over the length of the plasma column and the plasma is said to be inhomogeneous. Consequently, in the case of inhomogeneous high-density discharge plasmas, at large optical depths, the emergent spectra are characterised by the existence of self-reversed lines. The shape of a self-reversed line may contain copious information on the processes accompanying the discharge.

In order to make the connection between the emitted line and the plasma structure we have to consider some kind of equilibrium. An important case of equilibrium is the so-called local thermodynamic equilibrium (LTE), which implies that any volume element of the plasma can be described in terms of thermodynamic equilibrium, but that the conditions vary with the position in the plasma. The existence of LTE is equivalent to the existence of a local thermodynamic temperature ("equilibrium temperature"). Validity of the LTE-assumption simplifies the use of spectroscopic diagnostics and the performance of species-concentration calculations.

A plasma in a non-LTE state cannot be described correctly by any single temperature, in particular, when inhomogeneity and strong reabsorption of radiation are present. In this case the only practical way of investigation is to approximate the complex plasma state by simplified models.

In the following we will describe some of the aspects of the line emission spectroscopy methods. Special emphasis will be placed on the self-reversed lines for diagnostic purposes without the LTE-assumption.

The application examples will be largely confined to high-pressure arc light sources.

Physics and Applications of Pseudosparks
Edited by M. A. Gundersen and G. Schaefer
Plenum Press, New York, 1990

PLASMA IN LTE-STATE

The Equilibrium Laws

In a LTE plasma the electron collisions dominate the equilibrium processes determining the populations of excited levels. Therefore, electron temperature and electron density are often chosen as the leading parameters in characterizing the state of the plasma. The distribution of population among energy levels can be calculated using the Boltzmann and Saha laws along with the equation of state and the charge neutrality condition. Assuming only neutral atoms and singly charged ions, the equations can be written as follows : the Boltzmann law

$$n_p/n_q = (g_p/g_q) \exp [- (E_p - E_q)/kT_{el}] , \qquad (1)$$

the Saha equation

$$S(T_{el}, E_p) \frac{n_{el}n_i}{n_p} = \frac{2g_i}{g_p} \left(\frac{2\pi m l T_e}{h^2}\right)^{3/2} \exp \left(\frac{E_i E_p}{kT_{el}}\right) \qquad (2)$$

the equation of state

$$P = kT_{el} (n_{el} + n_i + n_o + \sum_{p=1}^{M} n_p) , \qquad (3)$$

the charge neutrality condition

$$n_{el} = n_i, \qquad (4)$$

where n_{el} = electron density; n_i = ion ground state density; n_o = neutral ground state density; n_p, n_q = density of levels p, q belonging to the same species; g_p, g_q = statistical weights of level p,q; E_p, E_q = excitation energy of levels p,q; E_i = ionization energy of the atom ground state; g_i = statistical weight of the ion ground state; T_{el} = electron temperature; and Σ = summation over all excited states, where M is the level with energy equal to the lowered ionization pontential. Further more, if we neglect the number of neutrals in excited states with respect to the neutral ground state density,

$$\sum_{p=1}^{m} n_p << n_o , \qquad (5)$$

from Eqs. (2)-(4), one obtains for the relation between n_{el} and T_{el}

$$n_{el}^2 = (p/kT_{el}) \cdot S (T_{el}, Ep) . \qquad (6)$$

The densities of the high lying levels ($E_i - E_p < kT_{el}$) are fairly independent of T_{el} and depend strongly on n_{el}, according to Eq. (2). On the opposite limit ($E_i - E_p >> kT_{el}$), n_p is very sensitive to T_{el}.

Electron Temperature and Density Determination

The electron temperature is invisible in an arc plasma because it is never determined explicitly. We determine an excitation temperature, rename it as electron temperature and then use it in the equations where electron temperature dominates.[7]

In LTE plasmas optically thin lines can be used for the construction of Boltzmann's diagram and temperature determination.[8,9] These lines are usually produced from transitions between high-energy levels and correction for reabsorption will be required. A

Boltzmann diagram of the excited state density distribution yields a straight line with slope -$1/kT_{el}$.

The electron density can be determined from Eq. (6). Since deviations from LTE population densities occur mainly for low lying excited states, the n_{el} value can be determined by taking the limit of Eq. (2) as g_p and E_p go to the ionization limit (g_∞ and E_∞, respectively). Rearranging Eq. (2) and assuming quasineutrality condition, gives[10]

$$\frac{n_\infty}{g_\infty} = \frac{n_{el}^2}{2g_i} \frac{h^3}{(2\pi mkT_{el})^{3/2}} \quad , \tag{7}$$

where n_∞/g_∞ is the level density per statistical weight at the ionization limit. The value of n_∞ may be determined experimentally by measuring absolute line intensities of highly excited states for which we assume that they are populated according to Saha law.

For determining the electron density in a plasma, a method based on the Stark broadening of spectral lines due to interaction of a radiating atom with surrounding electrons and ions can also be used. As a rule, lines sensitive to n_{el} and not subject to self-absorption as the hydrogen H-ß or the sodium 568.8 nm line,[12,13] for which the value of the Stark constant are known with sufficient accuracy, are used to determine the profiles and half-widths of the lines.

SPECTRAL LINE EMISSION FROM AN INHOMOGENEOUS LIGHT SOURCE

The shape of a spectral line with consideration of self-absorption and inhomogeneity along a line of sight of an axisymmetric radiating plasma (Fig.1) is described by the integral form of the equation of radiative transport in an absorbing medium:

$$I(v) = \int_{-R}^{+R} \varepsilon(v,r) \exp\left[- \int_{-R}^{+R} k(v,r')dr'\right]dr \quad , \tag{8}$$

where

$$\varepsilon(v,r) = (hv/en) \, A_{pq} \, n_p(r) \, P(v,r) \tag{9}$$

is the emission coefficient,

$$k(v,r) = (hv/c)B_{qp} \, n_q(v) \, P(v,r) \tag{10}$$

is the absorption coefficient, $\int_{r}^{R} K(v,r') \, dr'$ is the optical depth, and $P(v,r)$ is the local line profile with $\int P(v,r)dv=1$. The transition probability for emission A_{pq} is related to the line transition probability for absorption B_{qp} by

$$A_{pq}/B_{qp} = (g_p/g_q)(8\pi hv^3/c^3). \tag{11}$$

For an optically thin line $\int_{r}^{R} K(v,r')dr' \longrightarrow o$ within the line limits. Thus, Eq. (8) takes the form

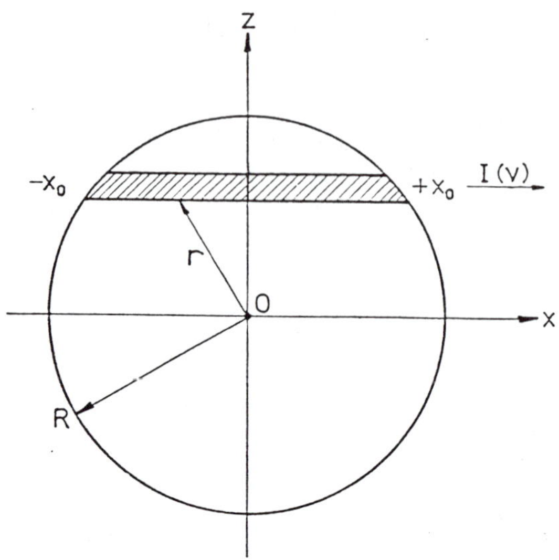

Fig. 1.   Illustration of the coordinate system.

$$I = \int\limits_{-R}^{+R} \varepsilon(v,r)dr \ , \tag{12}$$

In this case the total line intensity is given by

$$I = \int\limits_{-R}^{+R} J(r)dr \ , \tag{13}$$

where

$$J(r) = \int\limits_{line} \varepsilon(v,r)dv \tag{14}$$

is the local emissivity (=total local line intensity).

INHOMOGENEOUS, OPTICALLY THIN LTE PLASMAS

In an LTE arc plasma the equilibrium temperature can be determined from the absolute value of the total line intensity of an optically thin line.

For a line-of-sight through a radiating plasma of cylindrical symmetry (Fig.1) the emitted total intensity of an optically thin line I(z) is simply the sum of the local emissivities of all elements of the optical path. A set of side-on-measurements, that is, the intensity I(z) measured as a funtion of the coordinate z, provides us a set of integral equations,

$$I(z) = 2 \int\limits_{z}^{R} J(r) \ \frac{r \ dr}{(r^2-z^2)^{1/2}} \ , \tag{15}$$

which, after the so-called Abel inversion, gives the local emissivity,

$$J(r) = -\frac{1}{\pi} \int_r^R \frac{dI(z)}{dz} \frac{dz}{(z^2-r^2)^{1/2}} \cdot \tag{16}$$

$J(r)$ is related to $n_p(r)$,

$$J(r) = (h\nu/4n)A_{pq} g_p n_p(r) . \tag{17}$$

Thus, the radial profile of the population density of the upper state of the line-producing transition $p \to q$, can be determined from Eq.(17). For a plasma in LTE, $n_p(r)$ is related to $n_0(r)$ by Boltzmann's law according to Eq. (1). Then, if $n(r)$ or the pressure is known the local temperature can be deduced. In a closed burner arc discharge as in a high-pressure mercury arc lamp, knowledge of the line density N_0,

$$N_0 = \int_0^R 2\pi r n_0(r) dr , \tag{18}$$

i.e., the number of the ground state atoms in the arc per unit length, allows the determination of the temperature profile and the plasma pressure. When the temperature profile is known the measurement of the emissivity for a thin line can be used for the determination of the ground state density of the emitting element.

Figure 2 depicts the time-averaged intensity $I(z)$ of the optically thin 577-nm line of mercury, the local emissivity $J(r)$ obtained by Abel inversion, and the resulting equilibrium temperature profile $T^{eq}(r)$ for a 1.5-bar Hg-NaI arc lamp, with 1.8-cm internal diameter and 4.5-cm interelectrode distance, operated at ac (50 Hz, 360 W).

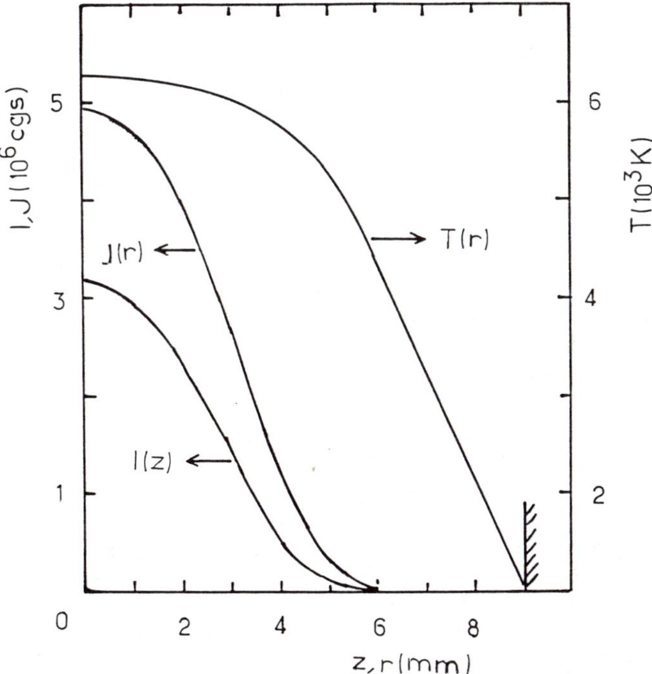

Fig. 2.   Time-averaged total intensity $I(z)$ and local emissivity $J(r)$ for the Hg 577-nm line, and temperature profile $T(r)$ in an ac high-pressure Hg-NaI arc discharge. From Ref. 14.

Figure 3 depicts I(z) and J(r) for the optically thin 616-nm line of sodium as well as the radial distribution of the sodium ground state atoms deduced from the emissivity of the 616 nm line and the temperature profile shown in Fig. 2.

INHOMOGENEOUS, OPTICALLY THICK PLASMAS

In optically thick plasmas self-reversed emission lines are usually considered (Fig. 4). In such a line the absorption coefficient given by Eq. (10), decreases with the frequency distance to the line center; consequently, the plasma from optically thick around the line center becomes optically thin in the far wings.

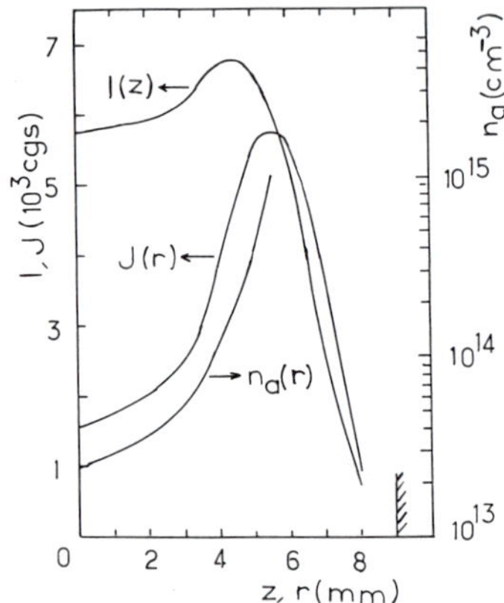

Fig. 3. Time-averaged total intensity I(z) and local emissivity J(r) for the Na 616-nm line, and sodium density distribution Na(r) in an ac high-pressure Hg-Na I arc discharge. From Ref. 14.

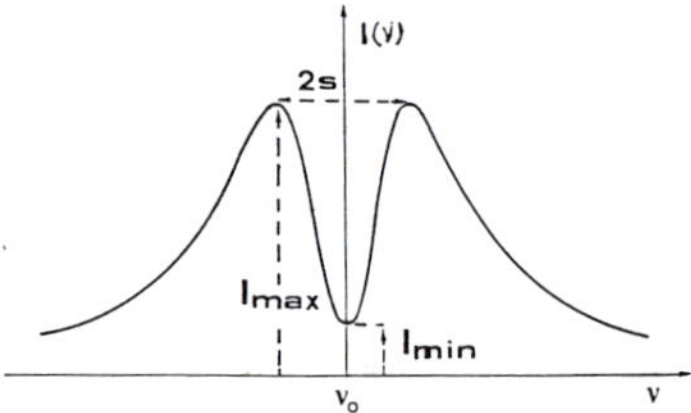

Fig. 4. Key parameters for a self-reversed emission line.

For spectral line diagnostics it is necessary to have suitable relations between the line intensity I(v) and the distributions of $\varepsilon(v)$ and K(V) within the plasma, according to Eqs. (8)-(10). This problem is too complicated for general solution, because it requires knowledge of the plasma structure (temperatures and densities) and of the line properties (broadening constants and transition probabilities).[4] Different authors have used agreement between calculated and measured spectral line shapes to validate model calculations or to estimate broadening constants.[15-20] Discharges excited in ac have also been used to separate the effect of the electrons from that of the atoms on the line broadening.[21,22] This was accomplished by taking advantage of the strong variation in the electron density during the arc current cycle. It has thus been possible to determine broadening constants and transitions probabilities for the mercury visible triplet lines by comparison between calculated and observed line shapes at different instants within the current cycle.

Determination of such constants was only possible under optically thin plasma conditions, where special experimental conditions and rigorous allowance for self-absorption and plasma inhomogeneity are required. Figure 5 illustrates the calculated growth and formation of the self-reversed H9-546.1 nm line along a diameter within a 3-bar, 50 Hz mercury discharge at the moment of maximum emission, i.e, about 5 ms after the voltage-zero crossing.[22]

Since the line shape depends in a complicated way on a large number of parameters, diagnostics based on a direct comparison of calculated and observed line shapes is extremely cumbersome and suitable only in particular cases. It is therefore preferred to find out characteristic quantities of the observed line shapes which are not sensitive to certain plasma and line profile parameters, but, at the same time, quite sensitive to other parameters; the intensity f.e., at the line self-reversal is determined primarily by axis temperature and is insensitive to radial temperature profile.

Many phenomenological models for line self-absorption in inhomogeneous plasmas have been proposed.[23-30] In particular, the Bartels and Cowan-Dieke models have found many applications in spectroscopic diagnostics of LTE plasmas.[31-37] A detailed discussion of those two models have been recently given in Ref.4.

DIAGNOSTICS BASED ON BARTELS' THEORY

According to Bartels' theory[24,25] the relation between the intensity of a self-reversed non resonant spectral line I(v) and the optical thickness of the plasma column T_o (V) can be approximately written in the form

$$I(v) = \frac{2hv^3}{c^2} \exp\left(-\frac{hv}{kT_m}\right) M\, Y\, (T_o, p), \tag{19}$$

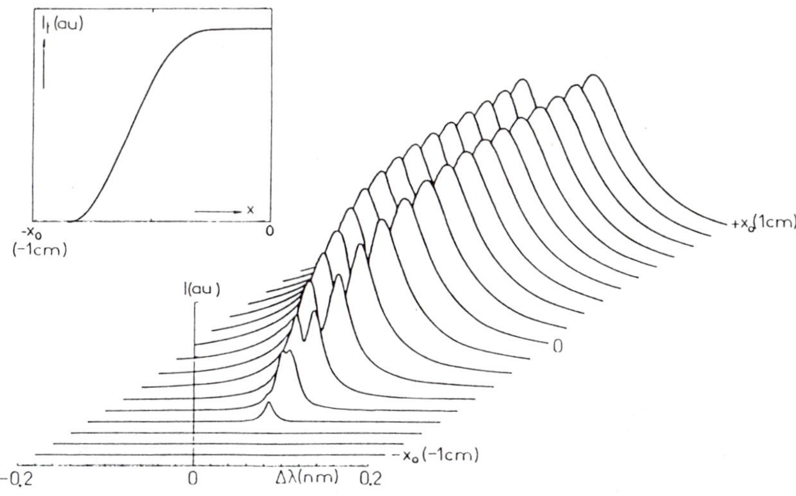

Fig. 5.   Spectral formation and total indensity growth of the Hg 546-nm line along a diameter of an ac high-pressure mercury discharge at the maximum emission phase. au: arbitrary units; I: spectral intensity distribution; I_t: total line intensity. From Ref. 22.

where T_{max} is the maximum temperature along a line-of-sight. Calculating M and Y, Bartels made the following basic assumptions: (i) the plasma is in LTE, (ii) the temperature profile is parabolic, (iii) the atomic pressure of the emitting element is constant. If $E_a/KT \gg 1$ the factor M can be represented by the equation

$$M = \begin{cases} (E_a/E_e)^{1/2} \text{ for pressure broadening,} \\ [(E_a + 0.5E_i)/(E_e + 0.5E_i)]^{1/2} \text{ for Stark broadening} \end{cases} \tag{20}$$

where E_e and E_a are the excitation energies of the lower and upper levels, and E_i the ionization energy of the atom. Reasonance lines or lines with a sufficiently low lower level are therefore excluded. The function $Y(T_0,p)$ in a limited region of $T_0(<10)$ can be represented with sufficient accuracy by a single-parameter family of curves

$$Y(T_0,p) = \exp(-\frac{T_0}{2} \{ \frac{T_0}{2} (1-p) + p\sinh (\frac{T_0}{2}) + \frac{1}{\sqrt{p}} \sinh (\frac{T_0\sqrt{p}}{2})\} \tag{21}$$

Neglecting the induced transitions, the parameter p can be approximated by the expression

$$p = \frac{6}{n} \text{ arc tan} \frac{M^2}{\sqrt{1 + M^2}}. \tag{22}$$

From the measured intensity distribution $I(v)/I_{max}$ it is possible to deduced the unknown $Y(\tau_0,p)$ distribution[38]

$$\frac{I(v)}{I_{max}} = \frac{Y(t_0,p)}{Y_{max}}, \tag{23}$$

where I_{max} is the intensity at the maximum of the self-reversed line, and Y_{max} is the maximum value of the function $Y(T_0,p)$. The Y_{max} value can be calculated with 0.4% accuracy by

$$Y_{max}(p) = \frac{2}{\sqrt{1-p}} \left(\frac{1 - \sqrt{p}}{1 + \sqrt{p}}\right)^{1/2 \sqrt{p}} \tag{24}$$

The corresponding optical thickness then becomes

$$T_0 = \frac{1}{\sqrt{p}} \ln \frac{1 + \sqrt{p}}{1 - \sqrt{p}}. \tag{25}$$

From the measurement of $I(v)/I_{max}$ and using Eqs.(20) and (21), the value of $Y(T_0,p)$ can be found. Then from the values of $Y(T_0,p)$ the distribution of $T_0(v)$ is determined within the limits of the line.[39]

Measurement of I_{max} allows determination of T_m by

$$I_{max} = \frac{2hv^3}{c^2} \exp \left(\frac{hv}{kT_m}\right) M Y_{max}. \tag{26}$$

The validity of Bartel's theory and the applicability of Eq.(26) for plasma temperature determination have been confirmed[41] using the H9 546.1 nm line. The mercury resonance 253.7 nm line has also been studied. By means of a numerical solution of the equation of radiative transfer it has been found that the value of MY_{max} is practically insensitive to both the radial temperature profile (linear, parabolic or cubic) and the line broadening (position independent or pressure broadening) for the 546.1 nm line, but it is strongly dependent on the line broadening mode for the resonance line 253.7 nm. Side-on-measurements of I_{max} for the 253.7 nm line emitted by a high pressure (5.3 bar) mercury discharge have been performed[42] and the exact value of MY has been determined by an iterative method assuming a dispersion and a quasistatic line profile for the blue and the red wing of the H9-253.7 nm line, respectively. Then, with the aid of the LTE-assumption the radial temperature profile has been determined (Fig.6).

Since, the wavelength distance between the two-side line maxima is primarily determined by species concentration, a method based on Bartel's model hasbeen proposed[43-45] for the determination of partial pressures in high-pressure metal halide discharges from the separation of the maxima of a non-resonant self-reversed spectral line. Recently this technique has been developed and applied in the case of resonant lines (T1-377.6 nm).[42]

Bartel's model, however, is only valid at low and intermediate values of optical depth; this precludes its application in the neighborhood of the self-reversal minimum.[46]

DIAGNOSTICS BASED ON COWAN-DIEKE'S THEORY

For the study of line self-absorption in inhomogeneous plasmas Cowan and Dieke[23] have considered that the shape and the intensity of a spectral line e $\leftrightarrow$ a (a<e) depends on

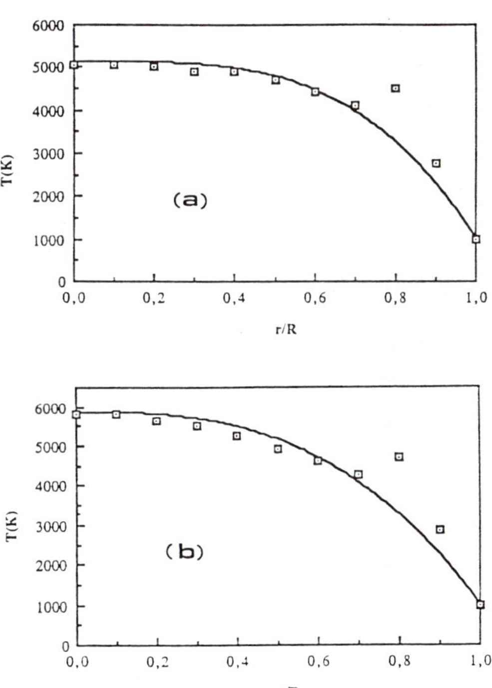

Fig. 6.   Radial temperature profile in a 5.4-bar ac mercury arc at the moments of minimum (a) and maximum (b) emission, using the optically thin Hg 577-nm (solid line) and the resonance Hg 253.8-nm (experimental points) lines. From Ref. 42.

i) the local line profile, which was assumed to be constant throughout the plasma column, viz.

$$P(u,r) = P(u), \quad u = v - v_0 \ , \tag{27}$$

ii) the optical depth at the line center, given by the absorption parameter

$$p = (hv/c)B_{ae}N_aP(o) \ , \tag{28}$$

and iii) the excitation function $E(y)$ defined by

$$E(y) = n_e(r)/n_a(r) \ , \tag{29}$$

with

$$Y = \int_r^R n_a(r')dr' \ , \tag{30}$$

$$n_a(r) = n_a(r)/N_a \ , \tag{31}$$

$$n_e(r) = n_e(r)/N_e \ , \tag{32}$$

$$2N_a = \int_{-R}^{+R} n_a(r)dr \ , \tag{33}$$

$$N_e = \int_{-R}^{R} n_e(r)dr \ . \tag{34}$$

Thus, expression (8) for the line intensity takes a simplified form

$$I(u) = I_0 P(u) \int_o^2 E(y) \exp(-\mu y)dy \ , \tag{35}$$

where I_0 is the total line intensity in the absence of self-absorption and

$$\mu = p \frac{P(u)}{P(o)} \tag{36}$$

is the optical depth at frequency u with $\mu(o)=p$. Eq.(35) may then be written in the form:

$$I(u) = SV(o) \cdot K\mu \ , \tag{37}$$

where

$$S_v(o) = \frac{2hv^3}{c^2} \frac{g_a}{g_o} \frac{n_e(o)}{n_a(o)} \tag{38}$$

is the "source function"at the arc axis and

$$K_\mu = \mu \int_o^2 \frac{E(y)}{E(l)} \exp(-\mu y)dy \ . \tag{39}$$

Thus, I_{max} is given by

$$I_{max} = Sv(o) \cdot K_q \ , \tag{40}$$

where $q = \mu(s)$ is the optical depth at the distance s from the line center, where the line self-reversal occurs. Differentation of Eq.(35) then gives as a condition for reversal

$$\int_{0}^{2} (1-qy) \frac{E(u)}{E(l)} \exp(-qy)dy = 0 \ . \tag{41}$$

Solving Eq.(41) the value of q is obtained if E(y) is known. The value of p can be obtained from

$$\frac{I_{max}}{I_{min}} = \frac{K_q}{K_p} \tag{42}$$

where I_{min} is the minimal intensity at the line center. According to Cowan-Dieke E(y) may be approximated by

$$E(y) = \frac{n}{2} \ X \ \begin{cases} y^{n-1} & 0 \le y \le 1 \\ \\ (2-y)^{n-1} & 1 \le y \le 2 \end{cases} \tag{43}$$

with
$$E(l) = n/2, \tag{44}$$

where n is the so-called inhomogeneity parameter. From the definition of the excitation function given by Eq.(29) and the approximations (43) and (44), one obtains that

$$n = 2[n_e(o)/n_a(o)] \ \ . \tag{45}$$

We can now write K_μ in an explicit form

$$K_\mu(n) = \mu \int_{0}^{1} y^{n-1} \exp(-\mu y) \{ 1+\exp [2\mu (y-1)] \} \ dy \tag{46}$$

Figure 7 gives the calculated dependence of q and K_q on the inhomogeneity parameter n, while Fig. 8 shows I_{max}/I_{min} as a function of n for different values of p. On the other hand, assuming that P(u) is given by a dispersion function

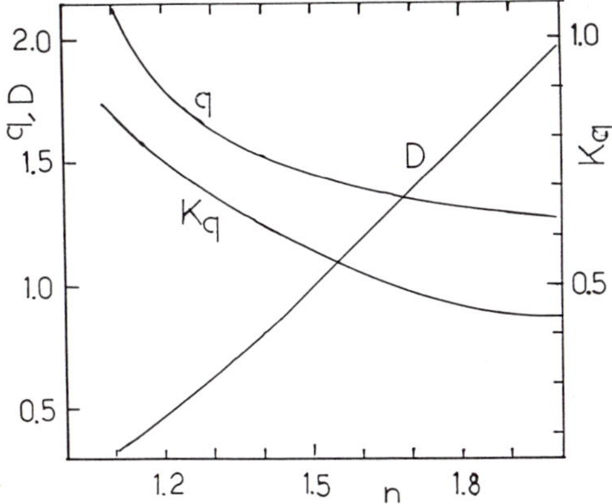

Fig. 7.   Calculated dependence of q, K_q and D on the inhomogeneity parameter n. From Ref. 34.

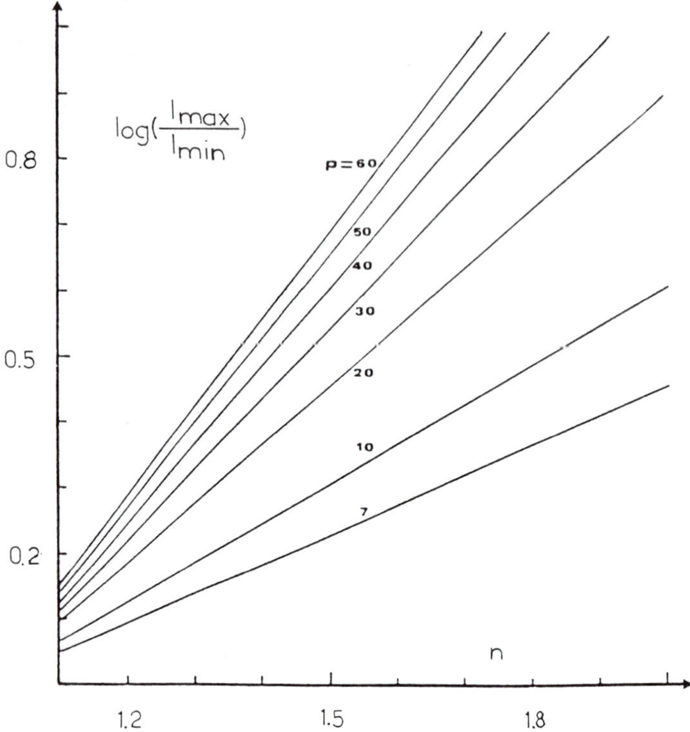

Fig. 8.   Calculated dependence of log (I_{max}/I_{min}) on the inhomogeneous parameter n, with line-center optical depth p as parameter. From Ref. 33.

$$P(u) = \frac{1}{\pi\delta} \cdot \frac{1}{(u/\delta)^2 + 1} \quad , \tag{47}$$

where δ is the HWHM of the line profile, one obtains that

$$\frac{q}{p} = \frac{1}{s_o^2 + 1} \quad . \tag{48}$$

Here $s_o(= s/\delta)$ is the frequency distance of the line reversal to the line center in units of δ.

The calculated dependence of $\log(I_{max}/I_{min})$ vs $\log(2s_o)$ for a given value of n is almost linear.[34] In Fig. 6 the slope D of the theoretically obtained quasi-straight lines $\log(I_{max}/I_{min})$ vs $\log(2s_o)$, in terms of the inhomogeneity parameter n is also given.Experimental data determining the ratio I_{max}/I_{min} and the wavelength separation 2s between the two side line maxima can be obtained by recording the spectral line shapes emitted at different lines of sight. The slope of the $\log(I_{max}/I_{min})$ vs $\log(2s)$ plot for these data (Fig.9) can be used to determine the inhomogeneity parameter n. From Eqs. (41) and

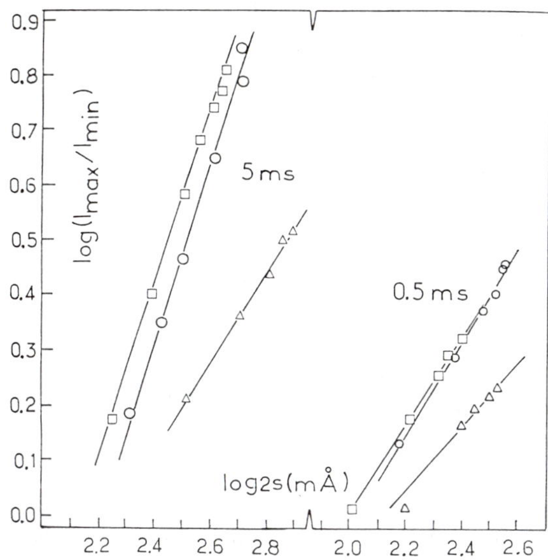

Fig. 9.   Experimental dependence of log (I_{max}/I_{min}) of log(2s) for the Hg 546.1 (Δ), 435.8 (0), and 404.7 nm ($\square$) lines emitted by a 3-bar mercury arc operating at 50 Hz, at the moments of maximum (5 ms) and minimum (0.5 ms) emission. Each consecutive point is displaced 1 mm from the arc axis or the previous point. From Ref. 50.

(43) the value of q is then calculated, while the value of p can be deduced from the measurement of I_{max}/I_{min} for a diametral scan, according to Eq.(42). For the same scan the measurement of the wavelength distance 2s can be used to determine the value of δ from Eq. (48).

Temperature Determination in LTE Arcs

In an arc plasma when LTE is valid at the arc core, the source function takes the form

$$S_v(0) = \frac{2hv^3}{c^2} \exp\left(- \frac{hv}{kT(o)} \right) .$$ (49)

Taking into account the radial plasma structure by the experimental determination of n, the arc axis temperature is derived from the measurement of I_{max} for the line emitted along an arc radius, according to Eq. (40).[34] Corrections concerning the radial line broadening profile have also been studied.[35] A combination of Bartel's and Cowan-Dieke's methods has also been used by Fishmann et al.[32] (see also references therein).

Pressure Determination in LTE Arcs

We define an "effective" pressure P_e for the emitting element with

$$P_e = \int_o^R P_e(r)f(r)dr / \int_o^R f(r)dr ,$$ (50)

with

$$f(r) = [Z(T) \cdot kT(r)]^{-1} \exp[-E_a/kT(r)] \, , \tag{51}$$

where $PE(r)$ is the radial pressure profile, and $Z(T)$ the partition function for the emitting atom. Knowing $p\delta$ and the temperature profile, the effective pressure is given[36] by

$$P_e = \frac{8\pi^2 v^2}{c^2} \frac{p\delta}{g_e A_{ae}} \Big/ \int_o^R f(r)dr \, . \tag{52}$$

This technique has been applied in the determination of the partial pressure of thallium in Hg-TlI arcs.[36] The value of n used in Ref. 36 is that obtained from Eq.(45) assuming LTE. Figure 10 gives the effective partial pressure of thallium against the arc axis partial pressure of thallium in a high-pressure (4.7 atm), ac (50 Hz) Hg-TlI arc discharge at the moment of minimum (0 ms) and maximum (5 ms) emission. The different thallium pressures have been obtained by varying the temperature minimum at the lower part of the discharge chamber.

Population Temperatures in Non-LTE Arcs

The Boltzmann law (1) referenced to the ground state population n_o give the population temperature T_j, $j=e,a$, of the j atomic level, viz.

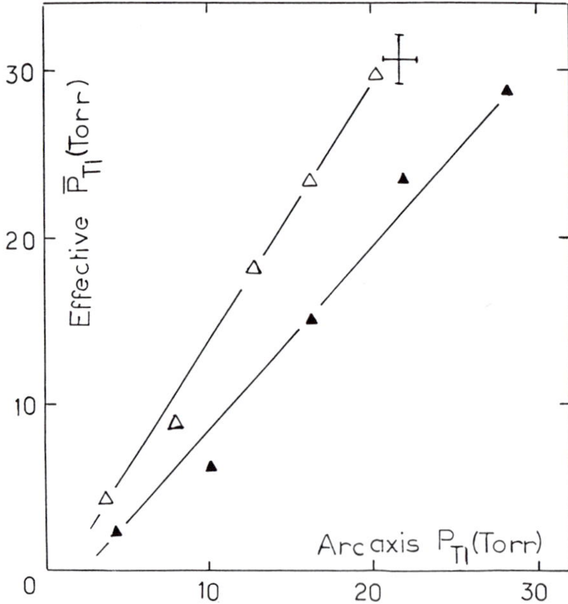

Fig. 10.   Effective partial pressure of thallium us the partial pressure at the arc axis of a 4.8-bar Hg-TlI arc operating at 50 Hz, at the moments of maximum (▲) and minimum (Δ) emission. From Ref. 36.

$$n_j/n_0 = (g_j/g_0) \exp(-E_j/kT_j) \tag{53}$$

In general,the population temperatures of the line-producing transition levels are not equal to each other, that is, LTE is not valid in the line. In this case, from Eqs. (38) and (53), the expression (40) is written[47]

$$I_{max} = \frac{2hv^3}{c^2} \exp\left(-\left(\frac{E_e}{kT_e} - \frac{E_a}{kT_a}\right)\right) K_q(n) \ . \tag{54}$$

Therefore, if T_e is known, T_a can be determined from the measurement of n and I_{max}.

When two self-reversed lines e↔a and e'↔a (a<e,e') with a common lower level[a] are present, T_e is related to T_e by

$$\frac{I_{max}}{I_{min}} = \frac{v^3}{v'^3} \exp\left(-\left(\frac{E_e}{kT_e} - \frac{E_e'}{kT_e'}\right)\right) \frac{K_q}{K_q'} \ . \tag{55}$$

An application is given here for the case of the mercury atom (Fig. 11) at the axis of a high-pressure (5.4 bar), ac mercury arc, at the moment of maximum emission. The self-reversed Hg 546-, 436-, 405- and 365- nm lines have been used. The equilibrium temperature T^{eq} (=5875 K) was obtained from the optically thin Hg 577- nm line assuming LTE. We consider that the temperature of the level $e=7^3S_1$ (upper level of the mercury visible triplet lines) is equal to T^{eq}, viz. $T_e = 5875$ K. Applying Eq. (54) the temperatures T_a of the 6^3P levels were found (Table 1). The 546 and 365 nm lines have a common lower level (6^3P_2). On the other hand 6^3D_3 (upper level of the 365 nm line) is very close to 6^3D_2 (upper level of the 577 nm line). Therefore, with $e'=6^3D_3$ and assuming that $T_{e'} = T(6^3D_2) = T^{eq} = 5875$ K, one obtains from Eq. (55), that $T_e \ T(7^3S_1) = 5815$ K. With this value for T_e a new set for the temperatures T_a of the 6^3P levels was obtained (Table 1). The results show that the population temperatures of the 6^3P levels are lower than those of the

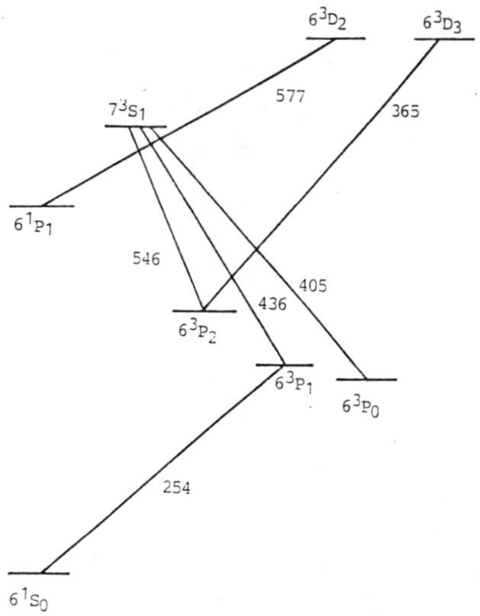

Fig. 11.   Partial term diagram of Hg (λ in nm).

Table 1. Experimentally determined valuesof the population temperatures T_a for a high-pressure H9 arc with $T_e = 5875$ K, determined from the 577 nm line (6^3D_2-6^1P_1), and $T_e = 5815$ K, determined from the lines 546.1 nm (7^3S_1 -6^3P_2) and 365.0 nm (6^3D_3 -6^3P_2). For comparison the valuesT$_m$ obtained by Bartels' method are also given.

Line (nm)	e	a	$T_e = 5875$ K	$T_e = 5815$ K	T_m
546.1	7^3S_1	6^3P_2	5750	5725	5955
435.8	7^3S_1	6^3P_1	5650	5570	5975
404.7	7^3S_1	6^3P_0	5665	5580	5980
365.0	6^3D_3	6^3P_2	5725		

upper levels at the arc axis. The 6^3P_1 resonance level is characterized by a lower temperature very close to that of the 6^3P_0.

Deviations From Excitation Equilibrium

Recently[48-52] an approach evolving from Cowan-Dieke's model has also been used in the study of deviations of the lower atomic levels from excitation equilibrium. For a dispersion line profile, rearranging of Eq. (28) gives

$$N_a = (cn/h\nu)B^{-1} p\delta . \qquad (56)$$

We introduce the degree of deviation from excitation equilibrium averaged along an arc radius $B_a = N_a/N_a^{eq}$, and the degree of deviation from excitation equilibrium in the arc axis $b_a = n_a(o)/n_a^{eq}(o)$. The superscript (eq) denotes the values under LTE conditions. The equilibrium values can be calculate from the equilibrium temperature profile $T^{eq}(r)$, obtained from optically thin lines, and Boltzmann's law. B_a and b_a are related by

$$b_a = (n^{eq}/n)B_a . \qquad (57)$$

Thus, B_a, b_a, N_a and $n_a(o)$ can be determined from the measurement of $p\delta$, n and T^{eq}.

In the following we present results concerning the metastable (6^3P_2 and 6^3P_0) and resonance (6^3P_1) mercury levels in high-pressure mercury arcs operating on 50 Hz (100 W/cm). The self-reversed Hg 546.1-, 435.8-, and 404.7-nm lines were used. The equilibrium values were obtained from Abel inversion of the optically thin Hg 577-nm line. Figure 12 shows the determined values of B_a and b_a in terms of the arc pressure at 0.5 ms and 5 ms after the voltage zero-crossing of the ac cycle. The time modulation of the temperature $T_a(o)$ for the 6^3P_1 level, is determined by the typical equation

$$\frac{n_a}{n_a^{eq}} = \exp\left(-\frac{E_a}{kT_a}\right) / \exp\left(-\frac{E_a}{kT^{eq}}\right) \qquad (58)$$

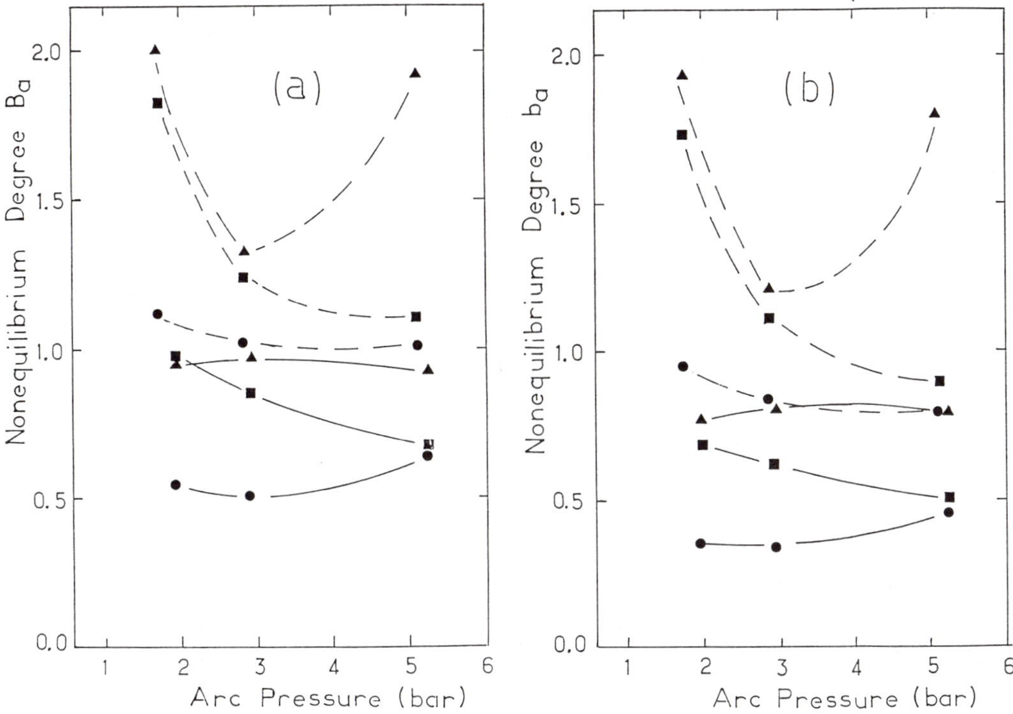

Fig. 12.   Degree of excitation nonequilibrium for the mercury triplet state against the arc pressure at the moments of maximum (solid lines) and minimum (dashed lines) emission. (a), degree of excitation non equilibrium averaged along an arc radi3s; (b), degree of excitation non equilibrium at the arc axis. ▲, 6^3P_2; ●, 6^3P_1; ■, 6^3P_0. From Ref. 52.

is given in Fig. 13 along with the modulation of $T^{eq}(0)$. The estimated experimental errors are $\Delta B_a/B_a = 8\%$, $\Delta b_a/b_a = 12\%$ at 5 ms, and $\Delta B_a/B_a = 12\%$, $\Delta b_a/b_a = 16\%$ at 0.5 ms.

<u>Discussion of the Experimental Results</u>

As seen in Fig.12, the population of the $6^3P_{2,0}$ metastable levels averaged along a radius and at the arc core is higher at 0.5 ms and lower at 5 ms with respect to the values of their population assuming equilibrium conditions. The relative population of the 6^3P_1 resonance level is always lower than those of the 6^3P levels.

The 6 P levels are associated with the ground state by inelastic collisions with the electrons and there is a tendency toward an equilibrium with the ground state, $n_a \sim n_0 \exp(-E_a/kT_{el})$. On the other hand the density of the 6^3D_2 level (upper level of the Hg 577 nm line) and, consequently, T_{eq}, is in relative equilibrium with the continuum, i.e., with the square of n_{el}.

From the hysteresis in the current-electric field behaviour of the plasma column[49] and assuming that current and field follow the n_{el} and T_{el} respectively, it follows that during the onser of ionization ($0<t<2ms$) there is n ionization equilibrium.[53] During this time a rapid increase of the population of the low lying levels takes place, because of the rapid increase in T_{el} while n_{el} and, consequently, T_{eq}, change slightly. This effect justifies

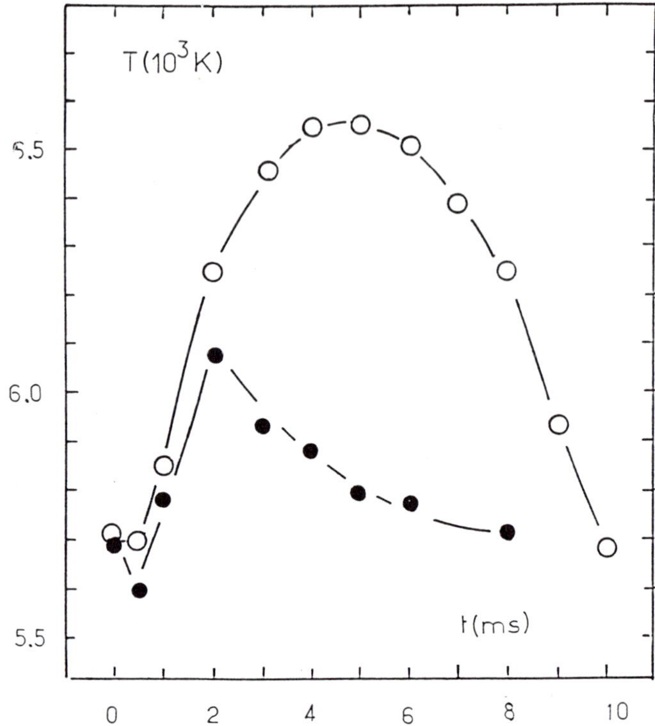

Fig.13.　Modulation of the equilibrium temperature $T^{eq}(0)$ and of the excitation temperature $T_a(\bullet)$ of the 6^3P_1 level, determined by Eq. (58) at the axis of an ac, 3-bar Hg arc. From Ref. 49.

the high value of the relative population of the 6^3P levels at 0.5 ms, where the value of T_{eq} is minimized.

About 5 ms the discharge current and T_{eq} pass through a maximum value, and T_{el} and n_{el} attain values which correspond to a steady-state situation, i.e., to a dc discharge operating at the same values of voltage and current. It has been found by an interferometric technique that in an ac mercury arc the atom kinetic temperature does not undergo a modulation over the current period and its value is much lower than the value of T_{eq} determined by the Hg 577 nm line at the moment of maximum arc current.[54]

These results reveal the existence of an overpopulation of the Hg ground state at 5 ms. Thus, the overpopulation of the ground state atoms seems to be the main nonequilibrium effect which causes the observed underpopulation of the levels (b_a, $B_a < 1$) at 5 ms.

At every moment the 6^3P levels are strongly correlated together via inelastic collisions with the electrons. Besides the 6^3P_1 resonance level is depopulated by radiative decay (emission of the 253.7 nm resonance line).[55] Thus the metastable levels play the role of population reservoirs for the resonance level which in its turn is unloaded into the ground state. This effect leads to a lowering in the relative population of the 6^3P levels and in particular of the 6^3P_1 level.

The absorption of the resonance radiation 253.7 nm arriving from the high temperature arc core is capable not only of compensating the local radiative losses at the arc periphery, but it can also cause an overpopulation of the 6^3P levels in this region. This explains why $B_a > b_a$.

As the pressure increases the electron-atom collision frequency increases too, so that b_a, $B_a \to 1$, as is the case for the $6^3P_{1,2}$ levels at 5 ms, and $6^3P_{1,0}$ levels at 0.5 ms (Fig. 12). The energy gap between 6^3P_1 and 6^3P_0 is 0.22 eV, and the atomic density two to three orders of magnitude greater than the electronic density. Therefore, the contribution of the atom-atom collision to the $6^3P_0 \to 6^3P_1$ transition should be considerable. This effect may explain the decrease of the 6^3P_0 population increasing the pressure.

The error in the determination of N_a introduced by the particular form of the excitation function [Eq.43] used in our model, was recently determined.[56] This has been done by solving Eq.(1) numerically on the basis of a simulated experiment. It has been found that for the case of a LTE Hg arc with parabolic or cubic temperature profile the model used gives N_a values that fall within 10% of the exact values.

We must, however, emphasize that the accuracy of the self-reversed line method for plasma diagnostics depends on the choice of the E(y) adn P(u) function.

In conclusion, we note that the interpretation of the experimental results given above are largely qualitative and to some extend speculative. A qualitative treatment of the variation of nonequilibrium with the pressure necessitates the solution of a system of equations governing the level populations including the equation for radiative transfer for the resonance radiation, and requires knowledge of the kinetic temperatures.

ACKNOWLEDGMENTS

This work was partially supported by the Greek Ministry of Research and Technology.

REFERENCES

1. H.R. Griem, "Plasma Spectroscopy" Mc Graw-Hill Company, New York (1964).

2. W. Locthe - Holtgreven, in "Plasma Diagnostics," W. Locthe-Holtgreven, ed., North-Holland P.C., Amsterdam (1968).

3. H.W. Drawin, in "Progress in Plasmas and Gas Electronics," Vol. 3, R. Rompe ed., Akademie-Verlag, Berlin (1975).

4. D. Karabourniotis, in "Radiative Processes in Discharge Plasmas," J.M. Proud, and L.H. Luessen, eds., Plenum Press, New York (1986).

5. J.F. Waymouth, in "Plasma Diagnostics," Vol. 1, O. Auciello, and D.L. Flamm, eds., Academic Press, Boston (1989).

6. I.J.M.M. Raaijmakers, P.W.J.M. Boumans, B. van der Sijde, and D.C. Schram, Spectrochim. Acta 38B, 697 (1983).

7. T.L. Eddy, J. Quant. Spectrosc. Radiat. Transfer 33, 197 (1985).

8. R. Assous, J. Quant. Spectrosc. Radiat. Transfer 23, 435 (1978).

9.   J. Bacri, and M. Lagreca, J. Phys. D : Appl. Phys. 16, 829 (1983).

10.   S. Nowak, J.A.M. van der Mullen, B. van der Sijde, and D.C. Schram, J. Quant. Spectrosc. Radiat. Transfer 41, 177 (1989).

11.   B.L. Caughlin, and M.W. Blades, Spectrochim. Acta 39B, 1583 (1984).

12.   L.A. Kirpichnikova, S.I. Krylova, L.A. Luizova, I.M. Nekrylova,V.A. Solyanikova, and A.D. Khakhaev, Opt. Spectrosc. 35, 470 (1973).

13.   M.A. Cappelli, and R.M. Measures, Applied Optics 26, 1058 (1987).

14.   D. Karabourniotis, A. Palladas, A. Tsakonas, Proc. 5th Int'l Symp. on Sci. Tech. Light Sources, York (1989).

15.   J.J. Damelincourt, D. Karabourniotis, M. Karabourniotis, and L. Scoarnec, J. Phys. D : Appl. Phys. 11, 2207 (1978).

16.   I.S. Fishman, M. Kh. Salakhov, E.V. Sarandaev, and P.S. Semin, Opt. Spectrosc. 51, 435 (1981).

17.   H.P. Stormberg, J. Appl. Phys. 51, 1963 (1980).

18.   H.P. Stormberg, and R. Schafer, J. Appl. Phys. 54, 4338 (1983).

19.   J.T. Dakin, and R.P. Gilliard, J.Appl. Phys. 60, 1281 (1986).

20.   J.T. Dakin, and R.P. Gilliard, J. Appl. Phys. 62, 79 (1987).

21.   D. Karabourniotis, These de Doctorat d' Etat, no 793, Universite P. Sabatier, Toulouse (1977).

22.   J.J. Damelincourt, M. Aubes, P. Fragnac, and D. Karabournioris, J. Appl. Phys. 54, 3087 (1983).

23.   R.D. Cowan, and G.H. Dieke, Rev. Mod. Phys. 20, 418 (1948).

24.   H. Bartels, Z. Phys. 127, 243 (1950).

25.   H. Bartels, Z. Phys. 128, 546 (1950).

26.   T. Tako, J. Phys.Soc. Japan 16, 2016 (1961)

27.   N.G. Preobrazhensky, Opt. Spectrosc. 17, 4 (1964).

28.   N.G. Preobrazhensky, O.B. Ravodina, and N.S. Terpugova, J. Appl. Spectrosc. 3, 151 (1965).

29.   N.G. Preobrazhensky, Opt. Spectrosc. 22, 95 (1967).

30.   N.G. Preobrazhensky, "Spectroscopy of an Optically Thick Plasma" Nauka, Novosibirsk (1971).

31.   W.J. van den Hoek, Philips J. Res. 38, 188 (1983).

32.   I.S. Fishman, G.G. Illin, and M. Kh. Salakhov, J. Phys.. D : Appl. Phys. 20, 728 (1987).

33. D. Karabourniotis, and J.J. Damelincourt, J. Appl. Phys. 53, 2965 (1982).

34. D. Karabourniotis, J. Phys. D : Appl. Phys. 16, 1267 (1983).

35. D. Karabourniotis, and C. Karras, J. Appl. Phys. 57, 4861 (1985).

36. D. Karabourniotis, S. Couris, J.J. Damelincourt, and M. Aubes, IEEE. Trans. Plasma Science PS 14, 325 (1986).

37. D. Karabourniotis, S. Couris, and J.J. Damelincourt, J. Quant. Spectrosc. Radiat. Transfer 38, 303 (1987).

38. D.C. Fromm, J. Seehawer, and W.J. Wagner, "Proc. Symp. High Temp. Metal Halide Chemistry", Vol. 1, D. Hildebrand, and D.D. Cubiciotti, eds., Proc. ECS (1978).

39. K.V. Aleksandrovich, and I.A. Berezin, Opt. Spectrosc. 43, 13 (1977).

40. G. Wesselink, D. de Mooy, and M.J.C. van Gemert, J. Phys. D: Appl. Phys. 6, L27 (1973).

41. J.J. de Groot, and A.G. Jack, J. Quant. Spectrosc. Radiat. Transfer 13, 615 (1973).

42. E. Drakakis, These, no 275, Univ. P. Sabatier, Toulouse (1988).

43. H.G. Kloss, and W. Funk, Beitraqe Plasma Phys. 13, 101 (1973).

44. H.G. Kloss, and W. Funk, in "Proc. 13th Int'l Conf. on Ioniz. Gases" Berlin (1977).

45. S. Couris, These 3^e cycle. no 2871, Univ. P. Sabatier, Toulouse (1983).

46. P.A. Vicharelli, and P.A. Reiser, Phys. Lett. A125, 326 (1987).

47. D. Karabourniotis, and E. Drakakis, "Proc. XIX Int'l. Conf. Phen. Ioniz. Cases," Belgrade (1989).

48. D. Karabourniotis, Optics Comm. 61, 38 (1987).

49. D. Karabourniotis, and S. Couris, Optics Comm. 65, 22 (1988).

50. D. Karabourniotis, and S. Couris, Optics Comm. 67, 214 (1988).

51. D. Karabourniotis, Optics Comm. 67, 218 (1988).

52. D. Karabourniotis, S. Couris, E. Drakakis, and J.J. Damelincourt, J. Appl. Phys. (in press).

53. L.M. Biberman, V.S. Vorobev, and I.T. Yakubov, Sov.Phys.Ups. 22, 411 (1979).

54. V.G. Vdovin, A.A. Pustoshkin, N.A. Vdovina, and A.A. Pritkov, "Proc. 3rd Int'l Symp. on Sci. Tech. Light Sources" p. 174, Toulouse (1983).

55. P. Mansbach, and J. Keck, Phys. Rev. 181, 275 (1969).

56. P. Vicharelli, and D. Karabourniotis, "Proc. 5th Int'l Symp. on Sci. Tech. Light Sources" York (1989).

AN ANALYSIS OF THE HIGH CURRENT GLOW DISCHARGE OPERATION OF THE BLT SWITCH

G. Kirkman-Amemiya, R. L. Liou, T.Y. Hsu and M. A. Gundersen

University of Southern California
Department of Electrical Engineering
Los Angeles, California 90089-0484

ABSTRACT

High current operation of the BLT switch is described. This paper includes a description of the switch structure and several optical triggering methods. Result of experimental observations of high current operation are reported including measurements of peak currents >80kA and rates of current rise >10^{11}A/sec. Observations of the discharge plasma using a streak camera and electrode surface analysis by electron microscope are used to describe the dense glow discharge and the self heated cathode of this switch. Two phases of the discharge are identified, a hollow cathode phase responsible for the fast buildup of the discharge plasma and a dense glow discharge phase with self heated thermionic cathode responsible for the high peak currents obtained.

INTRODUCTION

The Back of the cathode Light activated Thyratron or BLT switch[1,2] is an optically triggered version of the Pseudospark switch[2]. The BLT operates with peak currents and current rates of rise comparable to high pressure spark gaps but with the repetition rates and low electrode erosion characteristic of hydrogen thyratrons. Unlike the hydrogen thyratron the switch requires no cathode heater power and has a much simpler gridless structure, also the symmetric structure of the BLT allows it to conduct reverse currents without electrode damage. Unlike the Pseudospark switch the BLT is optically triggered allowing for complete electrical isolation of the triggering circuit. In this paper we discuss the BLT switch and several optical triggering methods, the electrical characteristics (peak current and rate of current rise) of the switch, discharge plasma and cathode emission observed in the switch and the formation of the discharge including streak camera observations of two phases of the pseudospark type discharge.

SWITCH STRUCTURE AND OPERATION

The switch structure (figure 1) is cylindrically symmetric consisting of two cup shaped electrodes with flat bottoms that face each other and are separated by about 3-5mm, each electrode has an aperture on the cylinder central axis of 3-5mm diameter. The electrodes are positioned within a cylindrical ceramic or glass insulator and the structure is filled with a low pressure gas usually hydrogen at about 0.2 torr. For this pressure and electrode separation the switch has a high breakdown voltage by operating on the left side of Paschen minimum[3].

Physics and Applications of Pseudosparks
Edited by M. A. Gundersen and G. Schaefer
Plenum Press, New York, 1990

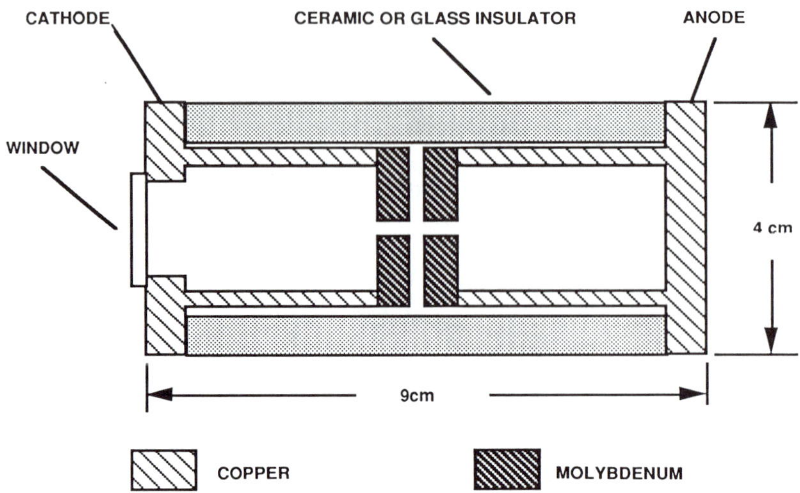

Figure 1.   Basic BLT switch structure. The switch is filled with low pressure gas placing the electrode gap on the left side of the paschen minimum. UV light enters through the window and is incident on the back side of the cathode initiating the discharge through photoemission. In the fiber optic triggered switch the window is replaced by an optical fiber feedthrough. The fiber then carries the light to the region of the cathode aperture.

In this region the mean free path between ionizing collisions for electrons released from the cathode is comparable to or larger than the electrode separation making and ionization avalanche and electrical breakdown highly unlikely. A higher breakdown voltage is obtained by decreasing the gas pressure or decreasing the electrode separation. The central apertures perturb the field giving a longer effective path for breakdown with a lower breakdown voltage. When electrons are added to the region of the cathode aperture an ionization avalanche can occur producing a homogeneous discharge plasma confined to this central region of the electrodes. A hollow cathode type discharge is formed that gives a rapid rise in ionization and current through the switch. The ions produced during these transient phases bombard the cathode producing secondary electron emission and heating the cathode surface to a temperature where thermionic emission is possible. The discharge continues in a high current dense glow discharge phase using the thermionic cathode produced at the front surface of the cathode electrode. The switch conducts at a current density of $\approx 10 kA/cm^2$ that continues for a time determined by the external circuit.

The operating voltage is similar to that of hydrogen thyratrons up to about 35kV for typical single gap devices however the peak currents and rate of current rise are significantly higher than thyratrons. The switch has operated at peak currents over 80kA with pulse lengths from 50nsec to several μsec. The rate of current rise is usually limited by the external circuit except if the switch is operated at a decreased gas pressure. Under normal operating conditions in a fast circuit this switch structure has been observed to switch peak currents >15kA with a rate of rise $>6 \times 10^{11}$ A/sec. The high peak currents and rise rates are important to several pulsed power applications such as lasers and modulators for accelerators that typically require magnetic pulse sharpening and compression to obtain the required pulse shape, in some applications replacing the thyratron by a BLT switch may eliminate the need for magnetic circuitry[4]. Also circuits that have poorly matched loads, such as the excimer laser discharge can produce reverse currents that are damaging to thyratrons due to arcing that can occur on reverse conduction, the symmetric structure of the BLT allows it to handle nearly 100% reverse currents without damage.

OPTICAL TRIGGERING

Only a small number ($\sim 10^9$) electrons are required to initiate this discharge. These electrons can be provided through photoemission using an unfocused uv light source incident on the back surface of the cathode. Measurements of delay and jitter[2] of the discharge formation show that with $\sim 1.5 \times 10^9$ starting electrons produced by photo emission the discharge is initiated with a delay of ~80nsec and jitter <1nsec for a 3mm electrode spacing at 0.2 torr H_2 and 10kV initial voltage. The values of delay and jitter increase when the number of starting electrons, gas pressure or applied voltage is decreased.

We have investigated several optical triggering methods including laser[1], flashlamp[4] and light guided by an optical fiber[5]. In each case a small amount of unfocused light is incident on the back surface of the cathode near the central aperture. The required starting electrons are produced by photoemission from the metal surface typically molybdenum, tungsten or nickel. At low power levels ionization within the gas medium by the triggering light is not significant.Through photoemission 1mJ of photons each having an energy about equal to the cathode work function can produce a number of starting electrons sufficient to initiate the discharge with a delay of ~100ns and jitter <1ns.

The number of photoelectrons released by a UV light pulse has been directly measured by collecting the electrons with a positively biased electrode. At light pulse energies from 1.5 to 6 mJ at a wavelength of 308 nm a quantum efficiency of approximately 2×10^{-7} was measured. This is roughly consistent with our expectations, since the work functions of the electrodes (app. 4.25 eV for molybdenum and 5 eV for nickel) are slightly greater than the energy per photon (4.025 eV). Shorter wavelengths produce improved quantum efficiency by several orders of magnitude. A light pulse at a wavelength of 222 nm (the energy of the photon is 5.58 eV) was observed to increase the quantum efficiency to $\sim 1 \times 10^{-4}$, an increase of approximately 700 over the quantum efficiency at a wavelength of 308 nm. This indicates that operation with low delay and jitter can be achieved with low light energies at shorter wavelengths. With 10 µJ at 222 nm, comparable results to the experiments with 4.4 mJ at 308 nm were achieved.

Laser Triggering

The first BLT[1] was triggered by an unfocussed XeCl laser of 308 nm wavelength and 10 nsec pulse length and pulse energy of 10 mJ. The light enters through a quartz window and is incident on the Mo or Ni cathode surface as in Fig. 1. The laser triggered switch operates with a typical delay ~ 100 ns and jitter ~ 1 ns when operated at ~ 0.2 Torr H_2 at voltages up to $\approx$ 35 KV. The repetition rate in these experiments was limited to 10 Hz by the test circuits used. Much higher repetition rates limited only by the laser used for triggering should be attainable.

Flashlamp Triggering

Since only modest amounts of unfocussed optical energy are required to initiate the discharge, the development of a simpler switch using a flashlamp as the light source was undertaken. The flashlamp triggered switch reported by Kirkman et al[4] is similar to the laser triggered switch with the window replaced by a small bulb type flashlamp such that the window of the flashlamp serves also as the window to the cathode backspace.

A reliable flashlamp triggered switch was designed using a small bulb type UV flashlamp as the light source. The bulb has an electrical to light energy conversion of about 14% with 39% of the light energy below 300 nm wavelength. The flashlamp circuit consists of a 0.5 µF capacitor charged to 600-1200 V that is discharged into the lamp yielding a UV output of about 15 mJ below 300 nm. The bulb is mounted directly behind the hollow cathode electrode such that the window of the bulb is also the window of the switch, allowing the maximum coupling of light into the hollow cathode. In initial tests we have switched 200 J with 0.15 J, an efficiency >1400.

The flashlamp has a large portion (~40%) of its output at wavelengths < 300 nm. A synthetic quartz window allows photons of wavelength as short as 160 nm to enter the cathode backspace and liberate photo-electrons to initiate the discharge. These short wavelength photons produce electrons very efficiently resulting in low jitter operation with a small amount of total light energy. Flashlamp triggered BLTs have operated at 10 -100 Hz with a typical delay of 250 nsec and jitter of ~ 20 ns at 0.2 Torr H_2 and at voltages up to ~ 38 Kv. The flashlamp triggered switch was also successfully tested as a thyratron replacement in a commercial excimer laser[4].

<u>Fiber-Optic Triggering</u>

Optical triggering using a fiber-optic to deliver a UV light pulse to the back surface of the cathode near the cathode aperture was first reported by Braun, et al.[5]. In this method a UV transmitting plastic coated silica or all silica fiber is used to bring light into the cathode back space and direct it to the cathode surface near the hollow cathode aperture. Using the fiber-optic with a XeCl-laser resulted in a small delay and jitter with a small amount of light incident. The best jitter and delay times were 0.4 ns and 78 ns respectively operating at a pressure of 0.2 Torr H_2 and 10 KV anode voltage using 4.4 mJ of XeCl laser light at 308 nm incident on the molybdenum cathode surface. This fiber-optic triggering method is particularly attractive in applications where it is desirable to trigger many switches simultaneously using one light source.

<u>Laser Window Cleaning</u>

The lifetimes of the optically triggered switches have been limited by coating of the windows by material sputtered from the electrode surfaces. Typically after about 10^5 discharges the transmission of the window is reduced to a point where low jitter triggering is no longer possible. We have found that this coating can be removed by laser ablation without having to remove the switch from service. Light of 308nm wavelength from a XeCl excimer laser is focussed to produce an energy density of ~100mJ/cm^2 incident on the coated window surface from the back side of the coating as in (figure 2). At sufficiently high energies the coating can be removed with a single 10nsec laser pulse. At an energy density of ~100mJ/cm^2 the coating is removed in a single pulse, at ~30mJ/cm^2 3-5 pulses are required while at < 5mJ/cm^2 little or no cleaning effect is observed. The clear area required for good triggering is typically only a fraction of a square centimeter and can therefore be cleaned with total laser energies <10mJ/pulse. We have also used this method to clean the optical surfaces of the optical fibers used for triggering.

A similar cleaning effect was observed[1] in the first laser triggered BLT switch where it was noted that the region of the cathode where the laser was incident showed the least change during operation while the other surfaces showed a distinct change in colour.

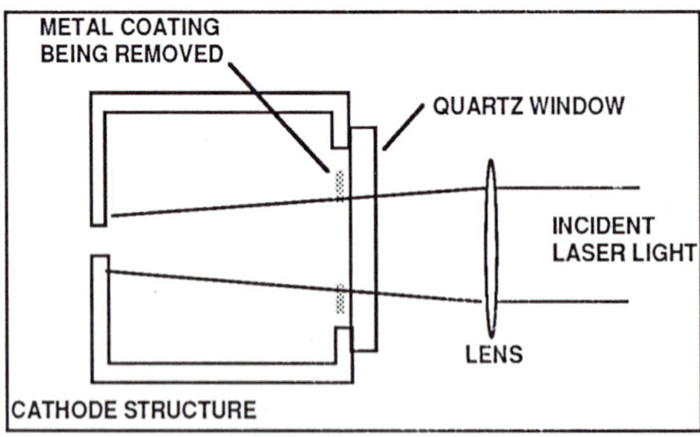

Figure 2.   Cleaning of BLT windows using laser ablation. Using 308nm wavelength light from a XeCl laser at an energy density of ~100mJ/cm^2 the coating is removed in a single pulse.

This was an interesting effect but was not investigated further because we knew of no useful properties of this effect. Recently we have determined the origin of this colour change on the cathode surface and offer an explanation for the clear areas based on laser ablation. An x-ray analysis of copper witness plates that were positioned in the cathode backspace during operation shows a significant amount of Molybdenum in the regions showing a colour change while the clear areas show no Molybdenum. The colour change on the cathode surfaces is due to a thin coating of sputtered electrode material that can be removed by the laser in the region of incidence leaving the clear areas observed.

This method of window cleaning makes long lifetime optically triggered switches possible. If the energy density on the window surface is sufficiently high the windows will remain clean and it is conceivable to produce BLT switches that are limited in lifetime only by degradation of the electrode structure which is slow and gas consumption.

HIGH CURRENT OPERATION

The rate of current rise and peak current handling ability in both forward and reverse directions is a significant issue in many pulsed power applications. The rate of current rise (dI/dt) and peak current in the BLT are normally limited by the external circuit and not the switch itself. Also the symmetric electrode structure allows the BLT to handle very high reverse currents without significant electrode degradation.

Recent experiments have shown that even for peak currents as high as 82 kA and dI/dt of 4×10^{11} A/sec the performance is not limited by the BLT switch. In these tests a flashlamp triggered switch was tested in a circuit whose load resistance could be changed to produce different peak currents. The circuit consisted of two $0.4 \mu F$ capacitors in parallel, a $1 m\Omega$ current viewing resistor , the flashlamp triggered switch and a carbon block load resistor of 1Ω to produce an overdamped discharge or 0.25Ω to give a near critical damping and for the highest currents the load resistors were replaced by copper conductors to produce a ringing discharge. The peak current and dI/dt show a linear dependence on initial voltage indicating that the current is determined by constant circuit parameters and not a switch discharge limit that would be indicated by a nonlinear behavior.

In nearly all cases the current is circuit limited and the switch operates in a diffuse discharge. Occasionally at high currents arc like behavior is observed. These arcs or sparks can be observed by eye or with camera as bright spots in the discharge plasma and at the cathode surfaces and are sometimes observed as dips in the current waveform or spikes on the voltage waveform. These defects however occur only for very short times compared to the total pulse length and the remainder of the current pulse shows a circuit dependent waveform. The reason for arc like behavior has not been carefully characterized and requires further study. The number of arc type events increases with increasing current however even at 82kA the current and voltage waveforms normally show show no symptoms of the arc like behavior.

DISCHARGE CHARACTER AND CATHODE EMISSION

It is desirable for plasma devices to operate with stable, homogeneous, and repeatable plasma parameters; in high power switches a glow discharge with large cross sectional area is desirable to minimize electrode damage and reduce switch impedance. Streak camera recordings in the radial direction show that in most cases a homogeneous discharge plasma exists[6]. The discharge is a stable high current glow discharge that uses a large area of the cathode, it has been referred to as a superdense glow discharge[8,9], it is not an arc type discharge that constricts to a small spot on the cathode resulting in non repeatable plasma conditions. Occasionally, arcing is observed to start on arbitrary locations in the vicinity of the cathode hole. However these arcs usually occur after the build-up of the homogeneous superdense glow and exist for a short time in comparison with the total length of the discharge. Thus it is straightforward to clearly distinguish between arcing and glow modes. These arc like events can be observed by eye as bright sparks at the cathode surface. The percentage of arcing events increases with increasing discharge current. When the electrodes are new many sparks are observed but after running for some time the electrodes become conditioned and operate at high currents with few sparks. During normal operation with well conditioned electrodes qualitative streak camera observations show the plasma extending homogeneously up to a radius of ~9 mm. Streak camera data have also been obtained for conditions where arcing occurs, and compared with the present data. A clear distinction between arcing and glow operation exists as evidenced from streak data.

Also observed is a uniform cathode surface etch pattern caused by plasma sputtering. The cathode surface shows a well defined outer limit of damage ~15 mm in diameter with a more intensely etched zone ~9 mm in diameter. Under closer examination[6] with a scanning electron microscope (SEM) it can be seen that large areas of the cathode surface have been melted. The melting is uniform over large areas $\sim 1 cm^2$ and not localized spots that would be characteristic of an arc or spark discharge. The melting indicates that the cathode surface has been heated at least to its melting temperature (2900 K for Mo and 3680 K for W) at such high temperatures Mo and W can produce a significant thermionic emission[7,8] producing the uniform glow discharge observed.

We have performed experiments using several molybdenum cathodes operated at several different peak currents. Analysis of the melted area shows that the melted area increases nearly linearly with increasing current. The melted cathode area increases to maintain a current density of 6 - 10kA/cm2 for total currents in the range of 3 - 10kA.

A temperature above the melting temperature of Mo (≈ 2900 K) has been obtained in these experiments. This high temperature can be achieved in a thin surface layer for a short time without a large amount of energy or temperature rise of the bulk cathode. Calculations show[8] that a power density of the order of 10-20 MW/cm^2 is necessary, on a time scale of 10^{-7} sec, to heat a thin surface layer (several μm) to the melting temperature. This high rate of energy deposition can be provided by ion bombardment of the surface during the initial phase of the discharge. During the initial closing phase of the discharge there can be an ion beam with energy comparable to the initial voltage on the gap. If we assume an ion (proton) beam of ~ 10 KeV (1/2 the typical initial voltage) and 2.5 KA/cm^2 (~25% of the peak current), then the power density related to this ion flux is ~25 MW/cm^2 which can heat the molybdenum surface to 3000°Kin ~ 30 nsec[8]. After the initial heating a much lower flux of ions is required to maintain this temperature. The total energy required to heat and maintain the temperature of the cathode surface is typically only a small fraction of the total energy switched.

In experiments using a tungsten cathode similar melting of the surface is observed. The melting point of tungsten is considerably higher than that of molybdenum while their thermal properties under pulsed conditions are similar therefore a significantly higher power density is required to melt the cathode surface during the current risetime.The observed melting of tungsten indicates that the higher power density is available and that a molybdenum cathode used in the same switch and circuit may reach a temperature much higher than its melting point. The emission from a liquid surface of the molybdenum cathode or the effects of molybdenum vapor above the cathode surface have not been characterized and should be investigated to more fully understand this switch and cathode.

At these high temperatures thermionic emission from the molybdenum surface is significant. Also the emission will be enhanced by the high field in the plasma sheath at the cathode surface. The sheath thickness and voltage during conduction are estimated to be 10^{-4}cm and 100 to 500V respectively giving a field at the cathode surface of 1-5x10^{6}V/cm. This field is still to low for field emission as described by the Fowler-Nordheim equation[9] but can significantly enhance thermionic emission. Murphy and Good[10] have considered the processes of field emission and thermionic emission in a common treatment and have calculated emission currents for different combinations of temperature and field. Their results indicate that for electric fields below 10^{7}V/cm the emission is well described by the Shottky emission equation[11] for field enhanced thermionic emission:

$$jth = \frac{4\,\pi\,m\,e\,(K_b\,T)^2}{h^3}\,\exp\,[(w-(e^3E)^{1/2})/(K_b\,T)]$$

where m and e are the electron mass and charge respectively, h is Plank constant, K_b is Boltzmann constant, w and T are the work function and surface temperature of the material respectively and E is the electric field at the surface. This equation is written in gaussian units giving the emission in statamperes/cm^2. The emission in A/cm^2 versus temperature for a surface with work function 4.3eV ($\approx$ work function of Mo or W) is shown in figure 3.

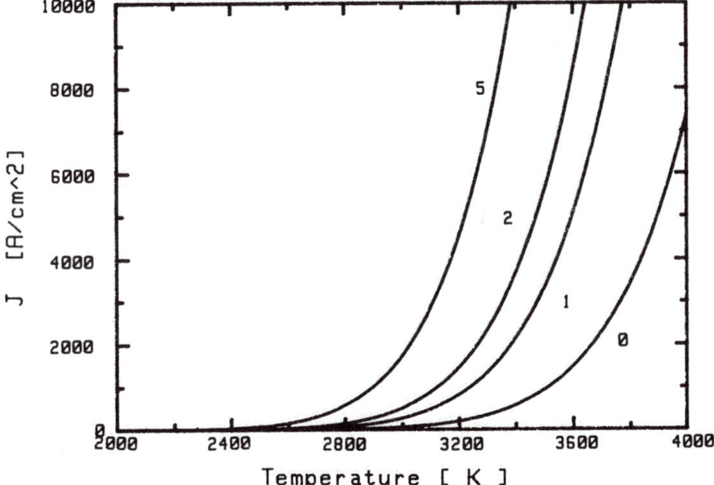

Figure 3.   Field enhanced thermionic emission from a metal surface with work function 4.3 eV (similar to Mo and W) for several values of electric field at the surface. Emission is shown for electric fields of 0, 1, 2, and 5 x10^{6}V/cm.

It can be seen that for the current densities observed in experiment a surface temperature greater than the melting point of molybdenum is required. It should also be noted that in the region of 3000-4000K the curves become very steep such that a small increase in temperature results in a large increase in emission. Although Molybdenum and Tungsten have high work functions there high melting points allow them to be among the best materials for thermionic cathodes.

These results suggest that cathodes may be designed employing this heating mechanism. By tailoring the area of ion beam heating it should be possible to optimize operating conditions according to pulse length, current rise time, voltage fall time, peak current, and electrode material. A detailed description of the cathode emission and heating processes is given by Hartmann et al. also published in these proceedings.

DISCHARGE FORMATION

The formation of the discharge has been studied using a streak camera and temporal analyzer system to give a time and space resolved measure of the light intensity emitted from the discharge plasma. The total light output resolved in space and time is shown in figures 4 and 5. Figure 4 shows the plasma for the first 60nsec of the discharge while figure 5 shows the light output from the center of the discharge versus time for the entire discharge along with the current pulse shown on the same time axis.

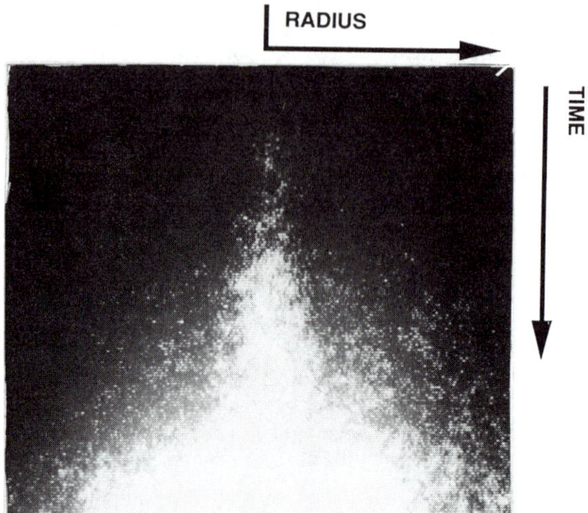

Figure 4(a).   Discharge formation showing two phases of the pseudospark discharge. Streak camera photo showing the radial distribution of light from the discharge plasma versus time.

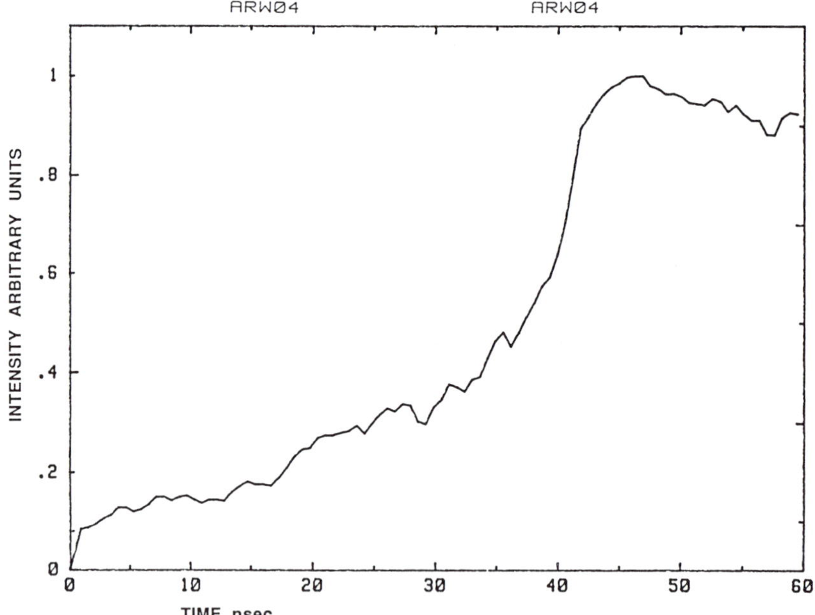

Figure 4(b).   Light intensity on the axis of the discharge versus time. The hollow cathode phase (0 to ~40nsec) is confined to the diameter of the central hole drawing current through the aperture. The dense glow discharge phase (times>40nsec) has a large diameter and draws current from the heated front surface of the cathode.

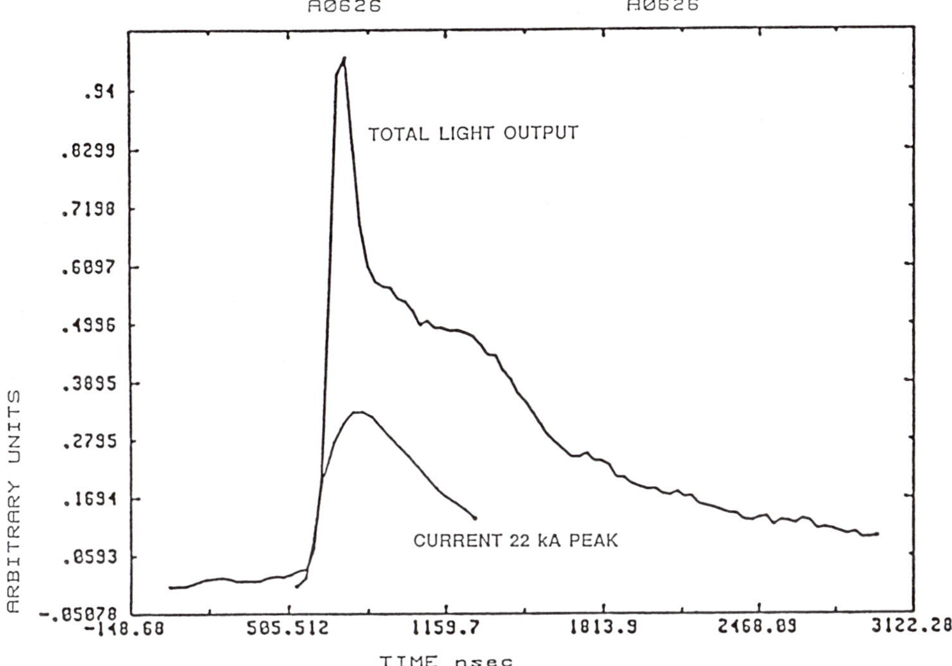

Figure 5.   Total light output vs time for the 22kA current pulse shown.

The light emitting plasma in the anode-cathode gap is observed to start on axis and remain confined for several 10's of nanoseconds to a diameter about equal to the cathode hole diameter. This indicates a hollow cathode phase of the discharge where current is being drawn from the space behind the cathode through the aperture and possibly from the inside of the cathode aperture. The light emitted during this phase is fairly low but monotonically increasing with time.

As the switch current rises the light emitting plasma suddenly shifts to a diameter of about 6 hole diameters and increases 5-10 times in intensity. If this is to be interpreted as an expanding plasma the expansion velocity measured would be $>10^8$cm/sec. We interpret this observation as a shift to the super dense glow discharge phase where high currents are being drawn from the front surface of the cathode that has been heated to a temperature suitable for high thermionic emission. This interpretation is also supported by the observation that the front surface of the cathode shows an intensely melted region of about the same diameter as the plasma observed in this phase.

After the peak of the switch current the light intensity falls rapidly to a lower value then decays with a time constant on the order of 1μsec. This rapid change may be due to a fast thermalization of the plasma when the voltage across the switch gap is lowered. During the current rise there is a high voltage across the switch producing fast electrons which can produce much light through direct ionization of H atoms and subsequent radiative recombination. After the current rise the voltage is lower and the plasma will thermalize on a time scale of nanoseconds due to the high collision frequency of electron-electron collisions which are very efficient in thermalization of the plasma. After this thermalization the plasma decays more slowly by recombination at the surfaces of the electrodes and walls of the switch.

These observations show two phases of the pseudospark type discharge, a hollow cathode phase utilizing the backspace and a dense glow discharge phase utilizing the front surface of the cathode. Observations behind the electrodes also show these stages of the discharge. Quantitative optical analysis behind the electrodes should be performed to better understand the formation of this fast discharge.

CONCLUSION

We have described the operation of the BLT switch at high peak current and current rates of rise. Several optical triggering method have been discussed. The results of observation of the discharge plasma with a streak camera and SEM analysis of the electrode surfaces indicate that the switch operates with a dense glow discharge with the high current density provided by a self heated thermionic cathode. An ion bombardment heating mechanism has been described which can explain the observed heating of the cathode surface. A calculation of the field enhanced thermionic emission from a molybdenum or tungsten surface heated to 3000-4000K shows that emission comparable to that observed in experiment is obtainable from these materials. Through time resolved observations of the discharge formation we have identified two phases of the pseudospark discharge. A hollow cathode discharge is observed that precedes the formation of the high current dense glow discharge phase. Future work should include studies of the cathode to understand the emission from a liquid surface and the effects of vaporization of the electrode material and of the discharge plasma to understand the conduction of the high current densities observed and the production of fast beam electrons.

ACKNOWLEDGEMENTS

The work presented here has been supported by the Air Force Office of Scientific Research and the Army Research Office.

REFERENCES

1) G. F. Kirkman and M. A. Gundersen, "A low pressure light initiated glow discharge switch for high power applications," Appl. Phys. Lett. **49**, 494 (1986).

2) K. Frank, E. Boggasch, J. Christiansen, A. Goertler, W. Hartmann, C. Kozlik, G. Kirkman, C. G. Braun, V. Dominic, M.A. Gundersen, H. Riege and G. Mechtersheimer, "High power pseudospark and BLT switches," IEEE Trans. Plasma Science, **16** (2), 317 (1988).

3) M. J. Schonhuber, "Breakdown of gases below paschen minimum: basic design data of high voltage equipment, IEEE Trans. Power Appar. Syst. PAS-88, 100 (1969).

4) G. Kirkman, W. Hartmann, and M. A. Gundersen, "A flashlamp triggered high power thyratron type switch," App. Phys. Lett. **52**, 613 (1988).

5) C. G. Braun, W. Hartmann, V. Dominic, G. Kirkman, M. Gundersen and G. McDuff, "Fiber optic triggered high-power low-pressure glow discharge switches," IEEE Trans. Electron Devices, **35**, (4), 559 (1988).

6) W. Hartmann, V. Dominic, G. F. Kirkman, and M. A. Gundersen, "Evidence for large area super-emission into a high current glow discharge," App. Phys. Lett. **53** (18), 1699 (1988).

7) W. Hartmann and M. A. Gundersen, "Origin of anomalous emission in super dense glow discharge," Phys. Rev. Lett. **60**, 2371 (1988).

8) W. Hartmann, V. Dominic, G. Kirkman, and M. A. Gundersen, "An analysis of the anomalous high current cathode emission in pseudo-spark and BLT switches," J. Appl. Phys. **65** (11), 4388 (1989).

9)   R. H. Fowler and L. W. Nordheim, "Electron emission in intense electric fields,"
       Proc. R. Soc. London Ser. A **119**, 173 (1928).

10)   E. L. Murphy and R. H. Good, "Thermionic emission, field emission and the
       transition region," Phys. Rev. **102**, 1464 (1956).

11)   W. Schottky, "Uber kalte und warme Elektronentladungen," Z. Phys. **14**, 63
       (1923).

LASER-INDUCED FLUORESCENCE MEASUREMENTS OF

NUMBER DENSITIES OF NEUTRAL AND IONIZED METAL ATOMS

Günter Lins

Siemens AG
Corporate Research and Development
D-8520 Erlangen, West Germany

ABSTRACT

Experiences are reported with the application of laser-induced fluorescence (LIF) to the measurement of atomic and ionic number densities during and after vacuum arcs. The investigations were undertaken to study the influence of neutral and charged particles on the switching properties of vacuum circuit breakers.

In the presence of a plasma, the process of LIF is strongly disturbed by collisional depopulation of the upper fluorescence level. In the case of copper, this results in sensitized fluorescence from the $4^2P_{1/2}$ level which is closely coupled by collisions to the optically pumped $4^2P_{3/2}$ level. Taking the ratio of populations of the upper levels of sensitized and line fluorescence as a measure of the collisional interaction, it is shown experimentally that after extinction of the metal vapour arc, the perturbation by collisions ceases sufficiently fast to allow LIF measurements of neutral and charged atoms.

Such measurements are presented for neutral copper and tungsten, and for singly ionized tungsten and chromium. Atomic and ionic densities between $10^{14}m^{-3}$ and $10^{19}m^{-3}$ were observed with local resolution generally of the order of 1 mm^3. The fast decay of the ion density (3 orders of magnitude within 15 μs for tungsten and 9 μs for chromium) after extinction of the arc can be well correlated to the recovery of vacuum switch gaps.

Measurements of the kind described could also be useful in understanding the role of metal atoms and ions in pseudosparks and related discharges.

1. INTRODUCTION

Laser-induced fluorescence (LIF) is a versatile diagnostic tool which, for example, has been used to measure ground state [1] and excited state [2] densities, excitation temperatures [3], atomic velocity distributions [4], spectral line profiles under the influence of electric field fluctuations [5], electron densities [6,7], and electron temperatures [7].

This paper describes measurements of number densities of neutral and singly ionized metal atoms during and after extinction of vacuum arcs. The purpose of this work was to gain insight into the relevance of neutral and charged particles to the recovery of dielectric strength of vacuum switch gaps after current zero of vacuum arcs. The kind of measurements to be reported should also be appropriate to aid the understanding of the role of sputtered cathode material in the operation of devices such as pseudospark switches and backlighted thyratrons.

After briefly summarizing the principles of LIF in the following section, we shall describe LIF experiments on vacuum arcs in section 3. The results will be presented in section 4. There we shall first discuss the influence of collisional transfer on LIF in the presence of a plasma, and afterwards treat the behaviour of densities of neutral and singly ionized tungsten, as well as singly ionized chromium after forced extinction of vacuum arcs. A summary will be given in section 5.

2. LASER-INDUCED FLUORESCENCE

The theory of laser-induced fluorescence rests on a number of prerequisites which are explicitly stated in [8]. In particular it is assumed that collisions are negligible, and, if total number densities are intended to be measured, that nearly all of the atoms are in their ground state. Then laser-induced fluorescence of metals such as copper, tungsten, and chromium or their singly charged ions can be described by a three-level model including the ground state 1, the laser-excited level 2, and a metastable level 3 into which, for the purpose of calculation, the lower levels of non-resonance transitions originating from level 2 are combined.

As an example, a partial energy level diagram of copper is given in Fig. 1. In order to explain the measurements concerning the effects of collisional transfer on LIF we have added two further levels, 4 and 5, which will be discussed later in this section. Irradiation of atoms, initially in the ground state 1, with laser light at 324.75 nm results in population of the excited level 2. By emission of light, the atoms can either return to the ground state or undergo a transition to the metastable state 3. The former process is referred to as resonance fluorescence, the latter is called line fluorescence. For the purpose of measurements we prefer line fluorescence, in the case of copper the transition $4^2P_{3/2}$ - $4^2D_{5/2}$, at a wavelength of 510.55 nm. This is well separated from the wavelength of the incident laser radiation, such that perturbations by scattering of laser light from windows and metal droplets in the vacuum gap can easily be avoided.

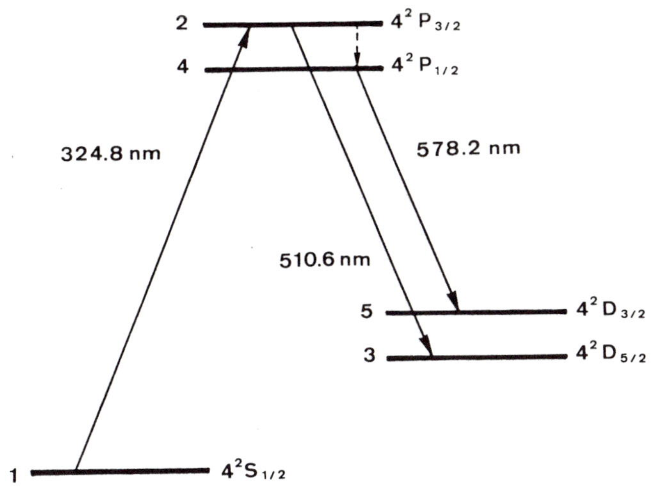

Fig. 1. Partial energy level diagram of copper. Levels $4^2P_{3/2}$ and $4^2P_{1/2}$ are closely coupled by collisions. Line fluorescence and sensitized fluorescence are observed at 510.6 nm and 578.2 nm, respectively. Note that the spacing of levels 2 and 4 is not drawn to scale. The difference in energy is only 0.03 eV [10].

The set of rate equations describing the process of LIF within the framework of the three-level model is explicitly stated in [8] and [9]. By solving the rate equations it can be shown that for sufficiently high intensity of the laser radiation the population density N_2 of level 2 tends towards an upper limit given by

$$N_{2sat} = N_{10} \frac{g_2}{g_1 + g_2} \tag{1}$$

which, except for the initial population density of the ground state, depends only on the the statistical weights g_1 and g_2 of the respective levels.

The saturated population density N_{2sat} will be closely approached, if the saturation parameter S, i.e. the ratio of the induced to the spontaneous rate of emission, is large as compared to unity.

$$S = \left(1 + \frac{g_1}{g_2}\right) B_{12} u_\nu \tau \gg 1 \tag{2}$$

Here B_{21} is the Einstein coefficient of induced emission, u_ν the spectral energy density and τ the radiative lifetime of the upper fluorescence level 2.

In case of saturation, N_{10} can be simply related to the peak value I_{23max} of the fluorescence intensity observed during the laser pulse by

$$I_{23max} = \frac{hc}{4\pi\lambda_{23}} A_{23} \frac{g_2}{g_1 + g_2} N_{10} l \tag{3}$$

where h is Planck's constant, λ_{23} the wavelength of the fluorescence transition, c the speed of light, A_{23} the transition probability, and l the dimension of the scattering volume in the direction of observation. If initially, i.e. before irradiation, all atoms are in their ground state 1, I_{23max} provides a measure of the total atomic number density.

The saturation parameter S is a useful criterion for saturation, only if the laser pulse is rectangular in time. For real laser pulses, with non-zero risetime and finite duration, the rate equations have to be solved numerically in order to predict whether saturation will occur. An example is given in Fig. 2 for the pulse shape produced by the flashlamp-pumped dye laser used for most of the experiments to be reported.

In a plasma environment, collisional rates of transition add to the radiative rates [9]. Furthermore, a significant fraction of the excitation energy contained in the laser-excited level 2 may be transferred to levels not included in the three-level model or to the plasma electrons by collisions during the excitation and fluorescence process. As a result, saturation may no longer be reached and the fluorescence intensity is reduced in an unpredictable way, such that (3) is no longer valid. A further complication arises from atoms no longer being in the ground state before irradiation.

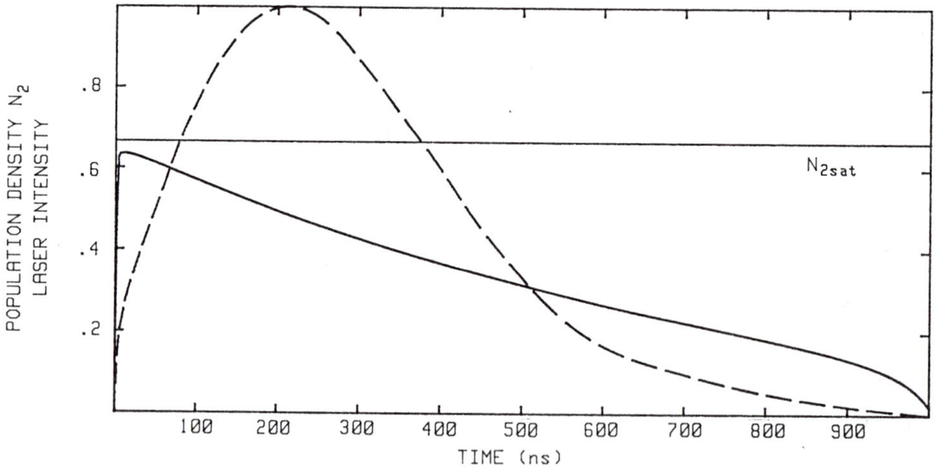

Fig. 2. Population density N_2, normalized to the total density N_{10}, (solid line) as obtained by numerically solving the rate equations [8, 9] for copper, according to Fig. 1, in the absence of collisions. N_2 as well as the maximum possible population density N_{2sat} are normalized to the total density N_{10}. The broken line shows the pulse shape of the flashlamp-pumped dye laser. As in the experiment, the peak power density in this example amounts to 30 kW/cm^2. N_2 closely approaches the saturated density N_{2sat}.

In the case of copper (Fig. 1) the upper fluorescence level $4^2P_{3/2}$ is in particular strongly coupled by collisions to the level $4^2P_{1/2}$ (labeled 4 in Fig. 1) which lies only 0.03 eV below [10]. Due to collisional population of the $4^2P_{1/2}$ level sensitized fluorescence $4^2P_{1/2} - 4^2D_{3/2}$ can be observed at a wavelength of 578.2 nm.

We take the ratio R of the population densities N_4 and N_2 of the upper levels of sensitized fluorescence, $4^2P_{1/2}$, and line fluorescence, $4^2P_{3/2}$, as a measure of the collisional perturbation. R is related to the intensities I_{45} and I_{23} of sensitized and line fluorescence by

$$R = \frac{N_4}{N_2} = \frac{I_{45}\lambda_{45}A_{23}}{I_{23}\lambda_{23}A_{45}} \qquad (4)$$

where λ_{45} and A_{45} are the wavelength and transition probability of the sensitized fluorescence transition. R is proportional to the electron density N_e [6]; however to measure N_e using R would require knowledge of the rate coefficient of collisional transfer from $4^2P_{3/2}$ to $4^2P_{1/2}$.

3. EXPERIMENTAL

3.1 Vacuum Arcs

An experimental setup typical of LIF measurements on vacuum arcs is shown in Fig. 3. A pair of either chromium-copper or tungsten-copper butt contacts, 40 mm or 28 mm in diameter, respectively (Fig. 4), was mounted in a stainless steel vacuum chamber fitted with quartz windows for optical access. The movable contact which was the anode had a slightly spherical surface to ascertain arc ignition in a well defined place. The pressure in the continuously pumped system was lower than 10^{-4} Pa.

Metal vapour arcs were drawn at an auxiliary current of 15 A. After the contacts had reached their final separation of 10 mm, a 50 Hz sinusoidal current was applied such that a peak current of 200 A was reached. At current crest, the current was forced to zero within 1.3 μs by application of a reverse voltage. A typical current waveform is shown in Fig. 5 together with an oscillogram of the contact separation.

3.2 Excitation of Fluorescence

To perform LIF experiments on neutral copper atoms a flashlamp-pumped dye laser was operated with a 6×10^{-5} molar solution of sulforhodamine 101 in methanol. Ultraviolet radiation at 324.75 nm obtained by second-harmonic generation in an ADP crystal was directed through the centre of the contact gap (Fig. 3) with the optical axis 5 mm above the cathode surface. Two slits of 2 mm width positioned near the entrance window of the vacuum vessel and oriented perpendicularly with respect to each other determined the cross-section of the beam. The power density in the observation volume amounted to 30 kW/cm^2, enough to come close to saturation in the absence of collisions (Fig. 2). The intensity profile in the direction of observation was measured with a diode array and found to be approximately rectangular.

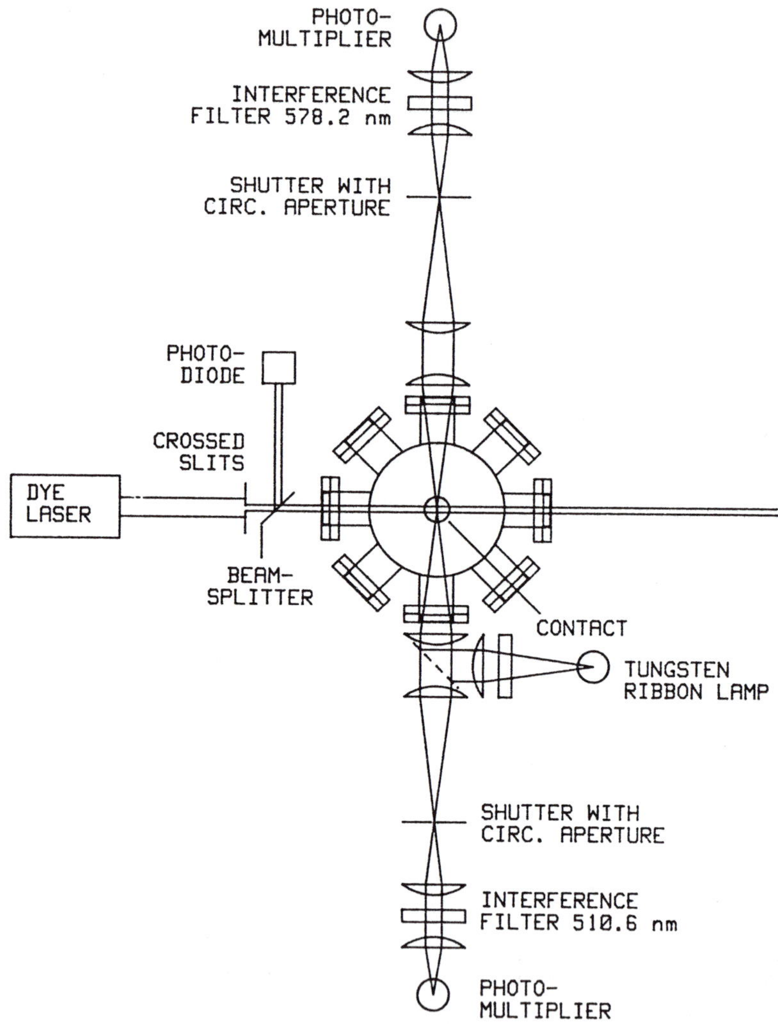

Fig. 3. Experimental setup for LIF on vacuum arcs. Sensitized and line fluorescence of copper were measured simultaneously. The setup for the rest of the experiments was different insofar as only one branch of the detection scheme was used and the laser radiation had to be focused into the contact gap.

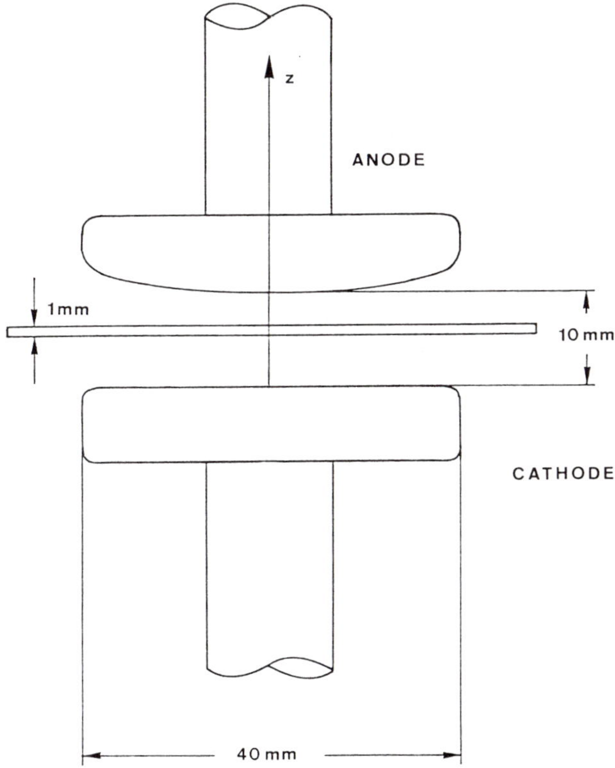

Fig. 4. Cross-section of contacts used in the experiments with chromium and copper. For the measurement of the chromium ion density the observation volume had a rectangular cross-section as shown. In all other cases its cross-section was circular, and 1 mm in diameter. Shape and separation of the tungsten-copper contacts used for experiments on tungsten atoms and ions were the same, except for the contact diameter which was 28 mm.

The same laser with a different crystal for second harmonic generation was used for measurements of the density of neutral (W I) and singly ionized (W II) tungsten. For the ions the laser was operated with a 4×10^{-5} molar solution of rhodamine 110 in methanol; for neutral tungsten a 5×10^{-5} molar solution of rhodamine 6G was used. The excitation and fluorescence wavelengths are given in Fig. 6. To obtain sufficient power density inside the observation volume (130 kW/cm^2), the laser radiation had to be focused into the contact gap. The beam diameter at the point of observation was 2 mm. Further details are given in [1].

Rise time and pulse duration of the flashlamp-pumped dye laser were 100 ns and 350 ns, respectively. The spectral bandwidth was 20 pm, enough to cover the Doppler width of the observed species completely.

For the excitation of fluorescence of singly ionized chromium (Cr II) a Nd:YAG-laser-pumped dye laser was used in connection with a Barium-Beta-Borate crystal for second-harmonic generation [12]. The dye laser was pumped with the third harmonic of the Nd:YAG laser at 355 nm and generated an energy of 5 to 7 mJ at a wavelength of 412.3 nm using a dye mixture of Quinolon 390 and Stilbene 420 dissolved in methanol. After frequency-doubling, pulses of 0.3 mJ were available with 13 pm bandwidth and 6 ns duration at the excitation wavelength of 206.15 nm. With an estimated power density of 3 MW/cm^2 of the focused radiation 80 per cent of the maximum population N_{2sat} of the upper fluorescence level was achieved as indicated by numerically solving the rate equations.

3.3 Detection of Fluorescence

Fluorescence was always observed perpendicularly to the incoming laser beam as shown in Fig. 3 for the example of copper. In that particular case two separate but nearly identical detection setups mounted opposite to each other received light from the same observation volume which was a cylinder whose length of 2 mm was given by the width of the laser beam. Its diameter of 1 mm was determined by a circular aperture inserted in each beam path and imaged into the centre of the contact gap with unit magnification. The apertures were part of fast mechanical shutters which protected the photomultipliers from arc radiation before current zero. Spectral selection was accomplished with interference filters of spectral width 1 nm (FWHM) centered at 510.6 nm for line fluorescence and 578.2 nm for sensitized fluoresence. A tungsten ribbon lamp was provided for the intensity calibration of the detection scheme. Transition probabilities for the evaluation of (3) were taken from [13].

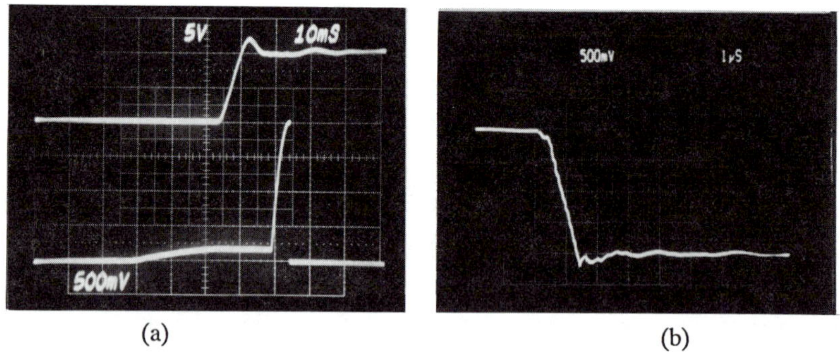

(a)   (b)

Fig. 5. (a) Contact separation (upper trace) and arc current(lower trace). The contacts are separated during an auxiliary current of 15 A. After they have reached their final separation, a 50-Hz sinusoidal half cycle is applied. At its crest, the current is forced to zero within 1.3 µs. One major division corresponds to 5 mm for contact separation, 50 A for the current, and 10 ms in time.

(b) Arc current near a forced current zero. One major division corresponds to 50 A for the current and 1 µs in time.

For the observation of LIF from tungsten atoms and ions only one branch of the detection scheme just described was used with interference filters centered at 302.5 nm for W I and 364.1 nm for W II. Transition probabilities were taken from [14] for neutral atoms and from [15] for ions.

In the case of ionized chromium the circular aperture was replaced with a rectangular one, imaged into the contact gap by a cylindrical lens, in order to collect as much light as possible. This resulted in a rectangular cross-section of the scattering volume which was 1 mm high, and extended 8 mm beyond the contact diameter on both sides (Fig. 4). The interference filter was replaced with a quarter-meter monochromator set to the fluorescence wavelength of 276.26 nm.

4. RESULTS AND DISCUSSION

<u>4.1 Influence of Collisional Transfer on Density Measurements</u>

The influence of collisional transfer on LIF was studied for the example of copper (Fig. 1). Sensitized fluorescence ($4^2P_{1/2}$- $4^2D_{3/2}$, $\lambda = 578.2$nm) and line fluorescence $4^2P_{3/2}$ - $4^2D_{5/2}$, $\lambda = 510.6$nm) were measured simultaneously for some microseconds before and after artificial current zeroes of vacuum arcs between chromium-copper contacts. Fig. 7 shows the ratio R of the population densities of levels $4^2P_{1/2}$ and $4^2P_{3/2}$ together with the copper vapour density N. R was evaluated according to (4) using the maximum values I_{45max} and I_{23max} of the respective intensities during the laser pulse. N was calculated from (3), i. e. based on the three-level model in the absence of collisions. (N is identical with N_{10} of (3); we drop the subscript for simplicity.)

R is constant and amounts to 7.2×10^{-3} until 1.3 μs before current zero, when the current is ramped down. Then it starts to decay exponentially with a decay time constant of 1.3 μs. At 4.5 μs after current

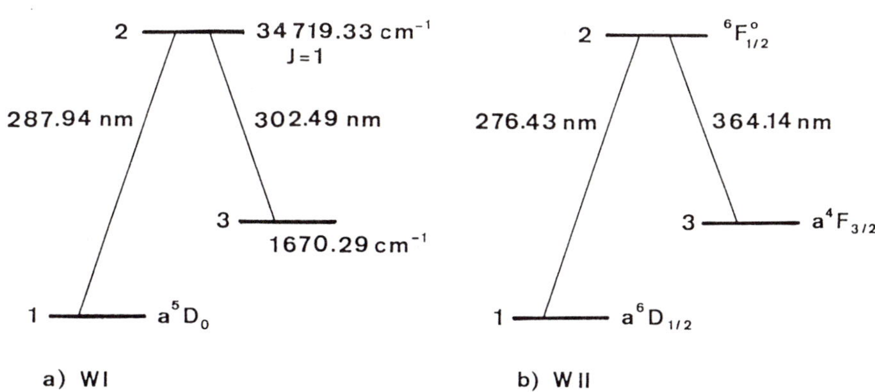

Fig. 6. Partial energy level diagrams of neutral (a) and singly ionized (b) tungsten. Term designations, wavelengths, and wavenumbers according to [14, 15, 25]

175

zero, R amounts to 10^{-4}. Taking R as a measure of the rate of the collisional perturbation of the upper fluorescence level, we conclude that collisional depopulation of the optically pumped $4^2P_{3/2}$ level is strong before, but rapidly reduced after current zero.

As a consequence, the three-level model fails if applied before current zero, and evaluation of fluorescence intensities using (3) yields vapour densities lower than the actual ones. Upon ramping down the current, the collisional interaction is continuously diminished due to decreasing electron density and temperature, and N rises by about a factor of ten. Between two to three microseconds after current zero, N reaches a maximum and subsequently decays as the copper vapour disperses.

One might object that the increase in copper vapour density near current zero is due to three-body recombination of copper ions. The corresponding cross-sections are not available in the literature. If the

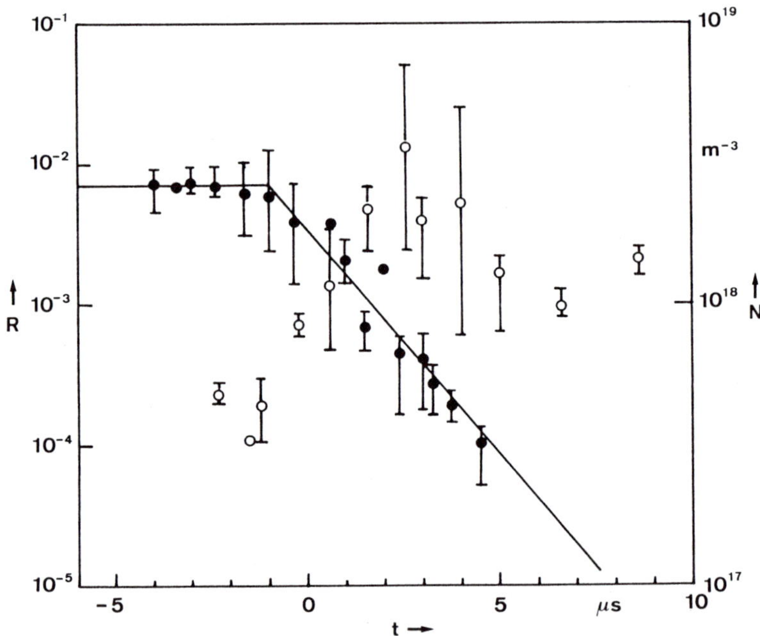

Fig. 7. Ratio R (dots) of population densities of levels $4^2P_{1/2}$ and $4^2P_{3/2}$ of copper, and copper vapour concentration N (circles) in the centre of the contact gap as inferred from the LIF signal at 510.6 nm according to (3). Each point represents an average of three to five measurements. Error bars indicate the highest and lowest values observed at the respective time. The time scale is chosen such that $t=0$ at the artificial current zero. The curve shows the constant value of R before ramping down the current, and its exponential decay. The decay time constant is 1.3 μs.

copper atoms are treated as hard spheres, 3.46×10^{-10} m in diameter [16], the mean free path for collisions of copper atoms among themselves is 0.5 m for the vapour density measured. The mean free path for electron impact ionization, which results if the maximum value of the strongly energy-dependent cross-section according to Lotz [17] is used, amounts to 5.3 m. Hence we conclude that three-body encounters are unlikely.

Under the experimental conditions described, laser-induced fluorescence cannot yield reliable density values before and up to about 3 µs after current zero. However, calculations based on models of Rich and Farrall [18] and Lins [19] indicate that the neutral copper vapour density in the centre of the contact gap after a rectangular current pulse decays by only 11 or 8 percent, respectively, within 5 µs after current zero, if the vapour temperature is chosen to be 3000 K. Both of these models tend to overestimate the rate of density decay, because they do not account for the contribution of metal droplets evaporating in flight after current zero. So the maximum value of N (Fig. 7) obtained by LIF measurements after current zero can be safely regarded as a good approximation to the actual vapour concentration shortly before current zero, i. e. during arcing.

The findings of the preceding paragraph were used to obtain a measurement of the copper vapour density in a metal vapour arc during a current half cycle, i. e. in the presence of a plasma. To do so, artifical current zeroes were forced at preselected times, and N was measured as described above. The maximum N observed after the respective artificial current zero was taken as the density value at the time within the half cycle when the reverse voltage was applied to induce the artificial current zero. Results and further details are given in [20].

<u>4.2 Density of Neutral and Singly Ionized Tungsten</u>

Vacuum switches are well-known for their fast dielectric recovery after current zero. For a long time it was generally accepted that this property was due to a rapid decay of the density of neutral metal vapour present in the contact gap after extinction of an arc. Measurements of several authors [1, 8, 21, 22] however showed that the decrease of the neutral metal vapour density after current zero of a vacuum arc proceeded much more slowly than dielectric recovery. Furthermore the observed vapour densities were found to be too small to cause breakdown by electron impact ionization. On the other hand it was concluded from post-arc current measurements [23] that after current zero substantial amounts of charge remained in the contact gap.

In this situation it seemed appealing to use laser-induced fluorescence to measure the densities of neutral and charged particles after current zero of vacuum arcs. Tungsten was chosen, because the resonance lines of singly ionized copper and chromium lie so far in the ultraviolet region of the spectrum that they cannot be reached with the flashlamp-pumped dye laser which was available for these experiments.

Fig. 8 shows the density of neutral and singly ionized tungsten atoms as measured in the centre of the contact gap, i. e. 5 mm above the cathode, from 7 μs before, till 20 μs after current zero of vacuum arcs as described in section 3.1. Before ramping down the current, i. e. during arcing, the tungsten ion density as inferred from the fluorescence intensity according to (3) is constant and amounts to about 5 x 10^{16}m^{-3}. As the current decreases, the ion density rises by nearly an order of magnitude within 2 μs, until shortly after current zero it starts to decay.

The increase of the ion density upon ramping down the current is attributed to the fact that during arcing the upper fluorescence level is strongly depopulated by electronic collisions as discussed in section 4.1. The same effect ought to be the reason why the fluorescence signal from the neutrals is so weak before current zero, that it disappears in the noise. Only after current zero, the neutral vapour density, again as evaluated according to (3), rises sharply. Since we know from section 4.1 that the collisional perturbation of the LIF process ceases shortly after current zero, we trust that we can rely on the densities obtained, after the maximum of the ion density has been reached.

After its maximum of 4 x 10^{17}m^{-3}, the ion density decays exponentially with a time constant of 2.6 μs, while the neutral density remains constant. The decay time constant for the decrease of the ion density comes close to the time scale typical of dielectric recovery of vacuum gaps [24]. Similar results

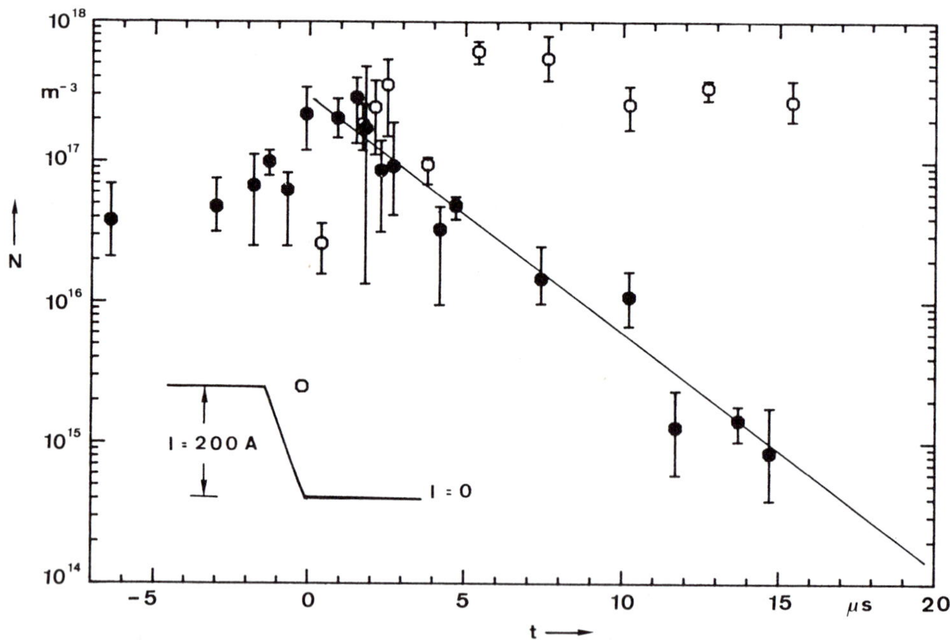

Fig. 8. Density of neutral (circles) and singly ionized (dots) tungsten in the centre of the contact gap. Averaging, error bars, and zero of time scale as in Figure 7. The time constant of exponential decay of the ion density (solid line) is 2.6 μs.

obtained at different distances from the cathode. The details are given in [11]. The long-term behaviour of the neutral-tungsten density is shown in Fig. 9. A density decay by three orders of magnitude which takes the ions only 20 μs lasts as long as 300 μs for the neutrals.

The ground states a^5D of W I and a^6D of W II are split into sublevels with energies of some tenths of an eV above the respective lowest levels [25] given in Fig. 6. Hence, even for a low temperature, a substantial fraction of the atoms or ions may not be in the ground state before irradiation with laser light, such that the densities inferred from LIF ought to be lower than the actual densities. However, since our statements are based on the development in time of densities rather than on absolute values, this should hardly affect their validity.

4.3 Density of Singly Ionized Chromium

The ion density measurements described in the preceding section were repeated with chromium ions. This was considered worthwhile, because, due to the high atomic mass of tungsten, the decay of the tungsten ion density was expected to be untypically long. Furthermore, chromium-copper nowadays has turned out to be the contact material of greatest technological importance to vacuum interrupters.

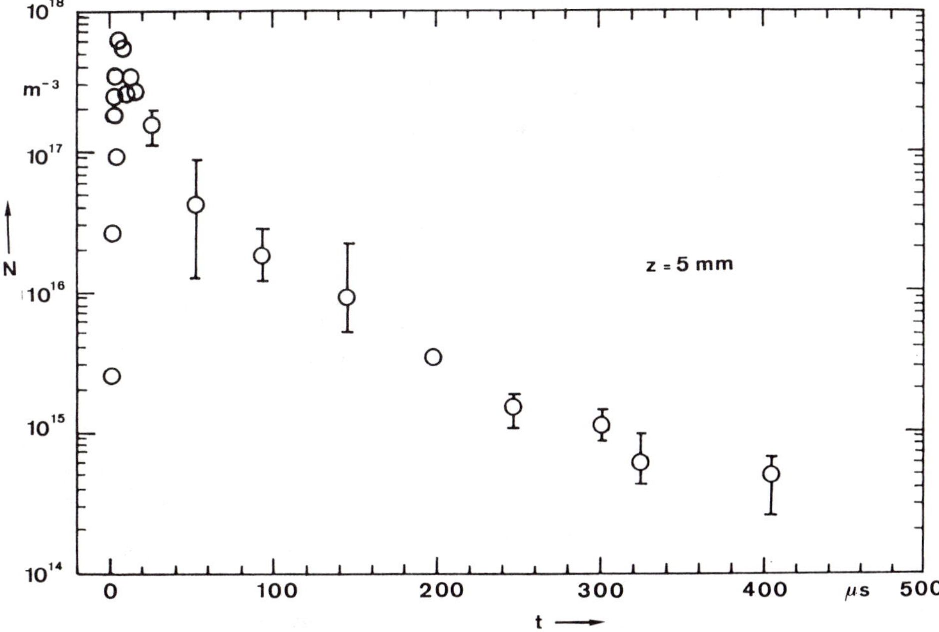

Fig. 9. Long-term behaviour of the neutral tungsten density in the centre of the contact gap. Averaging, error bars, and zero of time scale as in Fig. 7.

The result of a measurement 6 mm above the cathode is shown in Fig. 10. Since no calibration was made as yet, we give density data normalized to their maximum value.

As in the case of tungsten, the ion density is constant until the current decreases. The subsequent ion density increase, however, is less pronounced for chromium than for tungsten. The maximum density observed after current zero is only 40 per cent higher than its mean value before the current decline. The ion density decay is again exponential, but in this case characterized by a time constant of 1.3 μs, as opposed to tungsten where, presumedly due to the higher atomic mass, the time constant was 2.6 μs under conditions otherwise almost the same. The result is similar at a position 4 mm above the cathode, the decay time constant for this position being 1.1 μs.

We first note that the decay time constant for the chromium ion density comes very close to the time constant for dielectric recovery which was measured by [24] for the same contact material but smaller contact spacing, and found to be 1 μs in the early phase of recovery. This adds further evidence to the hypothesis that dielectric recovery of vacuum gaps after diffuse arcs is predominantly controlled by residual charge carriers rather than by neutrals.

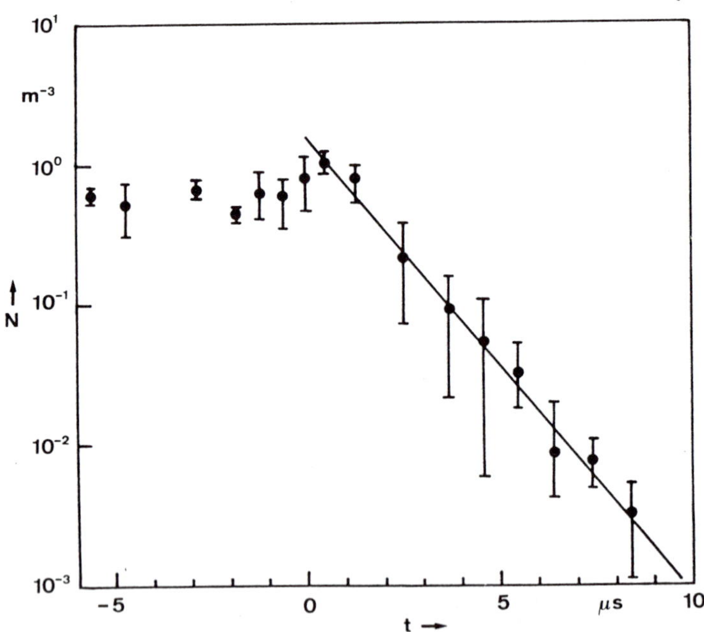

Fig. 10. Density of chromium ions, 6 mm above the cathode surface, normalized to its maximum value. Averaging, error bars, and zero of time scale as in Fig. 7. The time constant of the exponential decay (solid line) after current zero is 1.3 μs.

Secondly we observe that the decay time constants for the ratio R of section 4.1 and for the chromium ion density are practically equal. Both quantities were measured under the same conditions. According to [6], R is proportional to the electron density; the same should be true for the ion density, if the plasma obeys quasi-neutrality. This corroborates our view that the apparent increase in neutral atom as well as in ion density close to current zero is caused by a decrease of electron density and the ensuing disappearance of the perturbation of the upper fluorescence level.

5. SUMMARY AND CONCLUSIONS

Laser-induced fluorescence has been applied to study the behaviour of neutral and singly ionized atoms after forced extinction of vacuum arcs. By comparison of sensitized fluorescence with line fluorescence of copper it could be shown that in the presence of an arc plasma LIF is severely disturbed by collisional depopulation of the upper fluorescence level. By forcing artificial current zeroes, the collisional perturbation can be switched off rapidly, such that useful information concerning the neutral-vapour density in the arc plasma can be obtained by a measurement after current zero. This might be applied to LIF on pulsed plasmas with a rapid decay of the electron density at the end of their lifetime.

The ion density after extinction of vacuum arcs was found to decay on a time scale typical of the recovery of dielectric strength of vacuum gaps after low-current arcs. As anticipated from the difference in atomic mass, the density of chromium ions decays faster than the density of tungsten ions. While the ion density decreases by three orders of magnitude in less than 20 μs, the neutral-vapour density remains virtually constant during this time. It is concluded that for low-current vacuum arcs dielectric recovery is predominantly controlled by charge carriers rather than neutrals.

ACKNOWLEDGEMENTS

It is a pleasure to acknowledge many useful discussions with Dr. I. Paulus on the physics of vacuum arcs. I would also like to thank Mr. G. Grubmüller for maintenance and operation of much of the apparatus, Mrs. G. Brügmann for assistance in the evaluation and presentation of the data, and Mrs. B. Grüner for preparing the manuscript.

REFERENCES

[1] G. Lins, "Measurement of the neutral copper vapor density around current zero of a 500-A vacuum arc using laser-induced fluorescence", IEEE Trans. Plasma Sci. PS-13(6), 577(1985)

[2] D.A. Erwin and M.A. Gundersen, "Measurement of excited state densities during high-current operation of a hydrogen thyratron using laser-induced fluorescence", Appl. Phys. Lett. 48(26), 1773(1986)

[3] K.P. Selter and H.-J. Kunze, "Analysis of atomic beams produced by laser-induced ablation", Physica Scripta 25, 929(1982)

[4] B. Schweer, D. Rusbüldt, E. Hintz, J.B. Roberto, and W.R. Husinsky, "Measurement of the density and velocity distribution of neutral Fe in ISX-B by laser-fluorescence spectroscopy", J. Nucl. Mater. 93&94, 357(1980)

[5] U. Rebhan, "Investigation of the high-frequency Stark effect on lithium atoms by laser-induced fluorescence", J. Phys. B: At. Mol. Phys. 19, 3847(1986)

[6] K. Tsuchida, S. Miyake, K. Kadota, and J. Fujita, "Plasma electron density measurements by the laser- and collision-induced fluorescence method", Plasma Physics 2(9), 991(1983)

[7] K. Kadota, A. Pospieszczyk, P. Bogen, and E. Hintz, "Measurement of electron densities and temperatures in the plasma boundary combining neutral beam probes and laser-induced fluorescence", IEEE Trans. Plasma Sci. PS-12(4), 264(1984)

[8] B. Gellert, "Measurement of high copper vapour densities by laser-induced fluorescence", J. Phys. D: Appl. Phys. 21, 710(1988)

[9] H.C. Meng and H.-J. Kunze, "Investigation of the diffusion of impurity atoms in plasmas by laser-induced fluorescence", Phys. Fluids 22(6), 1082(1979)

[10] C.E. Moore, "A Multiplet Table of Astrophysical Interest", Nat. Stand. Ref. Data Ser., Nat. Bur. Stand. (U.S.), 40, Washington, D.C. 1972

[11] G. Lins, "Measurement of the tungsten ion concentration after forced extinction of a vacuum arc", Proc. XIIIth Int. Symp. Discharges and Electrical Insulation in Vacuum, Paris, 1988, vol. 1, pp. 208-210

[12] K. Kato, "Second-harmonic generation to 2048 A in ß-BaB$_2$O$_4$", IEEE J. Quantum Electronics QE-22(7), 1013(1986)

[13] A. Bielski, "A critical survey of atomic transition probabilities for Cu I", J. Quant. Spectrosc. Radiat. Transfer 15, 463 (1975)

[14] E.A. Den Hartog, D.W. Duquette, and J.E. Lawler, "Absolute transition probabilities in Ta I and W I", J. Opt. Soc. Am. B 4(1), 48(1987)

[15] M. Kwiatkowski, F. Naumann, K. Werner, and P. Zimmermann, "Measurements of radiative lifetimes in Ta II and W II", Physics Letters 103A (1,2), 49(1984)

[16] "Gmelins Handbuch der Anorganischen Chemie: Kupfer", Verlag Chemie, Weinheim, West-Germany, 1955, part A, vol. 2, p. 768

[17] W. Lotz, "Electron-impact ionization cross-sections and ionization rate coefficients for atoms and ions from scandium to zinc", Z. Physik 220, 466(1969)

[18] J.A. Rich and G.A. Farrall, "Vacuum arc recovery phenomena", Proc. IEEE 52, 1293(1964)

[19] G. Lins, "Evolution of copper vapor from the cathode of a diffuse vacuum arc", IEEE Trans. Plasma Sci. PS-15(5), 552(1987)

[20] G. Lins, "Measurement of the neutral copper vapour concentration during a vacuum arc by laser-induced fluorescence", Proc. XIXth Int. Conf. Phenomena in Ionized Gases, Belgrade, 1989, vol. 2, pp. 378-379

[21] J.E. Jenkins, J.C. Sherman, R. Webster, and S. Holmes, "Measurement of the neutral vapour density decay following the extinction of a high-current vacuum arc between copper electrodes", J. Phys. D: Appl. Phys. 8, L139 (1975)

[22] E. Hayeß, B. Jüttner, G. Lieder, H. Pursch, and L. Weixelbaum, "Measurements on the behaviour of neutral atom density in a diffuse vacuum arc by laser-induced fluorescence (LIF)", Proc. XIIIth Int. Symp. Discharges and Electrical Insulation in Vacuum, Paris, 1988, vol. 1, pp. 138-141

[23] E. Dullni, E. Schade, B. Gellert, "Dielectric recovery of vacuum arcs after strong anode spot activity", IEEE Trans. Plasma Sci. PS-15(5), 538(1987)

[24] G. Lins, I. Paulus, and F. Pohl, "Neutral copper vapour density and dielectric recovery after forced extinction of vacuum arcs", Proc. XIIIth Int. Symp. Discharges and Electrical Insulation in Vacuum, Paris, 1988, vol. 1, pp. 211-213

[25] C.E. Moore, "An Ultraviolet Multiplet Table", Circular of the National Bureau of Standards 488, Section 3, 4, and 5. Washington 1962

STREAMERS IN ATMOSPHERIC PRESSURE N_2:

EMPIRICAL RESULTS

P.F. Williams and F.E. Peterkin

Department of Electrical Engineering
University of Nebraska-Lincoln
Lincoln, NE 68588-0511

ABSTRACT

We have obtained streak and high speed shutter photographs
of streamers in a trigatron spark gap filled with N_2 at roughly
atmospheric pressure. The streamers are initiated in the high
field region surrounding the trigger pin, and propagate across
the main gap. We have observed streamers for main gap charging
voltages ranging from roughly 50% to 99% of V_{SB} (the static
self-breakdown voltage of the gap), and for both charging
polarities, producing cathode- and anode-directed streamers.
Streak photography allows us to measure the propagation velocity of the streamers. High speed shutter photography shows
the shape of the streamers. We observe a jump in the main gap
current when the streamer approaches the distant main gap electrode. From the magnitude of this jump and the streamer body
diameter determined from the shutter photographs, we can measure the free electron density in the streamer.

I. INTRODUCTION

The streamer mechanism for the electrical breakdown of
gases was first proposed more than 50 years ago by Raether[1]
and by Loeb and Meek.[2] Since that time the existence of
streamers has been firmly established, but many of the physical
properties of streamers have remained elusive. We report
empirical measurements of streamer properties in roughly atmospheric pressure N_2 which provide new information. These measurements are based on fast, high-sensitivity optical shutter
and streak photography, coupled with current measurements.
Such data have been obtained for pressures ranging from 200 to
900 Torr, and for charging voltages ranging from 99% of the
static self-break voltage for the gap, V_{SB}, to about 50% of
V_{SB}.

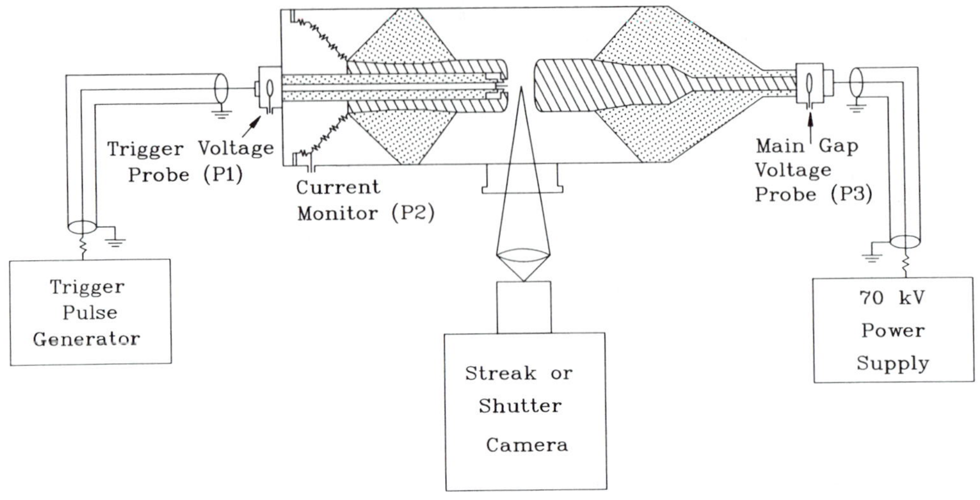

Fig. 1. Schematic diagram of the experimental setup used for the work discussed in this paper.

II.   EXPERIMENTAL APPARATUS

The experiments we discuss were carried out in a trigatron spark gap specially designed for good optical access, clean electrical diagnostics, and easy control of the fill gas pressure and composition. A trigatron spark gap is essentially an ordinary spark gap, except that there is a hole with a trigger pin electrode centered in and insulated from it in one electrode. Figure 1 shows a schematic drawing of the experimental setup. The details of the apparatus are described elsewhere.[3] Briefly, the cell was designed to operate with D.C. charging voltages up to 70 kV. The electrodes were shaped to provide a nearly constant field in the center of the gap, and with a 2.5 cm electrode separation the static self-breakdown voltage for 700 Torr N_2 was 63 kV. Care was taken to ensure that the cell appeared electrically as a constant-impedance, 50 Ω coaxial transmission line. The gap was charged through a 24 ns length of RG-218/U coaxial cable. The short cable was used to avoid swamping the sensitive cameras with the intense emission from the arc.

The cell was equipped with three electrical diagnostics, labelled P1, P2, and P3 in the figure. P1 and P3 were capacitively-coupled voltage probes monitoring the trigger voltage and the voltage on the charged main gap electrode, respectively. P2 was a low-inductance current-viewing resistor placed in series with the load resistor. Since voltage is applied to the gap through a 50 Ω coaxial cable, the voltage changes measured at P3 are also an accurate measure of the gap current at least for times less than one roundtrip transit time of the cable. We found this measure of current to be superior to that from P2 because P2 measures the total load current

including that induced by the trigger pulse, whereas P3 measures only the real and displacement current in the main gap.
All current measurements presented here are from P3. The
overall risetime of P3 (including the oscilloscope response)
was about 1 ns.

Optical diagnostics consisted of high speed, high sensitivity streak and shutter cameras. Both camera systems have
sufficient gain that single photo-electrons emitted from the
photocathode can be detected. The cameras viewed the gap
through a 4 inch, fused quartz window. The input optics for
the streak and shutter cameras were f/4 and f/6, respectively.
The optical and electrical diagnostics could be synchronized to
within about ±1 ns.

III. EXPERIMENTAL RESULTS

Figure 2 shows shutter photographs of the early stages of
electrical breakdown in the gap for charging voltage about 95%
of V_{SB}, and for the voltage polarity producing cathode-directed
streamers. The trigger pulse voltage was 10 kV, and of polarity opposite to that of the distant main gap electrode. As
discussed in the next section, we found that the properties of
the streamers were more nearly a function of the algebraic
difference between the charging and trigger voltages than of
just the charging voltage alone. In this sense, the gap was
about 11% overvolted. The photos in Figs. 2a) and 2b) show
streamers launched from the trigger pin in transit. We could
obtain only one shutter photo from each shot, so these photos
are each from different shots. There was 5-10 ns jitter in the
initiation time of the first streamer relative to the timing of
the trigger voltage pulse (due we believe to the 10-20 ns rise
time of the voltage pulse at the trigger pin tip). The times
listed in the figure are the times of shutter closing, measured
from the time when the leading streamer appeared in the gap as
estimated from streak photographs taken under the same conditions. The shutter time of the camera we used was about 5 ns,
so there is considerable motional blurring of the streamer
tips. The remaining photos show the early stages of the heating process leading to spark formation. The camera sensitivity
in 2c) was lower than in 2a) and 2b), and lower in 2d) than
2c).

Figure 3 shows a similar shutter photo of an anode-directed streamer, taken under the same conditions as Fig. 2,
except that the charging and trigger voltage polarities were
reversed. The shutter closed about 4 ns after the streamer
first appeared. This streamer appears more feather-like than
the cathode-directed streamer.

Figures 4 and 5 show streamers of both polarities taken
with lower charging voltage. Figure 4 shows a sequence of
shutter photographs taken with charging voltage of 30 kV,
corresponding to about 50% of V_{SB}, and with polarity producing
cathode-directed streamers. Including the trigger voltage as
discussed above, the gap was about 37% undervolted. The photos
in Figs. 4a) and 4b) show streamers in midgap, and the remaining photos in 4c)-4f) show the complex heating process leading

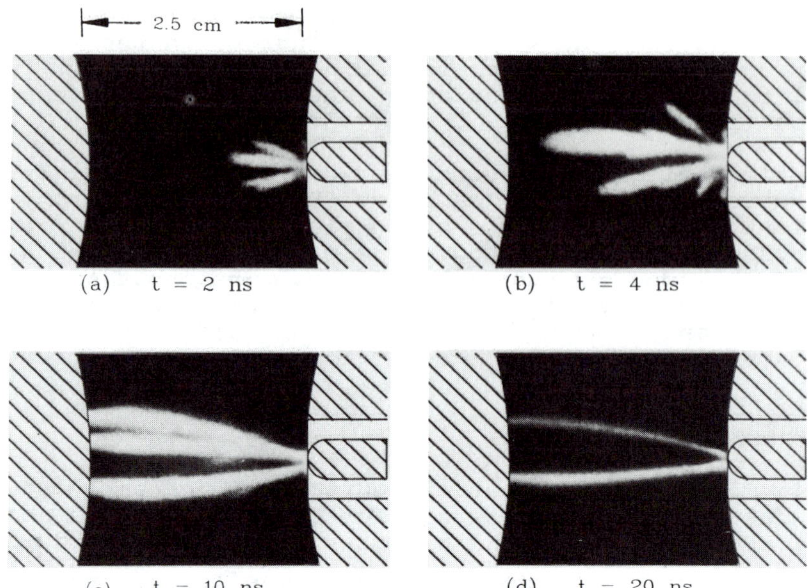

Fig. 2. Sequence of shutter photos showing the time development
of cathode-directed streamers in the main gap. The
streamers are pictured at various times after initiation. Due to the increasing intensity of the channels,
the image intensifier gain was lower for (c) and (d).
Conditions were: positive trigger, negative main gap
(+-) polarity, V_t = 10 kV, V_g = -60 kV, N_2 at 700 Torr,
2.5 cm gap separation, 4.76 mm diameter trigger pin
flush with the main electrode.

to spark channel formation. The streamers travel slower at
this voltage, and the shutter camera is able to resolve the
luminous tips of the streamers in Figs. 4a) and 4b). Figure 5
shows a photo of an anode-directed streamer, taken under similar conditions to Fig. 4, except that the voltage polarities
were reversed. The charging voltage in this case was 40 kV, or
about 66% of V_{SB}. Including the trigger voltage, the gap was
about 11% undervolted.

Figure 6 shows a streak photograph and synchronized
current trace of a cathode-directed streamer under the same
conditions as for Fig. 2. The time synchronization of the
streak and current trace is accurate to about ±1 ns. The
streak camera viewed a rectangular region about 1.2 mm wide,
centered on the gap axis. For the streak speed used, this

width corresponds to a time resolution of about 1 ns. The
streamer propagation speed increased as the streamer traversed
the gap, ranging from about 2×10^8 to more than 10^9 cm/sec.

IV. DISCUSSION

The variation of streamer properties with charging voltage
is of interest. Since the streamer tip is electrically connected to the tip of the trigger pin tip by the resistive
streamer channel, these properties might be better described as
functions of the algebraic difference between the potentials of
the trigger pin and the distant main gap electrode, rather than

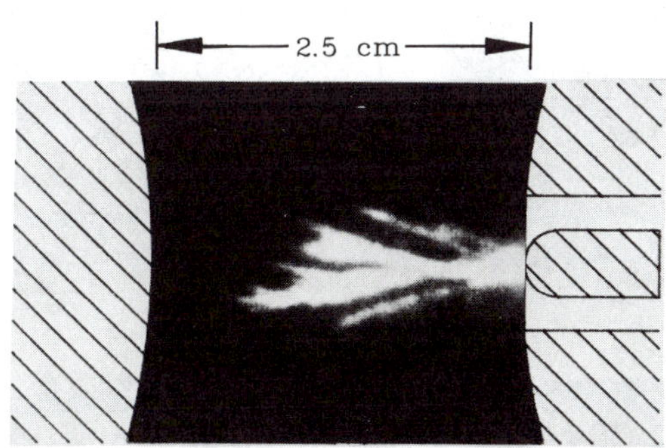

Fig. 3. Shutter photograph of anode-directed streamer in the
negative trigger, positive main gap (-+) polarity,
$V_t = -10$ kV, $V_g = 60$ kV. Other conditions were the
same as in Fig. 2.

simply the charging voltage. This suggestion has been made by
Martin.[4] To test this idea, we measured the streamer transit
time for a wide range of charging and trigger voltages and
plotted these data vs both charging voltage, and the difference
between charging and trigger voltage ($V_{charge} - V_{trigger}$). The
results are shown in Fig. 7. Although there is some experimental scatter, the data for different trigger voltages clearly
coalesce better when plotted against $|V_{charge} - V_{trigger}|$ than
just $|V_{charge}|$. The average streamer speed varied from about
1×10^8 to 5×10^8 cm/sec, depending on voltage magnitude and
polarity.

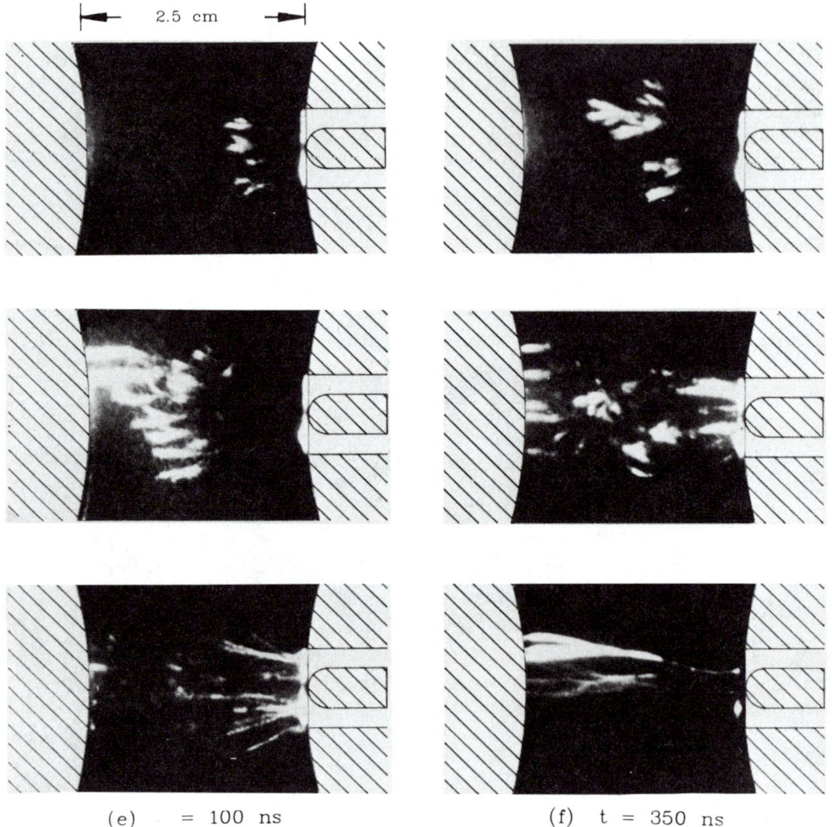

Fig. 4. Sequence of shutter photographs showing the time development of the streamer and arc discharge channels for 50% V_{SB} main gap charging with positive trigger, negative main gap (+-) polarity. The times indicate the approximate delay since streamers were first initiated. Voltages were $V_t = 10$ kV, $V_g = -30$ kV. Conditions were otherwise the same as in Fig. 2.

As the streak photograph in Fig. 6 shows, the instantane-
ous streamer velocity is by no means a constant as the streamer
traverses the gap. The acceleration as the streamer approaches
the distant main gap electrode is probably a result of the
increasing field just ahead of the streamer caused by the combined effects of the decreased field inside the streamer channel, and the requirement that the integrated field be a constant determined by the charging and trigger voltages.

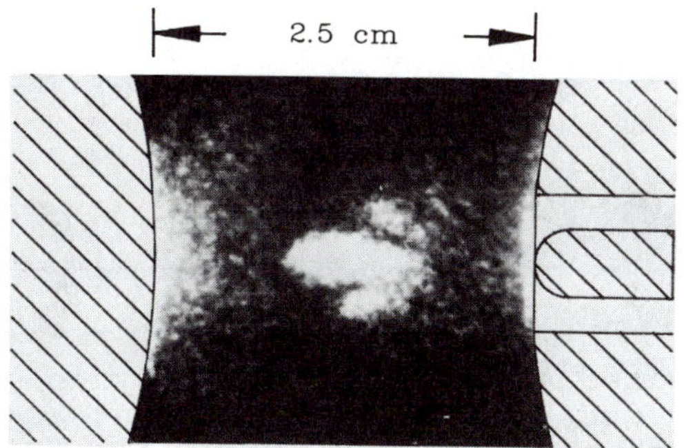

Fig. 5. Shutter photograph of the development of an anode-
directed streamer in the negative trigger, positive
main gap (-+) polarity for 66% V_{SB} main gap charging
voltage. Voltages were V_t = -10 kV, V_g = 40 kV, and
conditions were otherwise the same as in Fig. 2.

That the field is reduced inside the streamer channel is
verified by the streak photo. Consider the emission recorded
at a fixed time when the streamer is near midgap. The emission
is most intense in a small region, presumably corresponding to
the tip of the streamer. The emission is the result of
electron-impact excitation, and is a function of the local
electric field and the free electron density. Inside the
streamer there should be at least as high an electron density
as at the tip, but the emission is weaker, implying a lower
field there.

In most cases the gap current jumped simultaneously
(±1 ns) with the streamer arriving at the opposite main gap
electrode. When the streamer contacts this electrode, the
requirement of constant potential drop is inconsistent with

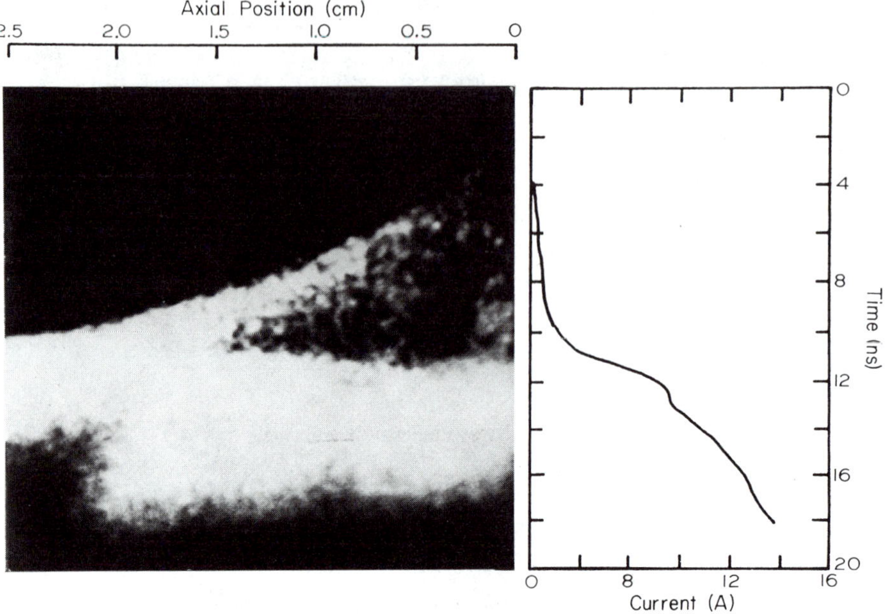

Fig. 6. Synchronized streak picture and main gap current trace,
obtained under the same conditions as Fig. 2. Synchronization is accurate to within ± 1 ns.

significant shielding of the main streamer body, and the field
inside the streamer must rise.[5] This effect is seen in the
electrical diagnostic as this current jump, and in the optical
diagnostic as a sudden increase in luminosity. Using the gap
current just after the jump and neglecting any voltage drop
across the electrode-plasma interfaces, we estimate the resistance of the streamer channel at this time to be somewhat
larger than 6 kΩ, and the average free-electron density in the
streamer channel to lie in the range 10^{14}-10^{15} cm^{-3}, in good
agreement with theoretical expectation.[6][7]

It is interesting that we found in all cases we looked at
that cathode-directed streamers travelled at least as fast as
anode-directed streamers (usually faster). This observation is
somewhat surprising. Wagner[8], for example, has published
streak photographs of streamer propagation in overvolted, uniform field gaps. In these experiments electron avalanches were
started by photoelectrons ejected at the cathode by U.V.
illumination. Streamers are formed at some point in midgap
where the free electron density in an avalanche becomes large
enough, and propagate in both directions. In N_2, as well as
other gases he reported on, the anode-directed streamers propagated faster than the cathode-directed. Other workers report
similar behavior.[9]

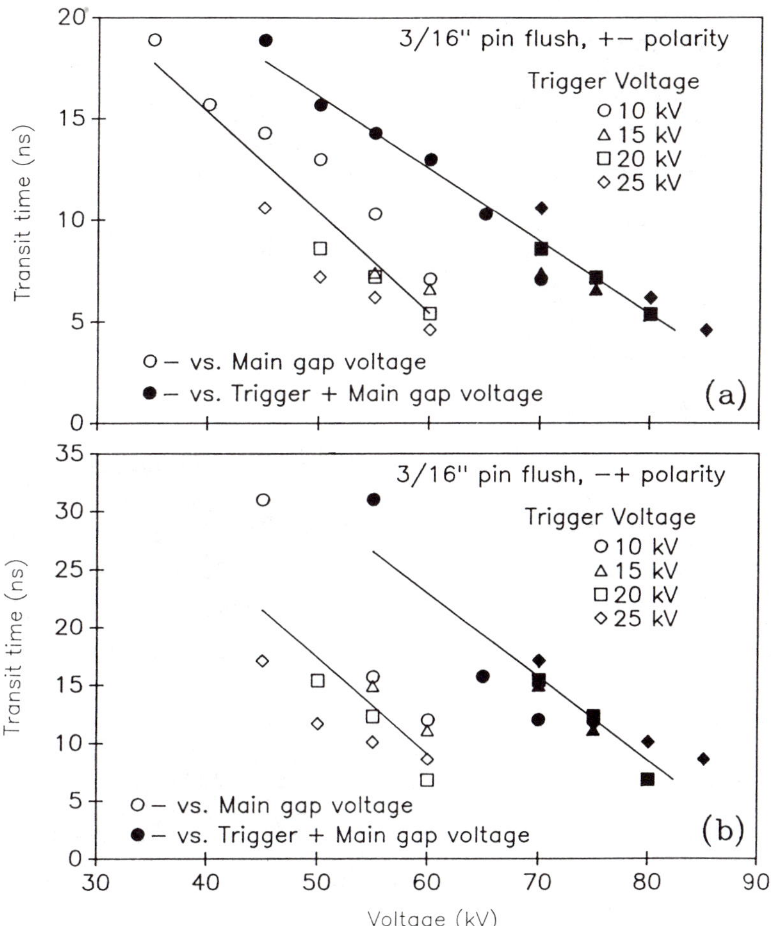

Fig. 7. Plot of streamer transit time vs. main gap voltage and vs. main gap plus trigger voltage, for a) positive trigger, negative main gap (+−), and b) negative trigger, positive main gap (−+) polarity. Conditions were otherwise the same as in Fig. 2.

A principal difference between our experiments and
Wagner's[8] is that his streamers were initiated in midgap and
were not anchored to any constant potential surface, whereas
ours were initiated very close to the trigger pin tip. The
potential of the streamers in Wagner's experiments floated
until one streamer propagated far enough to touch an electrode.
In our experiment, one end of the streamer was initiated nearly
in contact with the trigger pin tip. For the voltage polarity
producing cathode-directed streamers the electrical contact
between the streamer and the tip should be good because the
applied field drives the free electrons into the tip. The
potential of the streamer base is then tied to that of the pin
tip. For the opposite polarity, however, good electrical contact requires that electrons be emitted from the tip. In a
conventional glow discharge, such electron emission is accomplished through the creation of the cathode fall. In our case,
there may be insufficient time to set up the cathode fall
region, and it may be that during the propagation of the streamer the electrical contact to the pin tip is poor for this reason, leaving the potential of the base of the anode-directed
streamer floating, somewhat like the streamers initiated in
mid-gap. This effect would slow the propagation of our anode-
directed streamers, and might explain our unconventional observation of slower velocities for anode-directed streamers than
for cathode-directed.

The variation with charging voltage and polarity of the
streamer appearance in the shutter photographs is interesting.
At the highest charging voltage we investigated the cathode-
directed streamers propagate roughly along field lines, and
appear smooth with few features. Anode-directed streamers at
the same voltage, however, have a more feather-like appearance.
This difference may be due to the poorer electrical contact
with the trigger pin tip for and anode-directed streamer, as
discussed above.

For the lower charging voltage the cathode-directed streamer becomes considerable more bifurcated. This behavior probably reflects the increasingly important role played by the
space charge field of the streamer relative to that of the
applied field. The space charge field is subject to statistical variations in local charge density. As the role of the
applied field becomes less important, the streamer path becomes
more and more subject to these statistical variations. The
anode-directed streamer, on the other hand, becomes fatter and
more featureless as the charging voltage is decreased. This
behavior probably reflects the decay of the anode-directed
streamer into an anode-directed, drifting electron swarm. For
the lowest charging voltage at which we observed such streamers, the speed of the streamers was still several times an
electron drift velocity in the applied field, implying that
space-charge fields were still important.

REFERENCES

[1] H. Raether, Z. Phys. _112_, 164 (1939).

[2] L.B. Loeb and J.M. Meek, J. Appl. Phys. _11_, 438 (1940).

[3] F.E. Peterkin and P.F. Williams, Appl. Phys. Lett. _53_, 182 (1988); "Triggering in Trigatron Spark Gaps: A Fundamental Study," P.F. Williams and F.E. Peterkin, to appear in J. Appl. Physics, Nov. 1989.

[4] T.H. Martin in _Digest of Tech. Papers, 5th IEEE Pulsed Power Conf._, ed. by P.J. Turchi and M.F. Rose (IEEE, New York, 1985), p. 74.

[5] R.S. Sigmond, J. Appl. Phys. _56_, 1355 (1984).

[6] S.K. Dhali and P.F. Williams, Phys. Rev. A _31_, 1219 (1985); S.K. Dhali and P.F. Williams, J. Appl. Phys. _62_, 4696 (1987).

[7] C. Wu and E.E. Kunhardt, Phys. Rev. A _37_, 687 (1988).

[8] K.-H. Wagner, Z. Phys. _204_, 177 (1967).

[9] P. Stritzke, I. Sander, and H. Raether, J. Phys. D: Appl. Phys. _10_, 2285 (1977).

THE SOLUTION OF THE CONTINUITY EQUATIONS

IN IONIZATION AND PLASMA GROWTH

A. J. Davies and W. Niessen

Department of Physics, University College of Swansea
Singleton Park, Swansea, SA2 8PP, Great Britain

INTRODUCTION

In the study of the operation of hollow–cathode devices it is of great importance to gain an understanding of the basic physical mechanisms by which the plasma can grow in the hollow-cathode and the way in which it can influence the ionization and discharge behaviour in the main anode/cathode region. In such devices there is normally a very large variation in the electric field, E, in different parts of the device and this means that, at the moment, it is not possible to use any one method to simulate the plasma initiation and growth.

The various simulation methods may be divided into two main types depending upon the value of the parameter E/N where N is the gas number density. At low E/N the various species of charged particles come rapidly into a dynamic quasi–equilibrium with the neutral gas molecules and the plasma may be considered to be a Eulerian fluid with ionization and transport of the various particle species described by appropriate ionization and transport coefficients. At high E/N, and in regions where the electric field varies rapidly in space or time, equilibrium conditions no longer hold and Lagrangian methods are usually applied. Here the plasma is modelled as an assembly of interacting pseudo-particles and the distribution function traced by following the motion of individual particles. Statistical effects may be taken into account by using Monte Carlo or Boltzmann techniques.

In the present paper we will be concerned with situations where the fluid model can be used and will be discussing the various numerical techniques that can be employed to solve the continuity equations that govern the ionization development.

If we denote by ρ_i the particle density of the various charged particle species in the discharge and $\mathbf{u}_i$ their corresponding drift velocities, then the continuity equation describing the ionization development may be written

$$\frac{\partial \rho_i}{\partial t} = S_i - \nabla(\rho \mathbf{u}_i) \tag{1}$$

where the S_i are source terms due, for example, to ionization by electron or ion impact with neutral gas molecules. In situations where the charged particles are in local equilibrium, S_i and $\mathbf{u}_i$ will be functions of the local values of E/N.

At low particle densities the field distribution will be the Laplacian field determined by the particular electrode configurations employed. Sooner or later in the ionization development, however, space-charge effects become important and the field will be given by the solution of Poisson's equation

$$\text{div } \mathbf{E} = \rho/\varepsilon_0. \tag{2}$$

Since the drift velocities and source terms in (1) are functions of $\mathbf{E}$, the problem is highly non–linear and calls for particularly accurate numerical methods in the solution of the system of hyperbolic equations (1) in order to avoid spurious, non–physical, instabilities occurring.

The above system of equations must be solved subject to specified boundary and initial conditions. Since, in general, the source terms and drift velocities will be varying in time due to space-charge distortion of the applied electric field, Poisson's equation (2) must be solved repeatedly at each stage of the computations so that an accurate and rapid Poisson-solver is an absolute necessity.

In nearly all the work so far, the complexity of the problem has restricted the electrode geometries to fairly simple configurations so that the ionization development may either by modelled in one or two dimensions - the former case corresponding to plane parallel electrode geometries and the latter to systems having an axis of symmetry.

In describing the various numerical techniques that may be employed, for simplicity we will be considering the one–dimensional case but the methods chosen must be capable of being extended in a fairly straightforward manner to treat axially symmetric systems.

NUMERICAL SOLUTION OF THE CONTINUITY EQUATIONS

In solving equation (1), great care must be taken in the numerical approximation of the convective term which, in one dimension may be written $\partial(\rho u)/\partial x$, particularly in the vicinity of a rapid change in ρ due, for example, to shock or space-charge wave formation. Spurious ripples and numerical diffusion at such a front can lead to large numerical errors. It should be noted that even when the initial conditions are quite smooth, space-charge effects sooner or later introduce regions of high electric field with very rapidly changing particle densities so that the accurate treatment of such transitions is an essential pre-requisite of the method chosen.

In searching for the optimum numerical method, the following factors should be taken into consideration.

1. The conservation of particle densities as expressed by the continuity equation (1) must be mirrored in the corresponding numerical algorithm.

2. The spatial finite difference mesh must be flexible in order to allow greater accuracy to be obtained near electrode surfaces and in regions where electric fields and particle densities vary rapidly in space.

3. The method should be stable for all harmonic components of the solution.

4. The particle densities should be monotonic so that no spurious ripples or negative densities are generated.

5. The accuracy should be at least of second order accuracy in regions where the particle densities vary smoothly.

6. Discontinuities should be resolved without introducing any spurious oscillations.

THE HYBRID METHOD OF CHARACTERISTICS

In discussing the various algorithms we consider the one dimensional form of (1), that is

$$\frac{\partial \rho_i}{\partial t} = S_i - \frac{\partial}{\partial x}(\rho_i u_i) \, . \tag{3}$$

As previously mentioned some of the earliest work on the solution of the above type of equation in the field of gas discharges (see for example Davies *et al*, 1971, 1975; Kline, 1974; Bayle and Bayle, 1974; Yoshida and Tagashira, 1976) employed the Hybrid Method of Characteristics. Consider, for example, the continuity equation for electrons including a source term representing ionization by electron impact with neutral gas atoms which may be written

$$\frac{\partial \rho_e}{\partial t} = \alpha \rho_e u_e - \frac{\partial}{\partial x}(\rho_e u_e) \, , \tag{4}$$

where α is the number of ionizations per unit length undergone by an electron travelling in the direction of the electric field. This equation may be expressed in the alternative form

$$\frac{\partial \rho_e}{\partial t} + u_e \frac{\partial \rho_e}{\partial x} = \alpha \rho_e u_e - \rho_e \frac{\partial u_e}{\partial x} \tag{5}$$

and if we introduce the operator $D/Dt = \partial/\partial t + u_e \partial/\partial x$ (which denotes differentiation in a frame of reference moving with the electrons), then (5) becomes

$$\frac{D \rho_e}{Dt} = \alpha \rho_e u_e - \rho_e \frac{\partial u_e}{\partial x} \, . \tag{6}$$

If all the relevant fields, densities and coefficients are known at time T (Fig. 1), then the new densities at point P' at time $T+\Delta t$ may be found by integrating (6) along the Characteristic Curve C_e following the motion of the electrons to give

$$\rho_e(P') = \rho_e(Q)\exp\left[\int_{C_e}(\alpha u_e - \frac{\partial u_e}{\partial x})dt\right] \tag{7}$$

so that the accuracy of the method is essentially determined by the method of interpolation used to obtain the value of $\rho_e(Q)$ at the foot of the Characteristic and the way in which the integral of $(\alpha u_e - \partial u_e/\partial x)$ along C_e is determined.

The method may be made explicit by approximating the integral by

$$[\alpha u_e - \frac{\partial u_e}{\partial x}]_Q \cdot \Delta t$$

but the accuracy may be greatly improved by taking the average values of $(\alpha u_e - \partial u_e/\partial x)$ at the points Q and P' at the foot and top of the characteristic. Since this involves a knowledge of the relevant quantities at the future time $T+\Delta t$, the method becomes implicit and an iterative procedure must be used (Davies, 1971) to obtain successively better approximations to the integral. The overall accuracy obtained is critically dependent upon the interpolation procedure used as may be seen from the results obtained for the standard test problem described below using (a) first order and (b) second order interpolation.

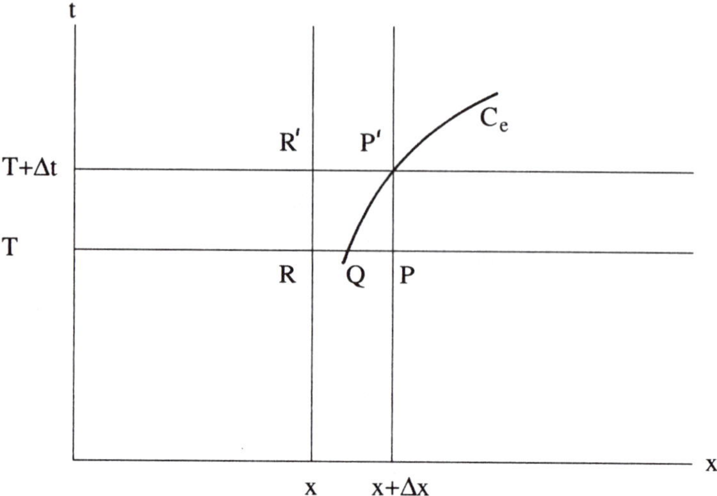

Figure 1. Characteristic curve C_e following the path of the electrons passing the point P' at the future time $T+\Delta t$. All quantities are known at time T.

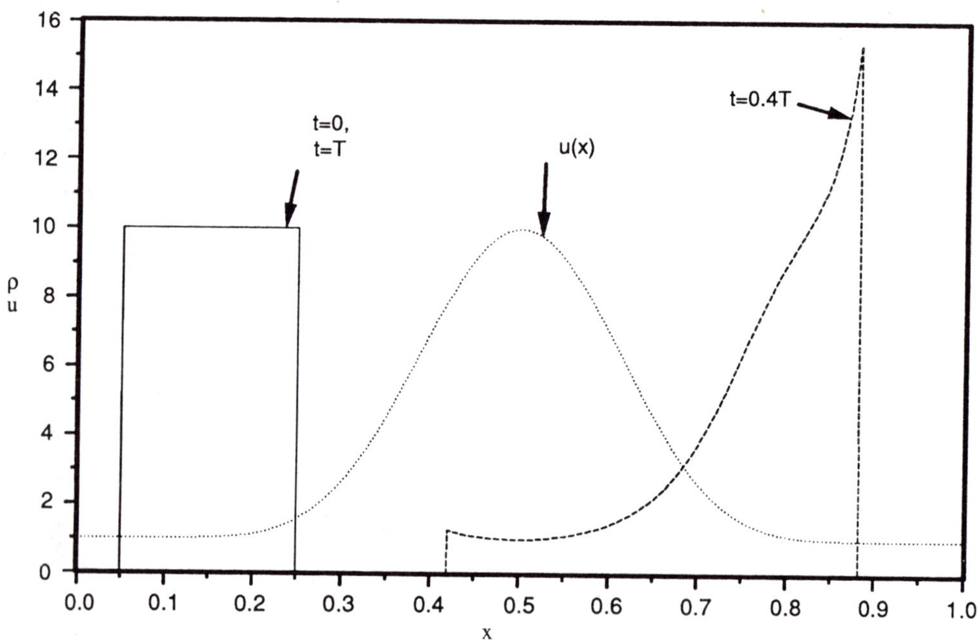

Figure 2. Test problem used to compare the various numerical algorithms.
ρ ——— initial density and distributions after periods T, 2T, 3T etc
ρ – – – accurate solution at time t = 0.4T.
u(x) · · · · · velocity field.

In the test problem the source term was omitted from (4) and the equation

$$\frac{\partial \rho}{\partial t} = - \frac{\partial (\rho u)}{\partial x} \qquad (8)$$

was solved on the x–interval [0,1] with a mesh consisting of 100 equal cells. The velocity field u was taken to be constant in time but vary with x according to the relation

$$u(x) = 1 + 9(\sin(\pi x))^8 \qquad (9)$$

which gives a peak value of u at x = 0.5 which is ten times greater than the values at the beginning and end of the interval (Fig. 2).

The initial distribution of ρ was

$$\rho(x,t = 0) = \begin{array}{ll} 10 & 0.05 \leq x \leq 0.25 \\ 0 & \text{elsewhere.} \end{array} \qquad (10)$$

The boundaries were periodic in the sense that any particle leaving the right hand boundary enters at the left so that, after a certain period T given by

$$T = \int_0^1 \frac{dx}{u(x)} \, ,$$

the true solution should again be identical with the initial distribution $\rho(x,t = 0)$. This is very useful for comparing the various numerical algorithms since the correct solution during the whole time interval does not have to be found. An approximation to the accurate solution at time t = T/4 was also obtained by following the characteristics of a very large number (500) of points.

We see from Fig. 3 that using a first order interpolation scheme introduces considerable numerical diffusion with the original rectangular distribution being completely lost after only one cycle. Furthermore the total number of particles (represented by the areas under the curves) is growing so that particle (and thus charge) conservation is not achieved.

As shown in Fig. 4, going to second order interpolation (with an undershoot-overshoot limiter) greatly improves the results with the degree of numerical diffusion being considerably reduced. Particle number is still not conserved, however, and the peak in the distribution has grown considerably. If the test cycle is repeated then the shape of the distribution does not change appreciably but the peak value grows without limit. In a practical situation this could lead to errors in the electric field computations since these are directly related to the particle densities.

Despite these limitations, provided it is used with care, the Hybrid Method of Characteristics have proved extremely successful in discharge simulation (Davies, 1986) and Morrow (1981) has shown that it compares favourably with several other schemes. Much of its success is due to the fact that in the early stages of discharge growth the applied field is not distorted by space charge so that, for plane parallel electrodes and a uniform mesh, the time increment Δt can be chosen so that the Characteristic Curve following the electrons intersects the line t = T in Fig. 1 exactly at at mesh point (i.e. QP = Δx) thus removing the necessity for interpolation. The condition for this,

$$u_e \frac{\Delta t}{\Delta x} = 1 \qquad (11)$$

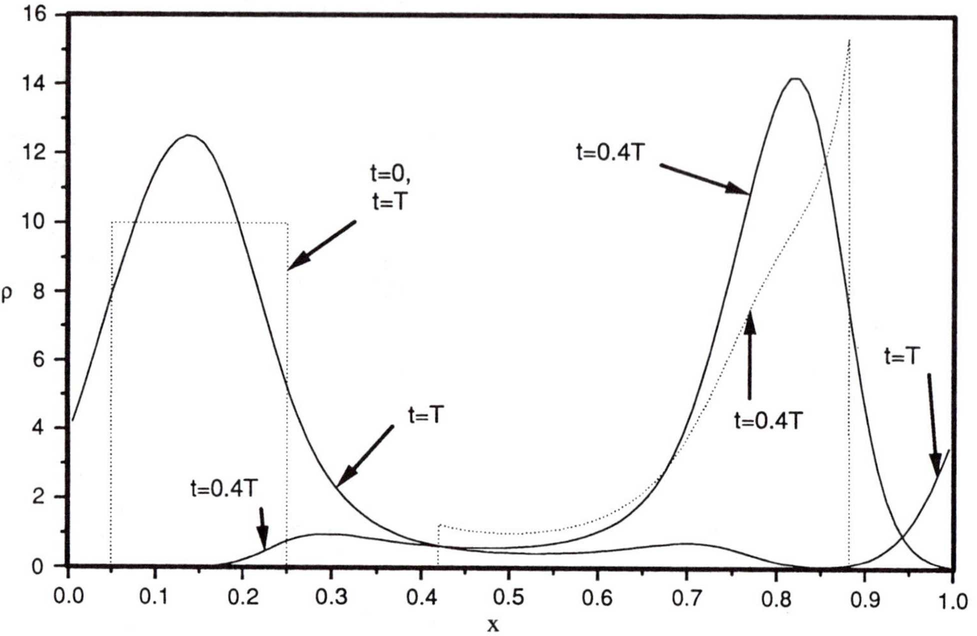

Figure 3.   Results obtained for the standard test problem using the Hybrid Method of Characteristics with first order interpolation in space.

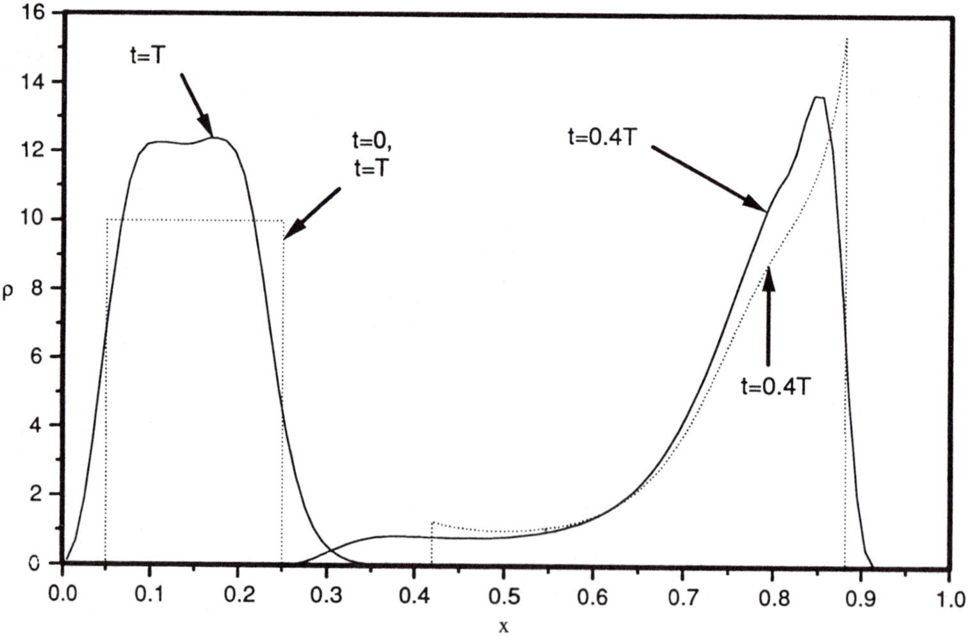

Figure 4.   Results obtained for the standard test problem using the Hybrid Method of Characteristics with second order interpolation in space.

is the limit of the Courant–Friedrichs stability criterion that applies to the other methods to be discussed later.

It should be pointed out, however, that as the field distortion becomes non–uniform due to space-charge distortion, (11) can no longer be satisfied at all mesh points and, in general, Δt must be chosen so that

$$u_e \frac{\Delta t}{\Delta x} \leq 1 \tag{12}$$

for all the mesh points, so that the Characteristic always intersects $t = T$ at less than one mesh point away.

For this reason, although the Hybrid Method of Characteristics is extremely versatile and is easily extended to two dimensions (Davies $et\ al$, 1977), care must be taken in its implementation. A companion paper (Mittag, 1989) in this proceedings describes a two dimensional treatment of the early stages of pseudo–spark development using this technique.

BASIC FINITE DIFFERENCE METHODS

In setting up a finite difference scheme great care must be taken if the conditions 1 to 6 mentioned above are to be satisfied.

Consider, for example, the continuity equation without source terms which, for the case of a constant velocity field may be written

$$\frac{\partial \rho}{\partial t} = - u \frac{\partial \rho}{\partial x} . \tag{13}$$

Suppose we choose a finite difference mesh

$$x_0, x_1, x_2, ---x_i --$$

of constant spacing Δx in the x direction and a constant time increment Δt so that the solutions are evaluated at times

$$t_0, t_1, t_2, ---t_n --.$$

A natural way to approximate the time differential in (13) is

$$\left. \frac{\partial \rho}{\partial t} \right|_{i,n} = \frac{\rho_i^{n+1} - \rho_i^n}{\Delta t} + O(\Delta t) \tag{14}$$

which is of first order accuracy in Δt and has the advantage that the ρ_i^{n+1} can be immediately evaluated in terms of known quantities at time n.

The spatial differential in (13) may be approximated by the second order formula

$$\left. \frac{\partial \rho}{\partial x} \right|_{i,n} = \frac{\rho_{i+1}^n - \rho_{i-1}^n}{2\Delta x} + O(\Delta x^2) \tag{15}$$

so that the resulting difference equation is

$$\frac{\rho_i^{n+1} - \rho_i^n}{\Delta t} = - u \left[\frac{\rho_{i+1}^n - \rho_{i-1}^n}{2\Delta x} \right] \tag{16}$$

and ρ_i^{n+1} is easily evaluated in terms of known quantities at time n.

The Forward Time Centred Centred Space (FTCS) scheme, however, can be shown to be highly unstable. If u is constant or slowly varying then the independent solutions satisfying the difference equation (16) are given by

$$\rho_i^n = \xi^n(k)\exp(jki\Delta x) \tag{17}$$

where k is the spatial wave number, $\xi(k)$ is a complex number depending on k, and $j = \sqrt{-1}$. For stability the amplitude $\xi(k)$ of the ρ_i^n should be less than or equal to one. In order to find $\xi(k)$ (17) is substituted back into (16) to give

$$\xi(k) = 1 - j\frac{u\Delta t}{\Delta x}\sin(k\Delta x) \tag{18}$$

which is always greater than 1 for all k so that the ρ_i^n grow exponentially with time and the solution is unconditionally unstable.

The above scheme may be stabilised by introducing the modification due to Lax who simply replaced ρ_i^n by the average $\tfrac{1}{2}(\rho_{i-1}^n + \rho_{i+1}^n)$ in (14) thus yielding the finite difference equation

$$\rho_i^{n+1} = \tfrac{1}{2}(\rho_{i-1}^n + \rho_{i+1}^n) - \frac{u\Delta t}{2\Delta x}(\rho_{i+1}^n + \rho_{i-1}^n) \tag{19}$$

and the corresponding $\xi(k)$'s are now

$$\xi(k) = \cos(k\Delta x) - j\frac{u\Delta t}{\Delta x}\sin(k\Delta x) \tag{20}$$

so that, for stability, the condition

$$\frac{u\Delta t}{\Delta x} \leq 1 \tag{21}$$

must be met which, again, is the Courant Friedrichs criterion that holds for many finite–difference schemes.

Although the Lax scheme is stable, the densities computed at a point i at the future time depend on values of the density at points i–1 and i+1 at the present time. Thus, for positive u, the variation of ρ_i^{n+1} is not only affected by a disturbance at (i–1,n) (as is to be expected from the direction of flow of the particles), but also by any disturbance that may occur at (i+1,n). Thus in physical situations where sharp fronts or disturbances may form it is preferable to consider using an "upwind" difference scheme which will take on different forms depending on the sign of u. For u positive we may set

$$\frac{\rho_i^{n+1} - \rho_i^n}{\Delta t} = - u_i^n \frac{\rho_i^n - \rho_{i-1}^n}{\Delta x} \tag{22}$$

while for u negative

$$\frac{\rho_i^{n+1} - \rho_i^n}{\Delta t} = - u_i^n \frac{\rho_{i+1}^n - \rho_i^n}{\Delta x} \tag{23}$$

where the stability is again determined by the Courant Friedrichs criterion (21). We see that for these choices, the Characteristic Curve passing through the point (i,n+1) always intersects between the two mesh points on the right hand sides of the above equations provided (21) is satisfied.

The above schemes, although better suited to following the propagation of sharp fronts, are only of first order and will introduce considerable numerical diffusion. The method cannot therefore be used directly but may be extended to a higher order accuracy by, for example, the MUSCL (Van Leer, 1979) scheme described later.

HIGH RESOLUTION INTEGRATION SCHEMES

Flux Corrected Transport (FCT) Schemes

One of the modern high resolution schemes that has been used extensively by Morrow (1981) and Kunhardt and Wu (1987) in the simulation of discharges is the FCT (Flux Corrected Transport) technique first developed by Boris and Book (1976).

Referring again to the continuity equation without source terms (8), then any finite difference formulation of (8) will be conservative if it is expressed in the form

$$\rho_i^{n+1} = \rho_i^n + \frac{\Delta t}{x_{i+1/2} - x_{i-1/2}} (F_{i-1/2} - F_{i+1/2}) \tag{24}$$

where the F's are the particle fluxes.

There is no time here to go into the many variations of the FCT technique but all the FCT algorithms conserve total particle number density, and are monotonic and stable. In essence they consist of first calculating a low order solution (involving numerical diffusion) ρ_0 of (8). Secondly a high order, anti–diffusive flux $A_{i\pm1/2}$ is calculated with the flux being limited by the factor $C_{i\pm1/2}$, whose absolute value is less than or equal to unity, in order not to produce new ripples.

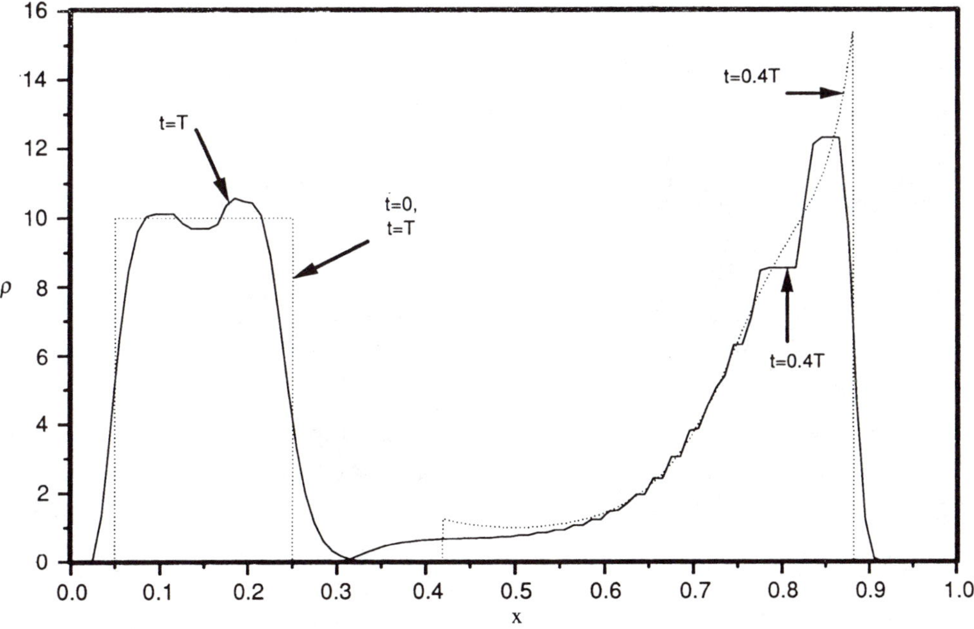

Figure 5. Results of the standard test problem using the Flux Corrected Transport Algorithm with the Zalesak (1979) peak preserver.

Lastly the corrected density ρ_c is calculated from

$$\rho_c = \rho_o + \frac{\Delta t}{x_{i+1/2} - x_{i-1/2}} (C_{i-1/2} A_{i-1/2} - C_{i+1/2} A_{i+1/2}) \tag{25}$$

Morrow (1981) has made detailed comparision of the FCT technique with other methods and also has investigated the effect of applying Zalesak's (1979) peak preserver and Fig. 5 shows the result of applying this method to the standard test problem.

We see here that, after one period T, a certain amount of diffusion has been introduced in the vicinity of the sharp edges of the square wave. The total particle number is conserved and the peak of the distribution has not grown. The curve at $t = 0.4T$, however, shows the well known staircase effect. This occurs because undamped ripples in the region where there is a steep slope may considerably distort the profile without actually creating any new maxima or minima. If, eventually, these ripples increase further then the flux limiter comes into play which has the effect of producing a "shelf" or "terrace" extending over a few mesh points. Book *et al* (1975) have shown that these terraces tend to occur where large scale ripples would form in less stable schemes.

If the test computations are extended further then the distributions at time 2T, 3T, etc are almost identical with those at $t = T$ with little sign of further diffusion or steepening of the profile.

<u>The Monotonic Upstream–Centred Scheme For Conservation Law (MUSCL)</u>

Recent work (Munz, 1988; Niessen, 1989) has shown that the so–called Monotonic Upstream-Centred Scheme for Conservation Law (MUSCL) has great potential in discharge simulation. This scheme, originally proposed by Van Leer (1979), has been used extensively for the study of fluid flow and is an extension of the "upwind" algorithm mentioned previously. It is of second order in space and in time and is able to cope with shock fronts without creating any artificial oscillations or introducing appreciable numerical diffusion.

We again consider the equation

$$\frac{\partial \rho_i}{\partial t} = - \frac{\partial}{\partial x} (\rho u)$$

and suppose that the region of interest has been divided into cells where the ith cell, centre x_i, is bounded by $x_{i-1/2}$ and $x_{i+1/2}$. The size, Δx_i of the ith cell is thus given by

$$\Delta x_i = x_{i+1/2} - x_{i-1/2} \tag{26}$$

and this may vary at different parts of the region considered. We define

$$\Delta x_{i+1/2} = x_{i+1} - x_i . \tag{27}$$

It is assumed that at time n the values of the density ρ_i^n at the centres of the cell and the values of the drift velocity $u_{i+1/2}^n$ at the cell boundaries are known. It is further assumed that the densities vary in a piecewise linear manner over any given cell so that

$$\rho^n(x) = \rho_i^n + (x - x_i)s_i^n \; ; \quad x_{i-1/2} \le x \le x_{i+1/2} \tag{28}$$

where the s_i^n are the slopes of the density distribution and it is the choice of these slopes that is a critical part of the algorithm influencing both the stability and degree of numerical diffusion. Munz (1988) has made a detailed comparison of the various slope formulas proposed by different workers and has discussed the way in which the different choices effect numerical diffusion and the preservation of peaks in the distribution.

Following Munz a whole range of possible slopes may be selected by the following formula

$$s(a_i, b_i) = \text{sign}(a_i) \max[\,|\text{minmod}(ka_i, b_i)|\,, |\text{minmod}(a_i, kb_i)|\,] \qquad (29)$$

where a_i and b_i are the right and left hand difference quotients

$$a_i = \frac{\rho_{i+1} - \rho_i}{\Delta x_{i+1/2}}, \quad b_i = \frac{\rho_i - \rho_{i-1}}{\Delta x_{i-1/2}}. \qquad (30)$$

The minmod function is defined by

$$\text{minmod}(a,b) = \begin{cases} a & \text{if } |a| \leq |b|, \ ab > 0 \\ b & \text{if } |a| > |b|, \ ab > 0 \\ 0 & \text{if } ab \leq 0 \end{cases} \qquad (31)$$

and effectively chooses the minimum of the right and left hand differences provided the differences are not of opposite sign in which case it is equated to zero.

In the slope formula (29) the right and left hand differences may be multiplied by a constant k. Taking k equal to zero, for example, gives a slope of zero so that the densities are taken to be constant over a cell.

For k = 1 the slope is equal to

$$s((a_i, b_i)) = |\text{minmod}(a_i, b_i)| \qquad ab > 0$$
$$= 0 \qquad ab \leq 0 \qquad (32)$$

so that it is simply taken to be the lower of the right and left hand differences. Introducing the constant k enables the slope to be increased but in all cases it is limited to the maximum of the right or left hand differences. In order to preserve second order accuracy in space and monotonicity, it can be shown that for a constant velocity field k must be restricted to the values

$$1 \leq k \leq \min\left[\frac{\Delta x_{i-1/2} + \Delta x_i}{\Delta x_i}, \ \frac{\Delta x_i + \Delta x_{i+1}}{\Delta x_i} \right]. \qquad (33)$$

For the case of a uniform mesh (33) reduces to

$$1 \leq k \leq 2 \qquad (33a)$$

and, provided, k does not exceed two the slope formula ensures that the values of the density at the boundaries of a cell lie between the density values at the centres of adjacent cells.

From (28) the values of ρ at the inner sides of the ith boundary are specified by

$$\rho_{i\pm}^n = \rho_i^n \pm \frac{\Delta x_i}{2} s_i^n \qquad (34)$$

so that the corresponding fluxes can be defined by

$$F_{i\pm}^n = \rho_{i\pm}^n u_{i\pm1/2}^n \qquad (35)$$

The values of ρ at these points after half a time step can be found from

$$\rho_{i\pm}^{n+1/2} = \rho_{i\pm}^{n} + \frac{\Delta t}{2\Delta x_i}(F_{i-}^{n} - F_{i+}^{n}) \tag{36}$$

Since, in general, the velocity field may be a function of the densities (through Poisson's equation for example), the values of $\rho_i^{n+1/2}$ must be found either from

$$\rho_i^{n+1/2} = \frac{1}{2}(\rho_{i-}^{n+1/2} + \rho_{i+}^{n+1/2}) \tag{37}$$

or by applying a first order upwind scheme

$$\rho_i^{n+1/2} = \rho_i^{n} + \frac{\Delta t}{2\Delta x_i}[g(F_{(i-1)+}^{n}, F_{i-}^{n}) - g(F_{i+}^{n}, F_{(i+1)-}^{n})] \tag{38}$$

where

$$g(a,b) = \begin{cases} a & \text{if } a \geq 0 \\ b & \text{if } b < 0 \end{cases}. \tag{39}$$

The values of $\rho_i^{n+1/2}$ can then be used to recalculate the velocity field at time $n+1/2$.

Finally the updated values of ρ after a complete time step are given by

$$\rho_i^{n+1} = \rho_i^{n} + \frac{\Delta t}{\Delta x_i}[g(F_{(i-1)+}^{n+1/2}, F_{i-}^{n+1/2}) - g(F_{i+}^{n+1/2}, F_{(i+1)-}^{n+1/2})] \tag{40}$$

where Δt has to satisfy the Courant Friedrichs criteria. Performing the time integration in two stages makes the method effectively second order in time.

The result of using the MUSCL scheme for different values of k for the test problem are shown in Figs. 6 to 8. Fig. 6 shows the results obtained for $k = 0$, corresponding to a first order scheme with the densities assumed constant over a mesh spacing. Here we see that there is a very large amount of numerical diffusion and after only a few periods the original rectangular distribution completely dissipates.

For $k = 1$ (Fig. 7) the method becomes second order with the peak being well preserved but still exhibits considerable numerical diffusion.

$k = 2$ (Fig. 8) corresponds to the so-called "superbee" scheme of Roe and Baines (1982) and we see that after one period the results are very similar to those produced by the FCT technique except that there is no sign of the staircase behaviour at $t = 0.4T$ observed with the FCT algorithm. After a large number of periods there is a tendency for the MUSCL scheme to induce pulse compression with the slopes at the edges of the distribution gradually becoming greater accompanied by an increase in the peak value and a corresponding decrease in the width.

In all the tests described above the Courant-Friedrichs number was chosen to be 0.01 using the largest value of u. Denoting the accurate density by ρ_{corr} we adopt as a measure of accuracy the absolute total error A define by

$$A = \sum_{i=1}^{n} |\rho_i - \rho_{corr}| \Delta x_i \tag{41}$$

Table 1 shows a comparison between the common methods together with the computing times found on a VAX 8700 computer although no great effort had been made to optimise the codes.

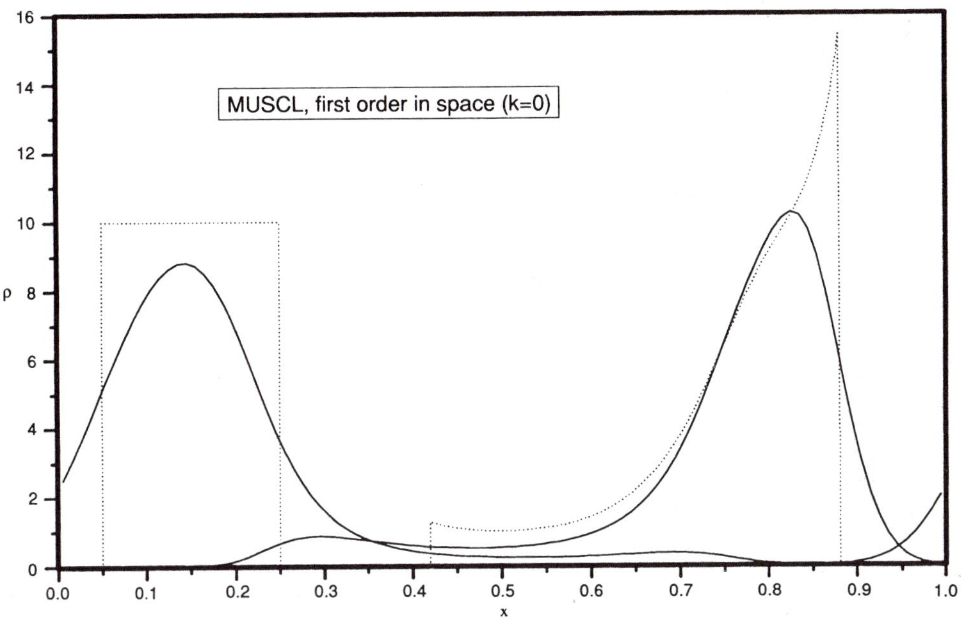

Figure 6. Results obtained from the standard test problem for the MUSCL scheme with the slope coefficient k = 0 corresponding to constant density over a cell. The method is of first order in this case.

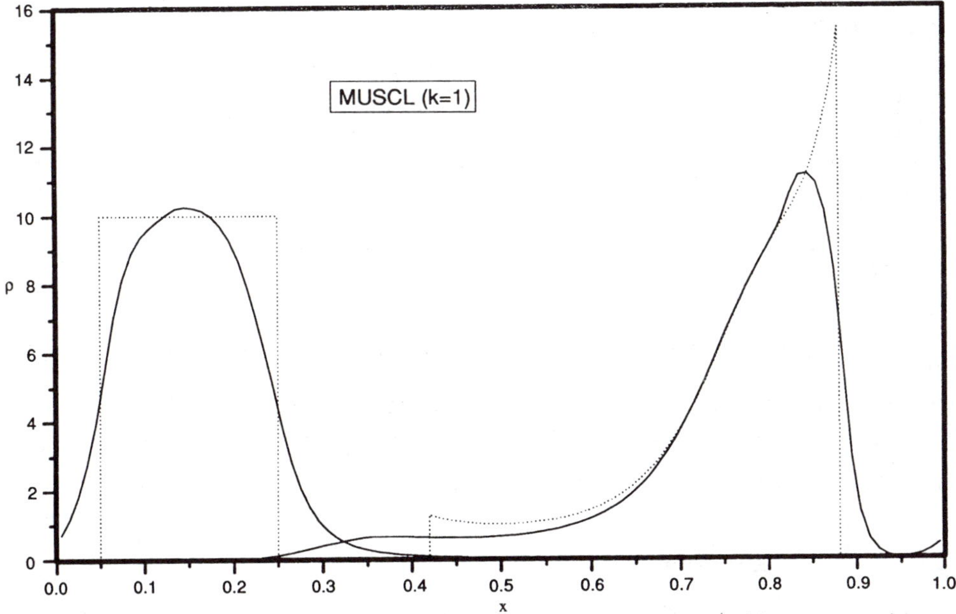

Figure 7. Results obtained from the standard test problem for a second order MUSCL scheme with slope coefficient k = 1.

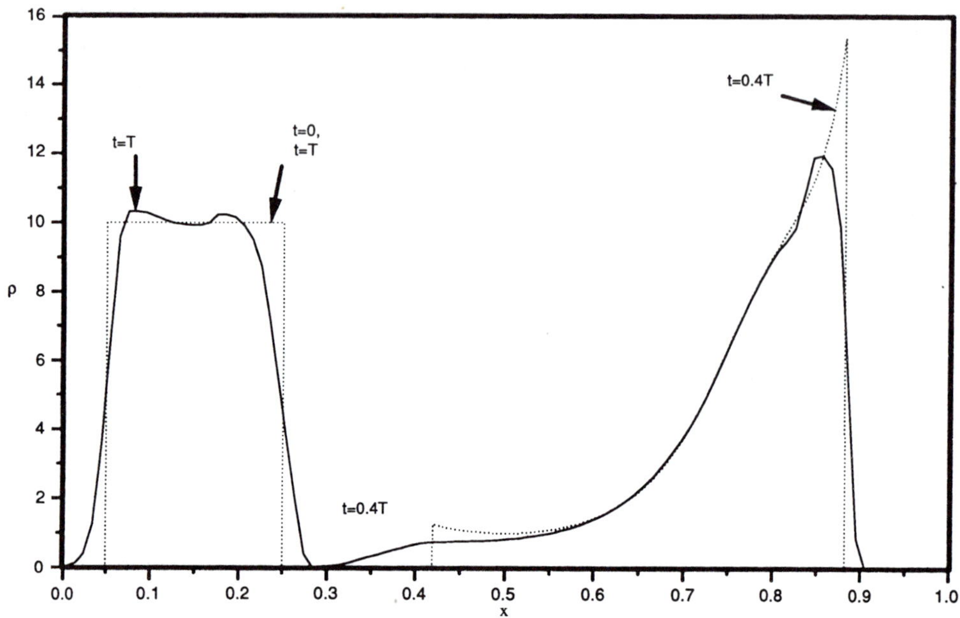

Figure 8. Results obtained from the standard test problem for a second order
MUSCL scheme with the slope coefficient k = 2 corresponding to the
"superbee" scheme of Roe and Baines (1982).

Table 1. Comparison of Various Integration Schemes for Standard Test Problem

MUSCL - Monotonic Upstream-Centred Scheme for Conservation Law
FCT - Flux Corrected Transport Scheme.
HMC - Hybrid Method of Characteristics.

Method	Total Error at $t = T$	CPU Time per step, s
MUSCL	0.26	6.6×10^{-4}
FCT	0.31	12.8×10^{-4}
HMC	0.64	11.6×10^{-4}

We see that the total errors in the FCT and MUSCL schemes are very similar
while the MSUCL scheme is about twice as fast as FCT with the Hybrid Method of
Characteristics lying inbetween. Note that these timings were obtained without taking
into account the source terms and including these would bring the timings of the
MUSCL and FCT schemes closer together because the evaluation and integration of the
source terms takes about the same time in either case.

Incorporation of source terms in high resolution schemes

If the source term in the continuity equation (1) is not too strong then it is possible
to use a second order Runge-Kutta scheme to evaluate it before or after accomplishing
the MUSCL or FCT step. In this scheme the densities, velocities and source terms are

all evaluated half a time step ahead of the present time prior to computing a complete solution a full time step ahead thus making the method of second order accuracy in time.

A more accurate method, however, is to include the source terms in the overall integration scheme. For example in the MUSCL scheme this is done by simply extending (36), (38) and (40) to

$$\rho_{i\pm}^{n+1/2} = \rho_{i\pm}^{n} + \frac{\Delta t}{2\Delta x_i}(F_{i-}^{n} - F_{i+}^{n}) + \frac{\Delta t}{2} S_{i\pm1/2}^{n} \tag{42}$$

$$\rho_{i}^{n+1/2} = \rho_{i}^{n} + \frac{\Delta t}{2\Delta x_i}[g(F_{(i-1)+}^{n}, F_{i-}^{n}) - g(F_{i+}^{n}, F_{(i+1)-}^{n})] + \frac{\Delta t}{2} S_{i}^{n} \tag{43}$$

$$\rho_{i}^{n+1} = \rho_{i}^{n} + \frac{\Delta t}{\Delta x_i}[g(F_{(i-1)+}^{n+1/2}, F_{i-}^{n+1/2}) - g(F_{i+}^{n+1/2}, F_{(i+1)-}^{n+1/2})] + \frac{\Delta t}{2} S_{i}^{n+1/2} \tag{44}$$

<u>An application of the MUSCL scheme with source terms</u>

As an example of the use of the MUSCL scheme, we consider the growth of ionization in a nitrogen discharge between plane parallel metal electrodes of separation d equal to 3 cm. The gas pressure is 26.6 kPa and the applied voltage is 26,400 V.

The continuity equations for the electrons and ions are

$$\frac{\partial \rho_e}{\partial t} = \alpha \rho_e u_e - \frac{\partial}{\partial x}(\rho_e u_e) \tag{45}$$

$$\frac{\partial \rho_p}{\partial t} = \alpha \rho_e u_e - \frac{\partial}{\partial x}(\rho_p u_p) \tag{46}$$

where the field dependence of the drift velocities, u_e and u_p, and ionization coefficient α are

$$u_e = A \frac{E}{N} \tag{47}$$

$$u_p = B(1 - C \frac{E}{N}) \frac{E}{N} \qquad \text{if } \frac{E}{N} \le 2.43 \times 10^{-19} \text{ Vm}^2$$

$$\quad\; = D\sqrt{\frac{E}{N}} - F\sqrt{\frac{N}{E}} \qquad \text{if } \frac{E}{N} > 2.43 \times 10^{-19} \text{ Vm}^2 \tag{48}$$

$$\frac{\alpha}{N} = G\exp(-H \frac{E}{N}) \tag{49}$$

with $\quad$ A = 9.56x10^{23} V^{-1}m^{-1}s^{-1}, $\quad$ B = 6.59x10^{21} V^{-1}m^{-1}s^{-1}, $\quad$ C = 1.32x10^{18} V^{-1}m^{-2}, D = 2.27x10^{12} V$^{-1/2}$s^{-1}, F = 7.61x10^{-18} Vm^{3}s^{-1}, G = 1.73x10^{-20} m^2, H = 7.9x10^{-19} Vm^2.

The discharge was taken to be of finite radius equal to 1.5 mm and was initiated by a Gaussian–shaped pulse of 400 electrons released at the cathode. The electric field was calculated by the method of disks (Davies and Evans, 1967) in which the cylindrical discharge is divided into thin disks and the field at any point found by summing the contribution of each disk. Since the discharge is bounded by metal electrodes, the effect of image charges induced in the electrodes have to be allowed for and sufficient accuracy was obtained by incorporating the effect of images within five gap lengths on either side of the electrodes.

Secondary emission of electrons at the cathode (x = 0) was calculated on the assumption that the rate of photon emission is proportional to the rate of ionization so that the electron current density at the cathode is given by

$$j(x = 0,t) = \gamma_{ph} \int_0^d \alpha \rho_e |u_e| dx \tag{50}$$

The secondary emission coefficient γ_{ph} was taken to be 6.2×10^{-4}.

In the early stages of the growth the particle densities are small so that the field remains uniform. Figs. 9 and 10 show the evolution of the electron density and total gap current as functions of time and at early times we see the initial and subsequent electron avalanches crossing the gap and the particle densities growing in time with space-charge distortion starting to become noticeable at about 1000 ns. The particle densities and variation of total current agree with the formal solutions of the continuity equation given by Davidson (1953, 1955, 1956, 1957) for the case of uniform electric fields.

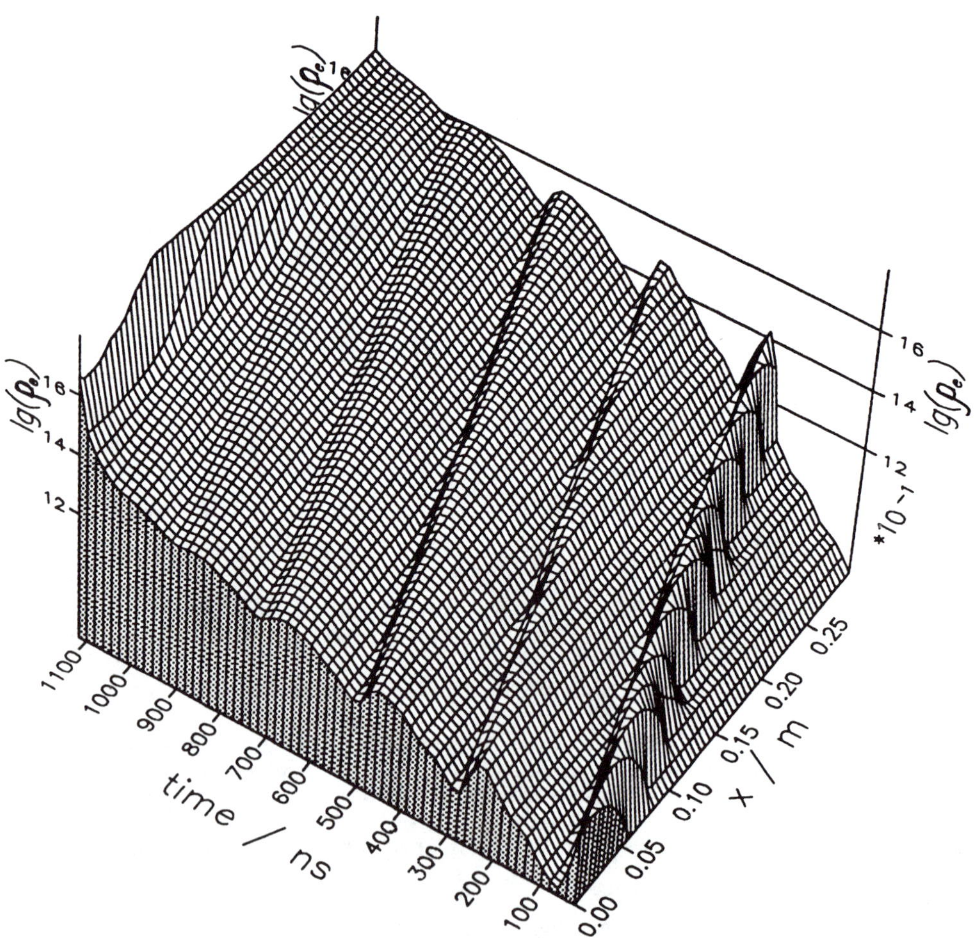

Figure 9. Spatio–temporal growth of electron density (in electrons per m^3) for a discharge between parallel plates in nitrogen initiated by a pulse of 400 electrons.

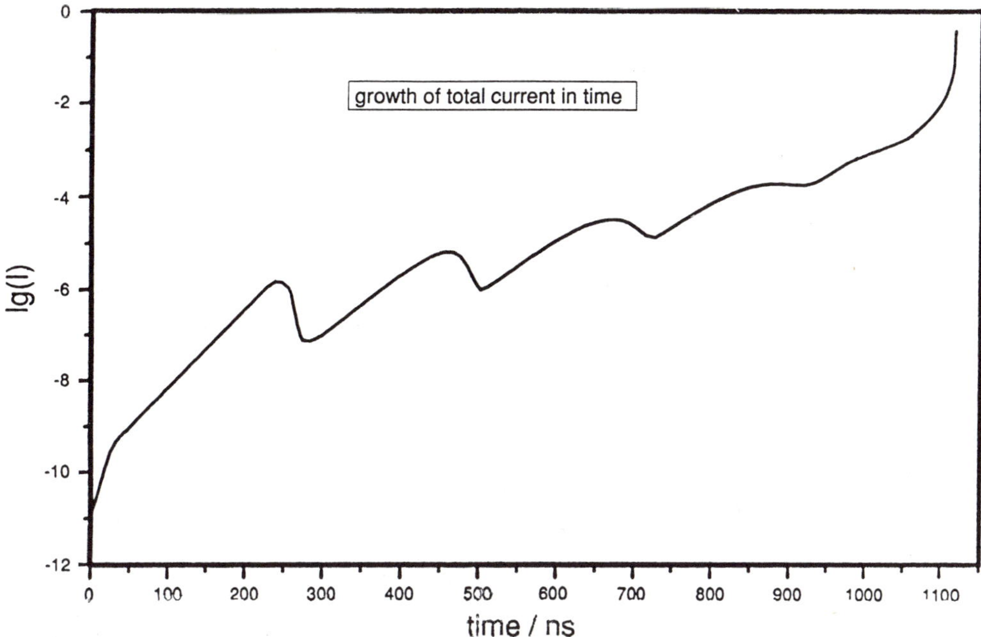

Figure 10. Growth of total current as a function of time for nitrogen discharge.

Once space-charge distortion becomes appreciable a sharp front starts to move towards the cathode as can be seen from the computed field distributions in Fig. 11. The magnitude of the source term is a measure of the light output from various points of the discharge and the maximum in the computed light ouput may be identified with the so-called "cathode-directed streamer" observed experimentally (Fig. 12).

In order to resolve the details of the propagation as the maximum moves towards the cathode a non-uniform graded mesh was chosen with the mesh spacing increasing exponentially from cathode to anode. In Figs. 11 and 12, 200 cells were used with the ratio of the mesh spacings at the cathode and anode being 1/4. In order to examine the behaviour when the field maximum actually arrives at the cathode an even more non-uniform mesh was used which again varied exponentially but had a ratio of 40/1 in the mesh sizes at the anode and cathode (Δx at the cathode was approximately .01 mm).

We see from Fig. 13 that the peak in the electric field immediately at the cathode is about 15 times the undistorted field and reaches a value almost high enough to produce field emission before falling to a very low value at about 0.1 mm. The light output shows a very sharp maximum propagating towards the cathode which then stabilises between 0.05 and 0.01 mm from the cathode surface.

The MUSCL scheme copes with the very rapid variation in the field and shows no sign of instability or staircase behaviour. Since all the discharge activity is in the immediate vicinity of the cathode and the field is so large, the Courant-Friedrichs condition enforces an ever decreasing value of Δt so that very little change in the particle densities is taking place in the interior of the gap.

The very high field, varying rapidly in space, also calls into question the whole validity of using an equilibrium model in this region. Several workers have attempted to obtain approximate models to treat the cathode-fall region in order to extend the computations further and follow the the discharge development at points in the gap remote from the cathode (Segur *et al*, 1983).

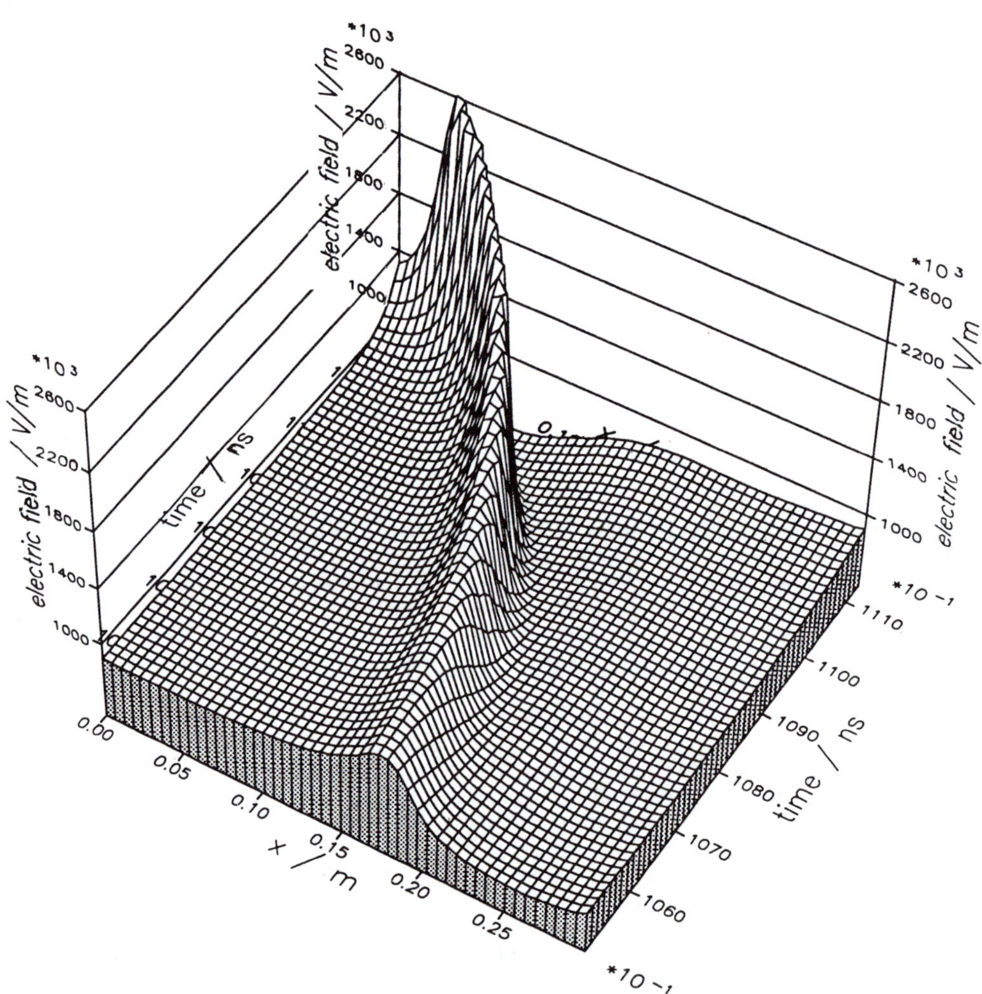

Figure 11. Variation of electric field as space-charge wave approaches cathode.

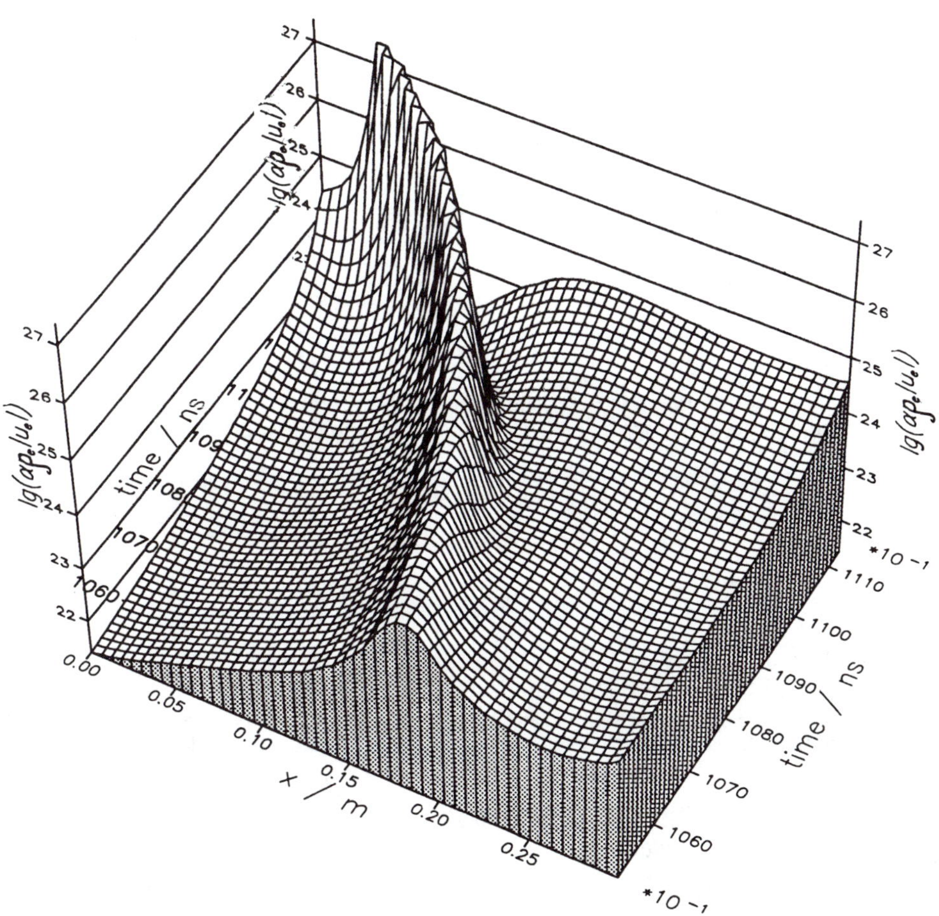

Figure 12. Magnitude of source term in nitrogen discharge as function of space and time. This is a measure of the light output from the discharge.

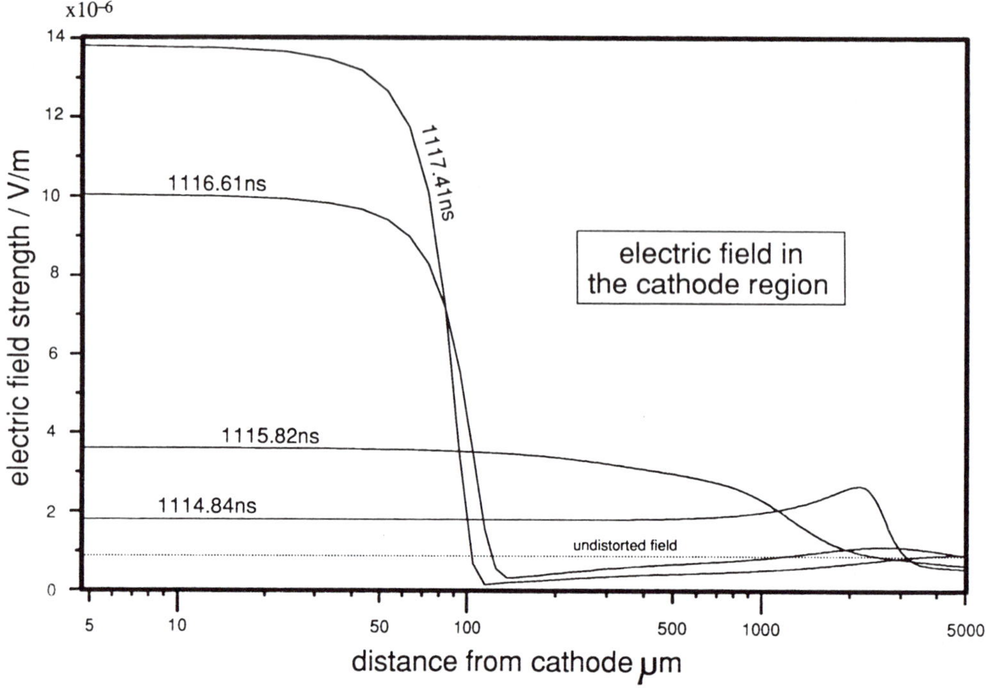

Figure 13. Variation of the electric field in the vicinity of the cathode after the space–charge wave has arrived at the cathode. Nitrogen discharge.

CONCLUSIONS

The Hybrid Method of Characteristics, FCT and MUSCL schemes have all been successfully used in discharge simulation. Although not conserving particle number the Hybrid Method of Characteristics is easily extended to two dimensions and has given a good qualitative description of the early stages of pseudospark development in the hollow cathode region (Mittag, 1989).

Both the FCT and MUSCL schemes are conservative and are of comparable accuracy, the MUSCL scheme being approximately twice as fast as FCT and being particularly promising for simulating discharge and plasma development. Both schemes are extendible in a fairly straightforward manner to two–dimensional axially symmetric systems having rectangular r–z grids, with the mesh spacing being capable of arbitrary gradation in both the r and z directions. Considerable work remains to be done, however, in developing conservative numerical algorithms for use in conjunction with two–dimensional curvilinear coordinate systems.

ACKNOWLEDGEMENTS

We gratefully acknowledge support from the Science and Engineering Research Council (Grant No. GR/D/79869) and the European Commission (Contract No. SCI–0055) which enabled this work to be carried out.

REFERENCES

Bayle, P., and Bayle, M, 1974, Simulation of secondary processes in breakdown in air, Z Phys, 226:275–281.

Book, D.L., Boris J.P., and Hain, K., 1975, Flux–corrected transport II: Generalizations of the method, J Comp Phys, 18:248–283.

Boris, J.P., and Book, D.L., 1976, Flux–corrected transport III: Minimal error FCT algorithms, J Comp Phys, 20:397–431.

Davidson, P.M., 1953, Appendix to "Formative time lags in the electrical breakdown of gases", Brit J Appl Phys, 4:170–175.

Davidson, P.M., 1955, Growth of current between parallel plates, Phys Rev, 99:1072–1074.

Davidson, P.M., 1956, Temporal growth of current between parallel plates, Phys Rev, 103:1897.

Davidson, P.M., 1957, Growth of current between parallel plates, Phys Rev, 106:1–2.

Davies, A.J., and Evans, C.J., 1967, Field distortion in gaseous discharges between parallel–plate electrodes, Proc IEE, 114:1547–1550.

Davies, A.J., Davies C.S., and Evans, C.J., 1971, Computer simulation of rapidly developing gaseous discharges, Proc IEE, 118:816–823.

Davies, A.J., Evans, C.J., and Woodison, P.M., 1975, Computation of current growth at high current densities, Proc IEE, 122:765–768.

Davies, A.J., Evans, C.J., and Townsend, P., 1977, Computation of axial and radial development of discharges between plane parallel electrodes, Proc IEE, 124:179–182.

Davies, A.J., 1986, Discharge simulation, Proc IEE, 133:217–240.

Kline, L.E., 1974, Calculations of discharge initiation in overvolted parallel-plane gaps, J Appl Phys, 45:2046–2054.

Kunhardt, E.E., and Wu J., 1987, Towards an accurate flux corrected transport algorithm, J Comp Phys, 68:127.

Mittag, K, 1989, A physical model of prebreakdown in the hollow cathode pseudospark discharge based on numerical simulations, see this proceedings.

Morrow, R., 1981, Numerical solution of hyperbolic equations for electron drift in strongly non–uniform electric fields, J Comp Phys, 43:1–15.

Munz, C.D., 1988, On the numerical dissipation of high resolution schemes for hyperbolic conservations laws, J Comp Phys, 77:18.

Niessen, W., 1989, Gas discharge simulation using the MUSCL scheme, submitted to J Comp Phys.

Roe, P.L., and Baines M.J., 1982, in "Proceedings of the 4th GAMM Conference on Numerical Methods in Fluid Mechanics," H. Viviand, ed., Vieweg, Braunschweig-Wiesbaden, p.281.

Segur, P., Yousfi, M., Boeuf, J.P, Marode, E., Davies, A.J., and Evans, J.G., 1983, The microscopic treatment of nonequilibrium regions in a weakly ionized gas, in "Electrical Breakdown and Discharges in Gases," Part A, L. Luessen, ed., Plenum, pp.331–394.

Van Leer, B., 1979, Towards the ultimate conservative difference scheme. V. A second-order sequel to Godunov's method, J. Comp Phys, 32:101–136.

Yoshida, K., and Tagashira, H., 1976, Computer simulation of a nitrogen discharge at high overvoltage, J Phys D, 9:491–505.

Zalesak, S.T., 1979, Fully multidimensional flux–corrected transport algorithms for fluids, J. Comp Phys, 31:335–362.

SCALING PARAMETERS FOR OPTICALLY TRIGGERED HOLLOW

CATHODE SWITCHES OBTAINED BY COMPUTER SIMULATION

Hoyoung Pak and Mark J. Kushner
University of Illinois
Department of Electrical and Computer Engineering
Gaseous Electronics Laboratory
607 E. Healey Street
Champaign, Illinois 61820
USA

ABSTRACT

Low pressure plasma switches are essential components of pulse power
devices. A new class of high current electrically isolatable switches, the
optically triggered pseudo-spark, is currently being developed for
applications where conventional thyratrons are inadequate. The analysis
and computational design of these switches is complicated by the fact that
the electron energy distribution in low pressure pulsed power plasma
devices is typically not in equilibrium with the local electric field. A
new computer model has been developed to describe electron transport for
these conditions and it has been applied to the optically triggered
pseudo-spark, or Back-Lit-Thyratron (BLT). The model uses two groups of
electrons described as the "bulk" and the "beam". The energy distribution
of the bulk electrons is nearly in equilibrium with the local electric
field while the beam represents those electrons whose energy is equal to
the local electric potential and which have not undergone collisions after
being emitted from the cathode. The model is used to investigate the
commutation phase of switching in the BLT.

I. INTRODUCTION

Low pressure plasma switches are critical components of pulse power
devices which require high holdoff voltages, high currents, and large
rates of current rise. The Back-Lit-Thyratron, or BLT, is a low pressure
(< 1.0 Torr) plasma switch that operates on the near side of the Paschen's
curve. (See Fig. 1.) The BLT is essentially an optically triggered
pseudo-spark and is currently being developed as a possible replacement
for conventional thyratrons and spark gaps. Performance specifications of
current > 50 kA, $dI/dt \geq 10^{12}$ A/s, current density ≥ 10's kA/cm^2, holdoff
voltage ≥ 40 kV and jitter ≤ 1 ns have thus far been demonstrated.[1-7]

In low pressure plasma devices having time varying electric
potentials, the electron energy distribution (EED) is typically not in
equilibrium with the local electric field. These nonequilibrium effects
have been difficult to include in multidimensional, time dependent models

Physics and Applications of Pseudosparks
Edited by M. A. Gundersen and G. Schaefer
Plenum Press, New York, 1990

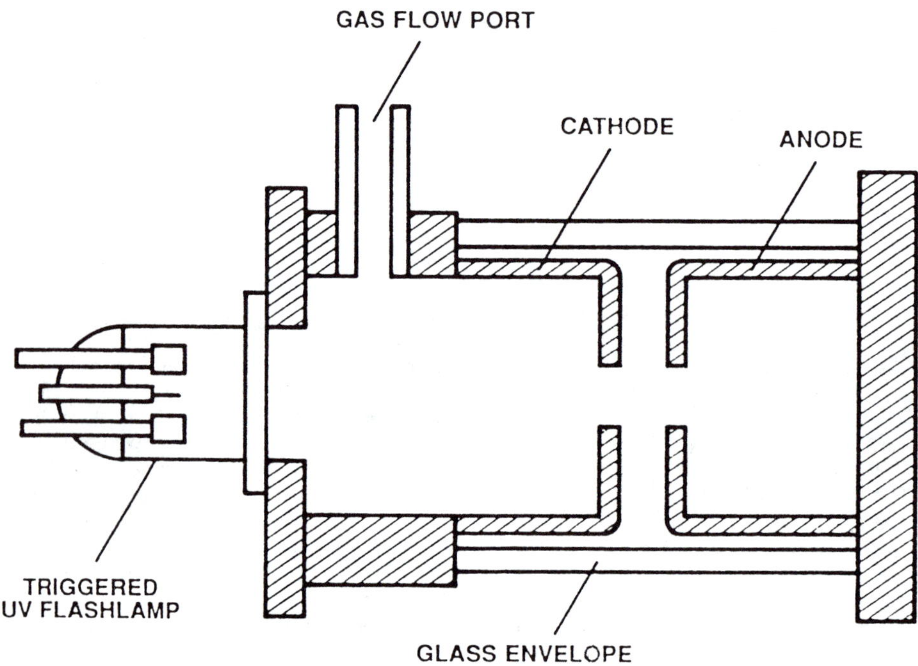

Fig. 1. Typical geometry for the optically triggered pseudo-spark, or Back-Lit-Thyratron, using a symmetric configuration. (From Ref. 1.)

of these devices due to the computational complexity of fully resolving Boltzmann's equation. In low pressures discharges, ≤ 0.1 Torr, Monte Carlo particle simulations may appropriately be used to model nonequilibrium electron transport. At higher gas pressures, ≥ 1.0 Torr, a fluid representation is an appropriate modeling technique. Fluid models typically assume that the EED is nearly in equilibrium with the local electric field thereby invoking the local field approximation (LFA), or they use an average energy representation which includes some transport effects. Large departures from the LFA, though, occur in the gas pressure range of interest to low pressure plasma switches, $0.1 < p < 1.0$ Torr. The application of fluid models to these switches is therefore questionable. The breakdown of the LFA in this pressure range results from the fact that there may exist a "beam" component to the EED, much like that in a cathode fall, which contains ballistic electrons whose trajectories are aligned along electric field lines.[8]

In this paper we describe a computer model that self-consistently simulates the performance of a BLT during the commutation phase of switching while including nonequilibrium aspects of the electron energy distribution resulting from ballistic electrons. To improve on the LFA description of electron transport a "beam-bulk" model for the EED has been developed. The beam-bulk model separates the EED into two groups. The first group, the bulk, represents low energy electrons for which the LFA applies. The transport of bulk electrons is therefore adequately represented by the fluid method used in our previous model of the BLT described in Ref. 9. This model is a 2-dimensional time dependent simulation in which the continuity equations for electrons and ions, and Poisson's equation are solved. The transport coefficients are obtained using the LFA. The second group in the beam-bulk model, the beam, represents electrons whose energy is equal to the local electric potential, whose trajectories are directed along electric field lines and who have survived from emission at the cathode without suffering collisions. Electrons in the beam component which have collisions "fall" into the bulk component of the distribution. The use of the beam-bulk model to simulate switching in the BLT is described below.

II. DESCRIPTION OF THE MODEL

In this section we will describe our model for the Back-Lit-Thyratron (BLT). The BLT is typically cylindrically symmetric with a hollow cathode and an anode which may be either planar or hollow. (See Fig. 1) The model uses finite differences to solve the continuity equations for bulk electrons and ions in 2-1/2 dimensions in either rectangular or cylindrical coordinates. A beam component to the EED is "overlayed" on the distribution of electrons and ions in the manner described below. The sequence of events in the simulation will first be briefly presented. As input, the user specifies the magnitude and time duration of the flux of triggering photons that are directed onto the inside surface of the cathode. The user also specifies the gas mixture, geometry, and holdoff voltage. The simulation begins by calculating the vacuum electric field and illuminating the inside of the cathode of the BLT which is holding off the voltage V_0. The photon flux generates secondary electrons at the surface of the cathode, which provide a source term for electrons. We modeled the case of using a laser as the photon source. The optical pulse was 8 ns (FWHM) having an incident fluence of a few tenths to a few mJ. The source term is included in the electron continuity equation when using only the bulk distribution to describe the electrons, or is "given" to the beam component in the beam-bulk model. The electron and ion continuity

equations are then integrated in time using the local field approximation
to obtain their transport coefficients. Contributions to the electron and
ion continuity equations from the beam are added to account for
ionizations and slowing beam electrons. The electric field is updated by
solving the time dependent form of Poisson's equation. Simultaneously to
integrating the continuity equations, the flux of photons resulting from
electron impact excitation of the gas and the flux of ions incident on the
cathode are also computed. These fluxes, coupled with the appropriate
secondary electron emission coefficients, generate a source of electrons
at the cathode surface which sustains the current. The self-generated
magnetic field is also computed and an effective magnetically driven drift
velocity is added to the electric field driven drift velocity for the bulk
electrons. The electron flux (bulk and beam) incident on the anode is
integrated to yield the total current. Switching of the BLT and the end
of commutation is denoted by a 10% decrease in voltage across the BLT
while in series with a 1 Ω load. The interval between triggering
conduction is the anode delay time.

The details of including a beam component in our model will now be
discussed. Photoelectrons are generated which initiate the electron
avalanche constituting the commutation phase of switch closure. After
triggering and during commutation, electrons continue to be emitted from
the cathode due and ion bombardment and plasma generated photons. All
electrons emitted from the cathode surface, either photoelectrically or by
ion bombardment, are initially classified as beam electrons in our model.
The beam electrons follow a trajectory parallel to the local electric

field, and have velocities given by the local electric potential, $\phi(\vec{r}, t)$;
$v = (\frac{2 \cdot q \cdot \phi}{m_e})^{1/2}$. Because the trajectories and velocities of the beam

electrons depend on the local electric potential the paths of the beam
electrons change as space charge or gradients in conductivity deform the
potential. Beam electrons that undergo collisions with the neutral gas
are taken out of the beam component and are placed into the bulk
component. Secondary electrons resulting from ionizing collisions by the
beam are also placed in the bulk distribution.

To model the trajectories of the beam electrons we work on a separate
grid from that used for transport of the bulk electrons, which is
cylindrical with azimuthal symmetry. The computational grid for the beam
electrons consists of a mesh parallel to the local electric field. A
beamlet, labeled i, is associated with each electric potential grid line

emanating from the cathode and the beamlet follows a path $\vec{s}_i$ on the grid.
The beam-grid overlaps the conventional computational grid upon which
transport for the bulk electrons and the electric potential are
calculated. The (r,z) position on the cylindrical grid underlying the

mesh point j along beam trajectory $\vec{s}_i$ is $\vec{r}(s_{ij})$. The electron flux in the

beamlet i at location s_{ij} is

$$\Gamma_i(s_{ij}) = A(s_{ij}) \cdot \Gamma_{io} \cdot \exp(-\int_{\vec{s}_i} N \cdot \sigma_T(\epsilon(s)) \cdot ds) \qquad (1)$$

where Γ_{io} is the flux of secondary electrons due to all processes at the cathode for trajectory i, N is the neutral gas density, σ_T is the total electron scattering cross section and the integral is along the path i to location s_{ij}. The energy at which the cross section is evaluated at path location s_{ij} is $\epsilon_{ij} = e\phi\,(\vec{r}(s_{ij}))$ where ϕ is the local electric potential evaluated on the cylindrical mesh. The integral in Eq. 1 accounts for the loss of electrons out of the beam by collisions as they traverse the path. $A(s_{ij})$ is a weighting function to account for changes in the cross sectional area of the flux tube of the beamlet, which will be discussed below. The use of σ_T in Eq. 1 emphasizes that our definition of beam electrons in based on momentum, and not necessarily energy.

Since beam electrons follow a trajectory parallel to the electric field and do not cross electric field lines, the total electron current in the beam between given field lines is conserved in the absence of collisions. The spacing between electric field lines, though, may vary as a function of position thereby resulting in a change in the cross sectional area between given field lines. This change in area results in a change in the magnitude of the electron beam flux, while conserving total current, trapped between field lines as the beam electrons travel from the cathode to anode. For example, in the BLT the spacing between electric field lines generally decreases while moving from the cathode and to the anode as all field lines from the inner surface of the cathode must pass through the cathode hole. The compression of the electric field lines result in an increase in the flux of the electron beam. The weighting function $A(s_{ij})$ in Eq. 1 accounts for the change in the magnitude beam flux due to this change in cross sectional area, and is essentially the ratio of the cross sectional area of the flux tube at the cathode to that of the local flux tube.

The time rate of change of the bulk electrons at location $\vec{r}_b$ due to collisions of the beam is

$$\left[\frac{\partial n_e(\vec{r}_b)}{\partial t}\right]_b =$$

$$\sum_{i,j} [N \cdot (\sigma_T(\epsilon_{ij}) + \sigma_I(\epsilon_{ij})) \cdot \Gamma_{ij} \cdot R_v(s_{ij})] \cdot \delta(\vec{r}_b - \vec{r}(s_{ij})) \qquad (2)$$

where the sum is over beamlets i and beam locations j, σ_I is the ionization cross section, and R_v is the ratio of the volume of the computational cell of the beam to the volume of the computational cell for bulk electrons. The contribution of beam collisions to the time rate of change in ion density is

$$\left[\frac{\partial n_I(\vec{r}_b)}{\partial t}\right]_b = \sum_{i,j} N \cdot \sigma_I(\epsilon_{ij}) \cdot \Gamma_{ij} \cdot R_v(s_{ij})\, \delta(\vec{r}_b - \vec{r}(s_{ij})) \qquad (3)$$

The transport of bulk electron and ions is described using conventional fluid equations as in our previous model.[9] For example, the time rates of change of the bulk electrons and ions are

$$\frac{\partial n_e}{\partial t} = -\nabla \cdot [\vec{v}_d\, n_e - D_e \vec{\nabla} n_e] + n_e k_I N \pm \sum_j n_e k_j(\bar{\epsilon}) N_j + [\frac{\partial n_e}{\partial t}]_b \qquad (4a)$$

$$\frac{\partial n_I}{\partial t} = -\nabla \cdot [\vec{v}_I\, n_I - D_I \vec{\nabla} n_I] + n_e k_I N \pm \sum_j n_I k_j(\bar{\epsilon}) N_j + [\frac{\partial n_I}{\partial t}]_b \qquad (4b)$$

where the subscripts e and I refer to electrons and ions, and the drift velocity $(\vec{v})$, diffusion coefficient (D), and ionization coefficient (k_I) are obtained using the LFA. The summation in Eq. 4 represents additional processes, such as recombination, which depend upon the local electron energy, $\bar{\epsilon}$, and which are not included in the net ionization coefficient, k_I. The total electron and ion densities are then used to obtain the local electric field by solution of the current conservation equation

$$\nabla \cdot j = - \nabla \cdot \sigma \nabla\phi = - \frac{\partial \rho}{\partial t} \qquad (5)$$

where j is the current density, σ is the plasma conductivity, and ρ is the charge density. Eq. 5 was solved using the method of successive-over-relaxation every $10^{-11} - 10^{-10}$ s during the simulation. In practice, due to the sharp gradients that may occur in conductivity, we usually smoothed the conductivity using a 2-dimensional digital filter before solving for the potential.

It is well known that high current arcs may pinch or constrict under the influence of their self-generated magnetic fields. As our interest is primarily with commutation when currents are small, we expect that pinching will not be a dominating effect. Therefore, we may approximate its effects on the bulk electrons by using a magnetically driven drift velocity. In this approximation, the azimuthal magnetic field, B_θ, is obtained by integrating the axial current density. The bulk electrons drifting in the axial electric field experience a Lorentz force in the radial direction. By assuming that bulk electron motion in the radial direction is mobility limited, one may compute an effective magnetically driven drift velocity, $v_r = \mu_e^2 E_z B_\theta$.[10]

III.   SIMULATION OF SWITCHING CHARACTERISTICS

A. Summary of Results Using the Bulk Model

In this section, we will review switching characteristics of the BLT obtained by assuming that the EED is composed of only the bulk component. Comparisons between the bulk and beam-bulk models will be made below. Computed electric potentials within the BLT at various times during commutation are shown in Fig. 2 using a planar anode. The gas pressure is 0.5 Torr H_2 and $V_0 = 10$ kV. In the vacuum configuration, as shown in Fig.

2a ($t = 0$), the anode potential penetrates through the cathode hole to a value of $\approx$ 1 kV. The penetration of the potential must be small in order to prevent long path breakdown extending from the inside surface of the cathode to the anode. The volume of positive potential, relative to the cathode, inside of and adjacent to the cathode hole resulting from potential penetration may be thought of as a virtual anode. The virtual anode effectively increases the electric field adjacent to the cathode surface from which electrons are emitted. Although at this time this region of positive potential results from geometry rather than from space charge, as does the virtual cathode in front of a thermionic cathode, the structures are functionally equivalent.

The virtual anode survives in its vacuum configuration for only 5-6 ns after triggering. During this time, the seed photoelectrons drift away from the cathode towards the virtual anode and bring with them negative space charge which neutralizes the virtual anode. This action reduces the penetration of the anode potential through the cathode hole as shown in Fig. 2b. If, however, this region retains a sufficiently high electric field, ionizations will occur, though at a reduced rate, adjacent to the cathode hole. The secondary electrons produced by gas phase ionizations in this region respond quickly to the electric field and pass through the cathode hole to the anode. The heavier ions, though, respond less quickly to the field and drift slowly to the cathode. The end result is that positive space charge builds up and neutralizes the negative space charge created by the photoelectrons. The virtual anode is regenerated giving the appearance of increased penetration of the anode potential through the cathode hole as shown in Fig. 2c. As electron avalanche proceeds the anode potential penetrates further through the cathode hole, as shown in Fig. 2d.

In parameterizing the operating characteristics of the BLT as a function of gas pressure, electrode separation and optical trigger intensity we found that switch closure (i.e., electron avalanche and voltage collapse) occured only when the virtual anode was regenerated in the manner described above at times $\leq$ 100 ns after illumination of the cathode. By inference, we suggest that regeneration of the virtual anode prior to electron avalanche and voltage collapse is a requisite for switch closure.

The bulk electron density as a function of position for various times during commutation is shown in Fig. 3. In Fig. 3a ($t = 4$ ns) electrons appear only near the inside surface of the cathode as they begin to avalanche after having been generated by photoemission. The electron density at this time is $\leq 10^9 \mathrm{cm}^{-3}$. As the electrons drift towards the cathode hole (Fig. 3b) and into a region of higher electric field, the electron avalanche which constitutes commutation begins to intensify (Fig. 3c). The onset of the conduction phase occurs approximately at 100-120 ns when the region of high electron density penetrates to the anode (Fig. 3d) At this time, the electron density is $\leq 10^{13}$ cm^{-3}. Note that as the current begins to increase, the electrons in the cathode-anode gap actually begin to spread transversely. The radial spreading of the electron swarm at later times results from generation of transverse components of the electric field due to negative space charge and gradients of conductivity in the gap.

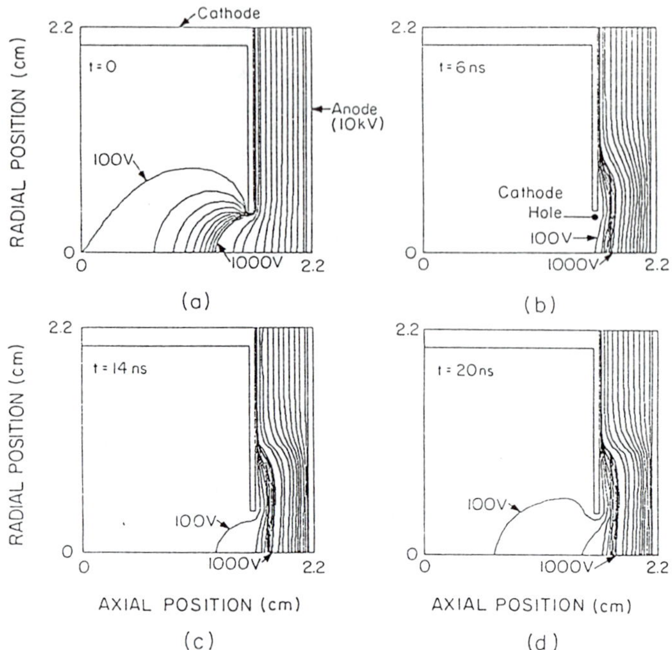

Fig. 2.
Electric potential for $V_0 = 10$ kV. (100 V contours for $V \leq 1$ kV, 1 kV contours for $V > 1$ kV). (a) t = 0. Vacuum configuration. (b) t = 6 ns. Negative space charge causes potential to recede. (c) t = 14 ns. Virtual anode is reforming. (d) t = 20 ns. The virtual anode is reformed.

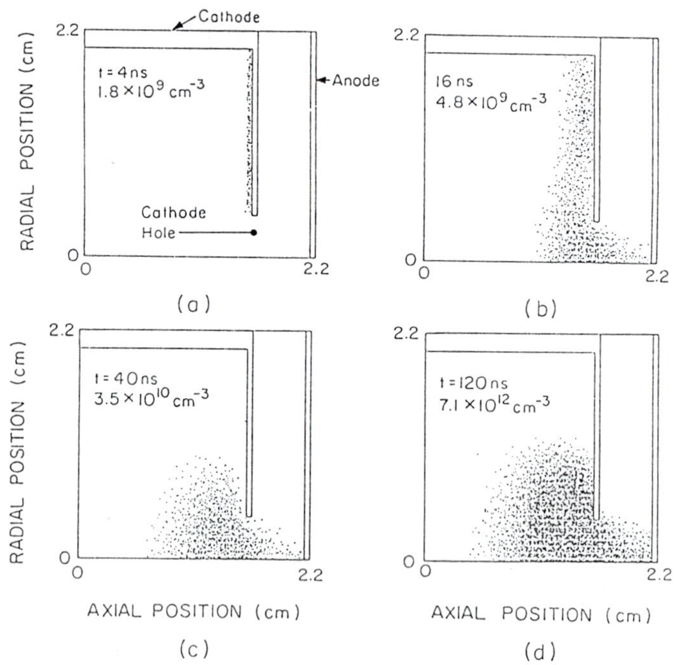

Fig. 3.
Electron density in BLT during commutation. The value noted is the maximum density at that time. (a) t = 4 ns. Photo-electrons are generated. (b) t = 16 ns. Electrons drift towards the cathode hole. (c) t = 40 ns. Electron avalanche begins. (d) t = 120 ns. End of commutation.

226

The volume which is deficient of electrons adjacent to the cathode is a region which resembles the cathode fall in a traditional glow discharge, somewhat mishapen due to the non-planar geometry. As shown in Fig. 4, this region is more heavily populated by ions which provide, by bombardment of the cathode, the sustaining current for the BLT during commutation.[11,12] Secondary electrons generated by this bombardment quickly accelerate away from the cathode. Note that the region of secondary electron generation by ion bombardment, as indicated by the spatial distribution of ion density, is within a few mm of the edge of the cathode hole, which agrees well with experimental observations of the current generation area.[6] It is not clear, though, that ion bombardment alone can sustain the observed peak currents in BLT's. Hartmann et. al.[6] suggest that field enhanced thermionic emission may be responsible for sustaining peak current.

The dependence of anode delay time on the diameter of the cathode hole is shown in Fig. 5. For sufficiently large cathode holes, the anode delay time is only weakly dependent on its diameter, decreasing slightly as the diameter increases. This occurs when the cathode hole diameter is significantly larger than the sheath thickness. The anode delay time increases significantly at smaller diameters until such time that closure cannot be obtained. This behavior results from reduced penetration of the anode potential through the cathode hole and overlapping of the sheaths at either side of the hole which effectively shield lower energy electrons away.

The dependence of anode delay time on holdoff voltage is shown in Fig. 6. For holdoff voltages greater than 10 kV, there is only a weak dependence of anode delay time on holdoff voltage. We expect this behavior because in hydrogen the electron transport properties (i.e., mobility and ionization coefficient) do not significantly increase in the electric fields provided by penetration of the anode potential through the cathode hole for voltages greater than $\approx$ 10 kV for this geometry.

B. Application of the Beam-Bulk Model

The beam-bulk model has been used to describe electron transport during the commutation phase of switching in the BLT when the holdoff voltage is high and non-equilibrium effects are likely to be important. The electrodes of the BLT we modeled consisted of a cylindrical hollow cathode with a symmetrical and opposing anode. The cathode was grounded and the anode was charged to 10 kV. The gas is 0.5 Torr H_2 unless otherwise stated. The anode-cathode separation was 0.5 cm, and the radius of the cathode hole was 0.5 cm.

The electron beam flux and the electron bulk flux 25 ns after cathode illumination are shown in Fig. 7 as a function of position. Since all electrons emitted from the cathode are initially in the beam the majority of the current is carried by the beam component close to the cathode. As one progresses away from the cathode, the electron current in the beam is depleted by collisions with the gas until current at the anode is carried largely by bulk electrons. At this time the fraction of current carried by the beam at the anode, η, is $\approx$2%. This fraction, though, is a function of time, as shown in Fig. 8. Early during triggering the EED consists virtually totally of the beam, and $\eta \simeq 1$. As secondary electrons are generated, and the bulk distribution begins to avalanche, η begins to decrease. By the time avalanche occurs and the switch closes, η reaches a nearly constant value.

The expression for the local flux of beam electrons, Eq. 1, implies that the magnitude of the beam flux away from the cathode will decrease at higher gas pressures due to a higher rate of collisions with the neutral

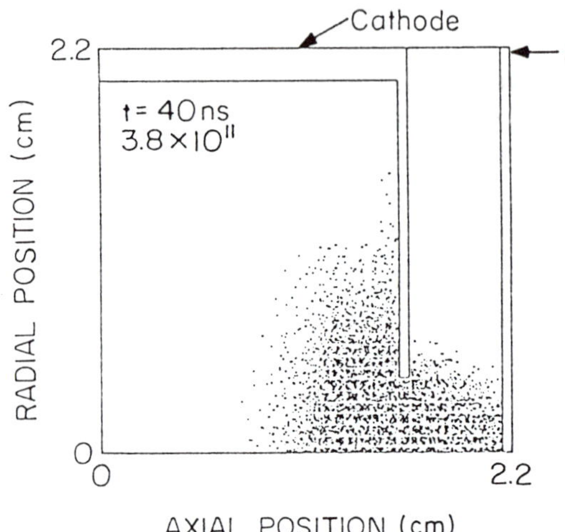

Fig. 4.
Ion density at t = 40 ns for conditions similar to Fig. 3. Ion bombardment of the cathode surface is sustains the discharge during commutation. Note that emission comes dominantly from within a few mm of the cathode hole. The resolution of the plot has been enhanced by a factor of 50 to show detail in the gap and near the cathode.

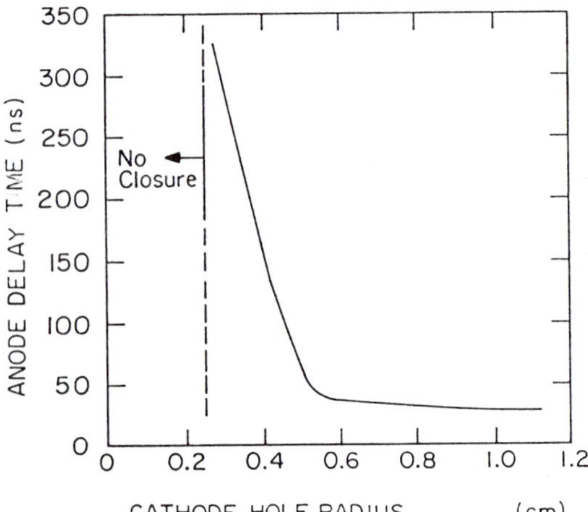

Fig. 5.
Anode delay time (ADT) as a function of cathode hole radius. The dependence of ADT on the cathode hole radius for small radii is a result of lower energy electrons being shielded away from the electrode gap by the cathode sheath.

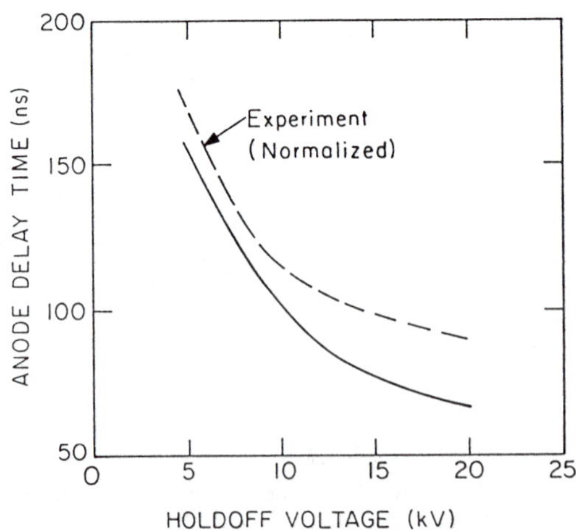

Fig. 6.
Anode delay time (ADT) as a function of holdoff voltage. At $V_0 > 10$ kV the ADT is relatively insensitive to holdoff voltage. The experimental results are from Ref. 13 and have been normalized to account for differences in trigger photon fluence.

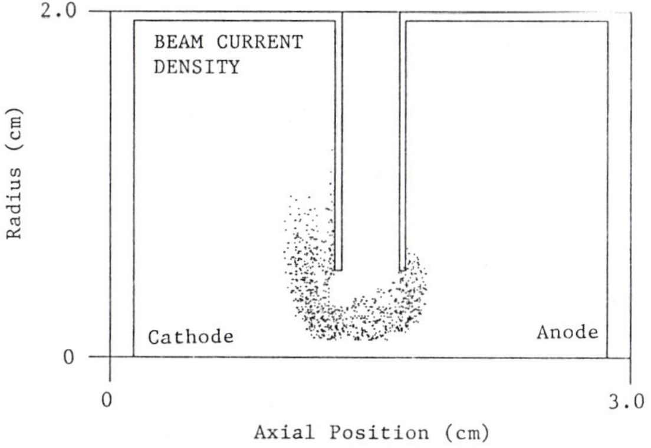

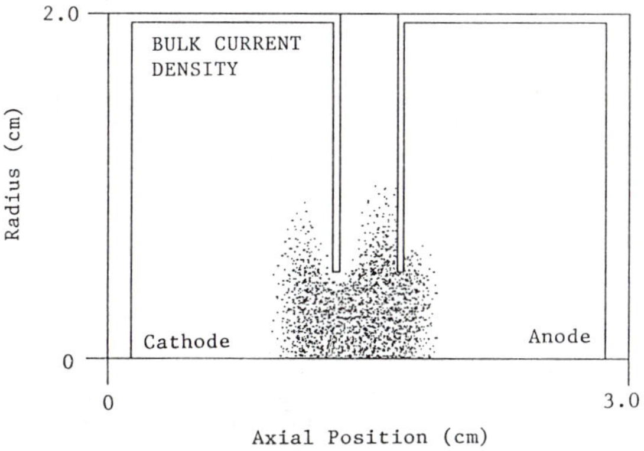

Fig. 7.
Relative current density carried by the (top) beam component and (bottom) the bulk component 25 ns after triggering the BLT. $P = 0.5$ Torr H_2 and $V_0 = 10$ kV. At this time, approximately 2% of the current at the anode is carried by the beam. Near the cathode, though, the majority of the current is carried by the beam.

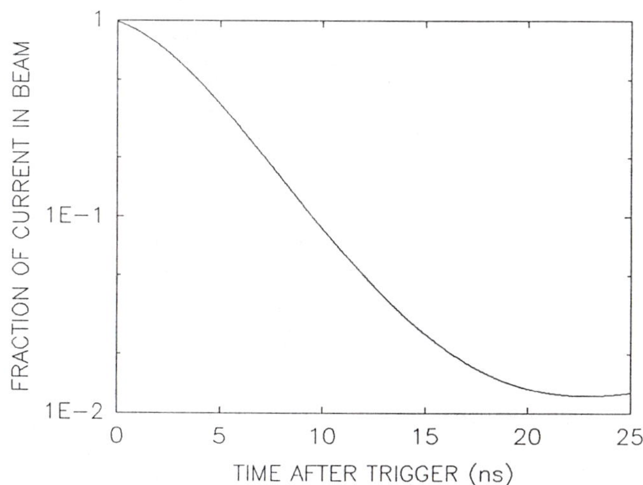

Fig. 8.
Fraction of the current at the anode carried by the beam component as a function of time after triggering. The contribution of the beam component reaches a constant value at the end of commutation.

gas, resulting in a lower penetration of beam electrons toward the anode. This pressure dependence is demonstrated by the fractional current carried by the beam at the anode at switch closure as a function of gas pressure, as shown in Fig. 9. At pressures nearing 1.0 Torr, the contribution of the beam electrons to the anode current is small whereas at pressures below 0.3 – 0.4 Torr a large contribution to the current is made by the beam. We have shown results for two values of the secondary emission coefficient, γ, resulting from ion bombardment of the cathode. The larger value is intended to be an approximation to conditions where the metal may have an oxide coating, which has high γ,[12] or when thermionic emission becomes important, which effectively has an infinite γ. Clearly, as γ increases, the importance of the beam component also increases, as less avalanching by the bulk distribution is required to sustain a given current.

The fraction of current carried by the beam at switch closure is shown in Fig. 10 as a function of hold-off voltage. For voltages > 6-7 kV, η is nearly constant. This effect results from the fact that the collision cross sections for electrons in H_2 decrease with increasing electron energy for ϵ > 100's eV. The mean free path, λ, therefore increases with increasing energy. When beam electrons accelerate above a critical energy where λ is greater than the additional distance to the anode, they undergo no further collisions. At this time, their contribution to the total current does not further decrease from collisional processes.

Although the fractional beam current is small for these conditions, the operation of the BLT is sensitive to the presence of beam electrons. The importance of the beam electrons with respect to switch closure time is shown in Fig. 11 as a function of pressure. Here we have computed the anode delay time using the beam-bulk model, and using the model without the beam component, which is essentially a fluid model using the LFA as described in Ref. 9. The differences in anode delay times obtained using the two methods are quantified by

$$\Delta t = 2 \cdot \frac{(\Delta t_{bb} - \Delta t_b)}{(\Delta t_{bb} + \Delta t_f)} \qquad (6)$$

where Δt_{bb} is the anode delay time of the beam-bulk model, and Δt_b is the anode delay time of the fluid model using only the bulk component. Typical anode delay times are 100's ns to a few μs at low pressure to 10's ns at high pressure. At lower pressures there is a large difference in the anode delay time between the two models, with the anode delay time computed with the beam-bulk model being larger. This is a consequence of the rate of ionization by beam electrons being lower than that by the bulk electrons at the local E/N in the vicinity of the cathode hole. This results from the ionization cross section corresponding to the plasma potential near the gap during commutation (100's eV – 1000's eV) being well down from the peak cross section, occuring at $\approx$ 100 eV. The differences between the two methods are small at pressures exceeding 0.5 – 1.0 Torr. At these pressures the fluid model using the LFA is an adequate representation.

IV. CONCLUDING REMARKS

In conclusion we have developed a beam-bulk model which uses two groups of electrons to describe the transport of electrons in an optically triggered psuedospark. Our results demonstrate that the importance of including nonequilibrium effects, as approximated by the beam-bulk model, at gas pressures less than 0.5 Torr. A fluid model using the LFA, though,

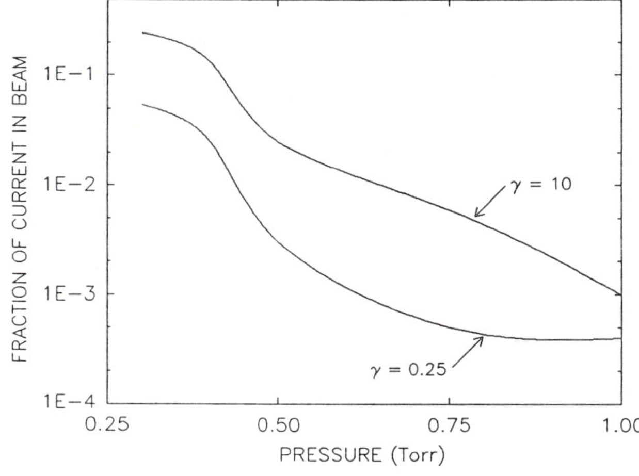

Fig. 9.
The fraction of current carried by the beam component at the anode as a function of gas pressure at the end of commutation. Results are shown for two values of the secondary emission coefficient at the cathode: $\gamma = 0.25$ and 10.

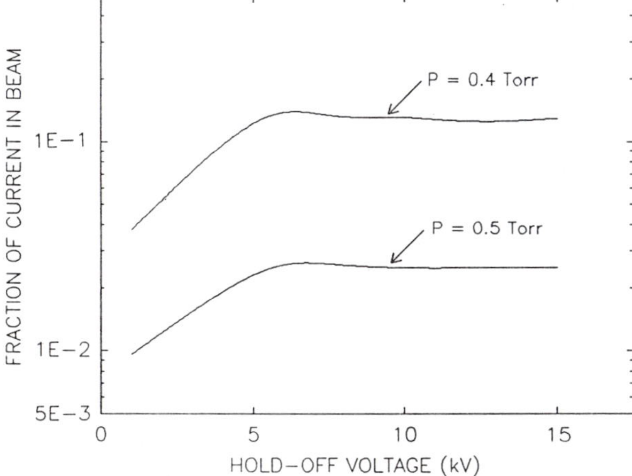

Fig. 10.
The fraction of current carried by the beam component at the anode as a function of hold-off voltage at the end of commutation. Results are shown for two values of gas pressure.

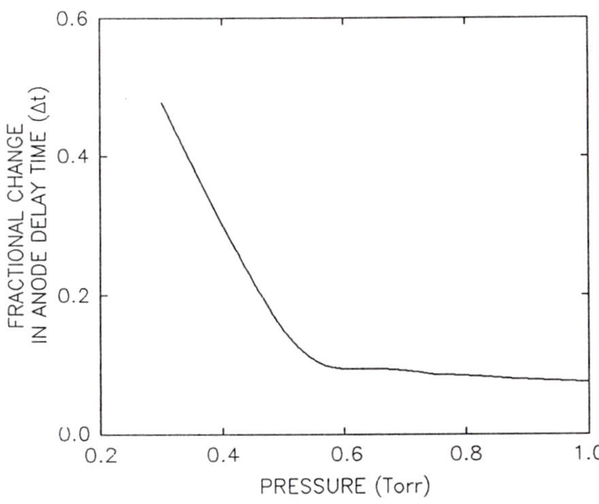

Fig. 11.
Fractional difference in switch closure time between the beam-bulk model (Δt_{bb}) and the single group bulk model (Δt_b) as a function of gas pressure. $\Delta t = [(\Delta t_{bb} - \Delta t_b)/(\Delta t_{bb} + \Delta t_b)/2]$. At P > 0.5 - 1.0 Torr, the single group model differs little from the beam-bulk model.

is fairly accurate at gas pressures exceeding 0.5 - 1.0 Torr. The
contribution of the beam component to the current at the anode increases
with decreasing gas pressure, increasing hold-off voltage (though not
dramatically) and decreases with time after trigger until the switch is
closed. The total current carried by the beam also decreases with distance
from the cathode as the beam is depleted of electrons by collisions.

V. ACKNOWLEDGEMENTS

This work was supported by the Texas Tech Research Foundation under
contract to the Defense Nuclear Agency through the direction of Dr. G.
McDuff and the National Science Foundation. The authors would like to
thank Prof. M. Gundersen and G. Kirkman for their comments.

References

1. G. Kirkman and M. Gundersen, Appl. Phys. Lett. $\underline{49}$, 494 (1986).
2. G. Kirkman, W. Hartmann, and M. Gundersen, Appl. Phys. Lett. $\underline{52}$, 613 (1988).
3. K. Frank, E. Boggasch, J. Christiansen, A. Geortler, W. Hartmann, C. Kozlik, G. Kirkman, C. Braun, V. Dominic, M. Gundersen, H. Riege, and G. Mechtersheimer, IEEE Trans. Plasma Sci. $\underline{16}$, 317, (1988).
4. G. Mechtersheimer, R. Kohler, T. Lasser, and R. Meyer, J. Phys. E: Sci. Instrum. $\underline{19}$, 466 (1986).
5. D. Bloess, I. Kamber, H. Riege, G. Bittner, V. Brueckner, J. Christiansen, K. Frank, W. Hartmann, N. Lieser, C. Scheltheiss, R. Seeboeck, and W. Seudtner, Nucl. Instrum Meth. $\underline{205}$, 172 (1983).
6. W. Hartmann, V. Dominic, G. Kirkman and M. Gundersen, Appl. Phys. Lett. $\underline{53}$, 1699 (1988).
7. C. Braun, V. Dominic, G. Kirkman, W. Hartmann, M. Gundersen, and G. McDuff, IEEE Trans. Electron Devices $\underline{35}$, 559 (1988).
8. R. J. Carman, J. Phys. D $\underline{22}$ 55 (1989).
9. H. Pak and M. Kushner in "Pulse Power for Laser II," Edited by T. R. Burkes, G. McDuff, Proc. SPIE 1046, pp. 64-71 (1989).
10. M. J. Kushner, R. D. Milroy and W. D. Kimura, J. Appl. Phys. $\underline{58}$, 2988 (1985).
11. B. Chapman, <u>Glow Discharge Processes</u> (Wiley, New York, 1980), pp. 82-95.
12. B. Szapiro, J. Rocca, and T. Prabhuram, Appl. Phys. Lett. $\underline{53}$, 358 (1988).

13. M. Gundersen, unpublished data.

A PHYSICAL MODEL OF PREBREAKDOWN IN THE

HOLLOW CATHODE PSEUDOSPARK DISCHARGE BASED ON NUMERICAL SIMULATIONS

Karl Mittag

Kernforschungszentrum Karlsruhe GmbH
Institut für Datenverarbeitung in der Technik
D 7500 Karlsruhe
Federal Republic of Germany

INTRODUCTION

The pseudospark discharge is a special form of hollow cathode discharge at pressures around 30 Pa, voltages between 1 and 400 kV, and typically 1 to 5 cm gap between anode and cathode. It promises applications in fast high power switches, in well pinched high intensity pulsed electron beams, as well as in point-like X-ray sources[1][2][3].

A comprehensive numerical simulation of the physics in pseudospark devices designed for such applications would have to take into account many primary effects on an atomic scale like elastic scattering, exitation, ionization, dissociation of molecules, attachment, detachment, charge exchange, and prompt and delayed emission of photons. In addition, secondary effects at the surfaces like the creation of electrons by photoeffect, or impinging slow and fast ions as well as fast neutral molecules would have to be considered, not forgetting backscattering. The motion of the charged particles would have to regard the externally applied electric and magnetic fields as well as those produced by the charges and currents in the later stages of the discharge. Also, the complex geometrical set-up together with the effect of surface layers on dielectric and metal materials play an important role in real pseudospark diodes.

From experiments it is known that the buildup of the pseudospark discharge is well separated into a slow, low current predischarge, and a very fast, high current main discharge. The former is governed by the motion of charged particles in the electric field together with the buildup of charge carriers mainly in the hollow cathode region. On the other hand, the latter is dominated by the combined action of electric and self-magnetic fields in the previously created plasma. The task to simulate such a complex phenomenon is beyond the possibilities of present-day computers. In this work we limit our model to the most essential features of the prebreakdown phase[4][5]. The geometry of the model diode can be seen in Fig. 1. It is a single gap diode with a hole at the center of the cathode allowing for a hollow cathode region at the outer side of the cathode.

For the parameter ranges of voltage and pressure at which pseudospark discharges typically are observed, the ratio of the electric field E to the background density N, E/N, generally is well above $5 \times 10^{-18} \, Vm^2$ in the gap region between anode and cathode. At these high E/N- values the charge

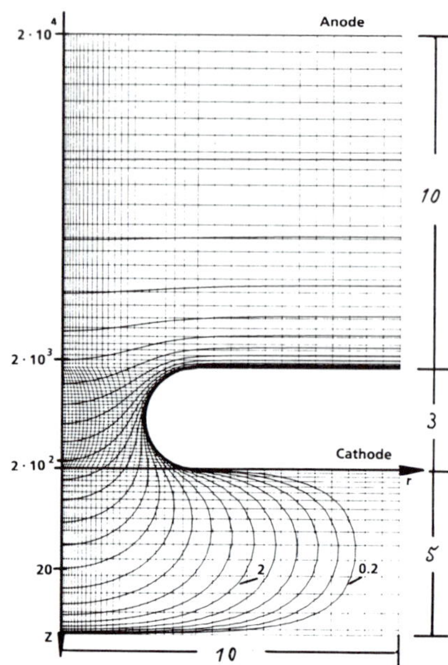

Fig. 1. Pseudospark geometry (mm), numerical mesh, and
equipotential lines (Volt)

carriers are continuously accelerated. Their velocity depends on the electric
field along their whole path. Since the cross sections for the various types
of collisions are strongly velocity dependent, a Monte Carlo type of simulation is appropriate for this region[5].

On the other hand, in the hollow cathode region the E/N-values typically
are below $5 \times 10^{-18} Vm^2$. For this parameter range the charge carriers lose
the energy, they pick up in the electric field, again in frequent elastic
and inelastic collosions with the neutral background gas molecules. As a
result, the charged particles drift along the electric field lines. Their
average drift velocity only depends on the value of the local electric field
at the position under consideration. For this equilibrium situation a fluid
model describing the drift motion is applied in the model presented below.

This model comprises the time dependent continuity equations for ions
and electrons, describing where the particles come from, where they drift
to, as well as the generation of new ones by ionization in the discharge
volume, and of secondaries e.g. due to photo effect at the boundaries. In
addition, the equations of motion are replaced by the relation between drift
velocity and electric field, obtained by analytical fits to experimental
data. (Thus also the Lorentz force equation is implicitely included in this
model, justifying to talk about attraction of electrons by a positive charge,
later on in the text). Also, the physical effects determining the source
terms in the continuity equations - such as ionization, exitation, secondary
electron emission due to photo effect or ion impact at the walls - are input
to the model as functions fitted to experimental data. Further, Poisson's
equation is solved in the whole interior of the model diode, taking the
change of the potential due to space charge into account.

This coupled set of differential equations is solved simultaneously,
and iteratively as a function of time in a two-dimensional, rotational symmetric r-z-geometry. A computer program written according to this swarm model

for an axially symmetric discharge in a plane parallel plate electrode geometry[6][7] founded the basis for our work.

FLUID MODEL FOR THE HOLLOW CATHODE REGION

The Basic Equations

The basic equations comprising this model[6] are given below for the case of only two fluids, namely electrons (e), and positive ions (p). The program can handle negative ions as well, and even more fluid components could be added, in principle. The equations are solved in a loop in the sequence they are written below, iterating for each time step until the most relevant quantities - potential ϕ and electron density ρ_e - differ not more than preset values from one iteration to the next. At the start of the simulation an initial electron and positive ion distribution has to be assumed.

Space charge:
$$\rho = \rho_p - \rho_e$$

Poissons equation:
$$\Delta\phi = \frac{\rho}{\epsilon_0}$$

Electric field:
$$\vec{E} = -\nabla\phi$$

Equations of motion:
$$\vec{w}_e = f_e(E,N)\vec{E} = \text{electron drift velocity}$$
$$\vec{w}_p = f_p(E,N)\vec{E} = \text{positive ion velocity}$$

Swarm paramters:
$$a = f_a(E,N) = \frac{d\rho_e}{\rho_e ds} \quad \text{due to ionization}$$
$$a_{ex} = f_{aex}(E,N)$$

Boundary Condition for secondary electrons:
$$\rho_e w_e = \nu_p \rho_p w_p + \Gamma_\phi \int_0^g \int_0^R g(z', r', r) a_{ex} \rho_e w_e r' dr' dz'$$

Continuity equations:
$$\frac{\partial \rho_e}{\partial t} = a\rho_e w_e - \nabla(\rho_e \vec{w}_e)$$
$$\frac{\partial \rho_p}{\partial t} = a\rho_e w_e - \nabla(\rho_p \vec{w}_p)$$

In these equations a is the first Townsend cefficient describing the relative increase of electron density due to ioniztion by electron impact on a given path length ds along the electric field lines. a_{ex} is similarly defined and yields the number of photons due to exitation by electron impact. ν_p is the secondary coefficient for electron generation by ions impinging on the surface. Γ_ϕ is a similar coefficient for photo effect. $a_{ex}\rho_e w_e$ gives the photon flux density at a position (z',r') in the hollow cathode region. $g(z',r',r)$ relates to the probability that a photon originating at that position reaches a point at the surface. The discharge volume extends to the radius R, and the axial extent of the hollow cathode is g. Additional equations, solved simultaneously as well, consider the decrease of the gap voltage due to discharging of an external capacitor by the discharge current in the gap.

The continuity equations are solved for the hollow cathode region only, using a non-equidistant, rectangular mesh of 38 x 17 lines in rotational symmetric in r-z-geometry (Fig. 1). The hybrid method of characteristics is used to solve these differential equations, choosing for each fluid a coordinate system which moves along with it[6]. For example, the continuity equation for the positive ions thus is transformed into the following integral equation, which follows the positive ions from the time t_1 at the location (z_1, r_1) to the time t_2 at the location (z_2, r_2):

$$\rho_p(t_2) = \rho_p(t_1) \, \exp\left(- \int_{t_1}^{t_2} \nabla \vec{w}_p \, dt \right) + \int_{t_1}^{t_2} a\rho_e w_e \, \exp\left(- \int_{t}^{t_2} \nabla \vec{w}_p(t') dt' \right) dt$$

The divergence terms describe deviations from the behavior of an incompressible fluid, e.g. the attraction or repulsion by the Coulomb force, or a non-linear dependence of the drift velocity on the electric field. $a\rho_e w_e dt$ gives the increase of ion density due to ionization.

Numerically the calculation of $\nabla \vec{w}_e$ and $\nabla \vec{w}_p$ - both functions of the electric field - requires special care to minimize errors. As the derivatives of the electric field components vary rapidly close to the hole, the problem was transformed by making use of Maxwell's equation

$$\nabla \vec{E} = \rho/\epsilon_0$$

and

$$\nabla \vec{w} = \nabla(f(E,N)\vec{E}) = \frac{df}{dE} \nabla E \cdot \vec{E} + f \frac{\rho}{\epsilon_0}$$

df/dE is calculated analytically from the given function f(E). The numerical calculation of ∇E remains. However, since the magnitude of $\vec{E}$, E, varies much more smoothly across the hollow cathode region than the electric field components, the accuracy has been improved considerably by this transformation.

In order to be economic both in computing time and storage the mesh was chosen finest close to the cathode hole, where the electric field and the ionization are largest. From here the mesh spacing gradually increases towards the outer radial and axial boundaries.

Poisson's equation is solved in all of the interior of the geometry shown in Fig. 1 by means of finite difference methods using boundary fitted coordinates[8]. Thus the electric field derived from the potential takes into account the distortion caused by the hole in the cathode, as well as the field changes due to space charge effects.

Further details about the numerical methods are given elsewhere[5][6].

The Boundary Condition at the Hole

A principle problem arises from the fact that the physics in the hollow cathode region and in the gap between anode and cathode is quite different. The E/N-values under consideration in this study are chosen such that in the hollow cathode region just no run-away conditions occur for the electrons, and the fluid model can be applied. As a consequence E/N is chosen to be about $4 \times 10^{-18} \, Vm^2$ on axis at the hole boundary. Inside the hole and in the diode gap E/N is much higher than this value, and the continuous acceleration of the electrons could be treated by two-dimensional Monte-Carlo methods. A major problem would be to couple these completely different simulation methods together at the hole boundary. At present this has not been achieved.

Physically these two regions are coupled by particle and photon fluxes crossing the boundary. For the treatment of the hollow cathode region the photon flux coming from the gap is neglected, and the positive ion flux is estimated from the following arguments.

The positive ion flux has to be continuous at the boundary. Then, assuming initially at the start of the simulation a uniform positive ion density in all of the hollow cathode region including on the hole boundary, this density had to remain constant in time, if no additional sources for positive ions were included, and space charge influence could still be neglected. The reason for this is that the positive ions move like an incompressible fluid, since their drift velocity taken from fits to experimental data is proportional to the electric field.

However, new positive ions are created by primary ionization by electrons. If these new-born ions are added to the original ones a density gradient builds up, because ionization is the larger the higher the electric field is. The positive ion density then increases from the hollow cathode walls towards the hole. This density gradient at the hole boundary relates to an additional ion flux entering the hole. In previous publications[4] [5] on this subject the positive ion flux at the hole boundary was estimated by extrapolating the ion density into the hole region, adding ions created by ionizing electrons leaving the hole as well. By this trick the simulation inside the hollow cathode region can be performed without any additional parameter depending on the physics in the diode gap. An obvious shortcoming of this approach is that once the density gradient at the hole boundary is established it will cause the positive ion flux to continue to increase in time due to a purely numerical feed back mechanism, even if the electron flux leaving the hole would be constant in time or die out. Thus, most probably this boundary model (1) overestimates the positive ion flux coming from the gap at the final stages of the simulation.

In model (2) this problem is avoided by neglecting that part of the incoming ion flux which is driven by the density gradient at the boundary. Instead it is assumed that at each time step the density inside the hole equals the density at the boundary. Otherwise model (2) is identical to model (1). Proceeding in this manner the continuity at the boundary is insured. However, the positive ion flux entering from the gap most probably is underestimated for a discharge which builds up in time, because the electric field increases towards the gap, and therefore also the ionization. From this follows that in reality more ions will be created in the hole than at the boundary. The real physical solution will lie between model (1) and model (2). In case of an electron density decreasing in time model (2) yields a positive ion density constant in time.

A third possibility to tackle this problem is to introduce a parameter which sums up the physics inside the diode gap (model (3)). Such a parameter can be based on the idea that the flux of positive ions entering the hollow cathode region through the hole certainly is related to the electron flux leaving the hole, for example, because the ionization by electrons is proportional to the electron flux. Such an assumption is supported by time dependent one-dimensional Monte-Carlo simulations for pre-breakdown of a discharge in a plane parallel plate geometry[9] which showed that the positive ion flux at the cathode is roughly an order of magnitude less then the electron flux at the anode. Exept for transients at the beginning of this simulation the ratio of these two fluxes turned out to be constant in time. An obvious disadvantage of model (3) is that it creates an unreal discontinuity at the boundary, because inside the hollow cathode region ions are created by ionization. Another shortcoming is a too fast response in ion density when the electron density is changing, neglecting that the ion drift velocity is two orders of magnitude smaller than the electron drift velocity.

These three models were compared for the following discharge parameters: gap voltage 70 kV, pressure 470 Pa (3.5 Torr), swarm parameters as given by[10], photon secondary coeffiecient 0.02, no secondary electrons due to ions. As a result the current density at the hole boundary is shown as function of radius and time for these three models in Fig. 2 to 4. It is seen that in all three cases the qualitative growth behavior is the same. After the input electron distribution has left the hollow cathode area the current density grows exponentially in time. The growth is identical for the different models as long as the positive space charge in the vicinity of the axis does not change the electric field noticable. Once this happens the current density grows overexponentially, and an extremely well focused electron beam is formed around the axis.

The time required for the onset of a pseudospark to develop is 102 ns for model (1), 133 ns for model (2), and 174 ns for model (3). Depending on the different incoming ion fluxes it takes more or less time for a critical positive space charge to be reached (note the different time scales in the figures).

For the parameters chosen model (3) definitely underestimates the ion flux, because the ion density created by ionization inside the hollow cathode turned out to be appreciably larger than that resulting from the incoming flux. As a consequence, for model (3) the ion density at a fixed time even decreases towards the hole in the vicinity of the boundary, which certainly is unreal in this region of the discharge. Of course this shortcoming could be overcome by enlarging the free parameter describing the ratio of ion to electron flux at the boundary until the discontinuity disappears. This approach leads to model (2), however.

It is noticed that models (1) and (2) give rather similar results. This is explained by the fact that the ion density is caused mainly by ionization in the hollow cathode and hole regions, and not by ions coming from the gap, as the slow ions on axis move only 1.5 mm from the hole boundary inside the hollow cathode during 100 ns. As long as the simulation takes into account properly the growth of ion density caused by ionization, additional incoming slow ion flux is of less relevance. These arguments lead to the choice of model (2) as a standard for further simulations, and give confidence that the results presented below are qualitatively correct, even though the discharge physics in the gap region is neglected.

Another problem at the hole boundary is related to the circumstance that the electric field is calculated in all of the discharge region, whereas the space charge term entering Poisson's equation is only known inside the hollow cathode region. In order to avoid numerical difficulties arising from a discontinuity in space charge density at this boundary, the space charge density is extrapolated into the hole region, decreasing exponentially in distance, typically a factor of 10^{12} across the disk. Changing this decay factor by orders of magnitude had only a minor effect on the results. This extrapolation is based on the fact that because the electric field is increasing towards the anode, also the particle velocities are increasing, hence there densities are decreasing, assuming a constant particle flux density in this argument.

<u>Accuracy Tests Performed on the Numerical Code</u>

An integral accuracy test of the electric field solver for the hollow cathode region can be based on the idea that for zero space charge the divergence of the electric field vanishes. Hence, making use of Gauss' theorem, the total electric flux entering the hollow cathode region must be equal to the total electric flux reaching the walls inside this region. A numerical integration of the two related surface integrals yielded that the flux entering the hole agrees with that reaching the walls to within 2%.

The next accuracy test was concerned with the time dependent motion of the electrons from the metal walls inside the hollow cathode region towards

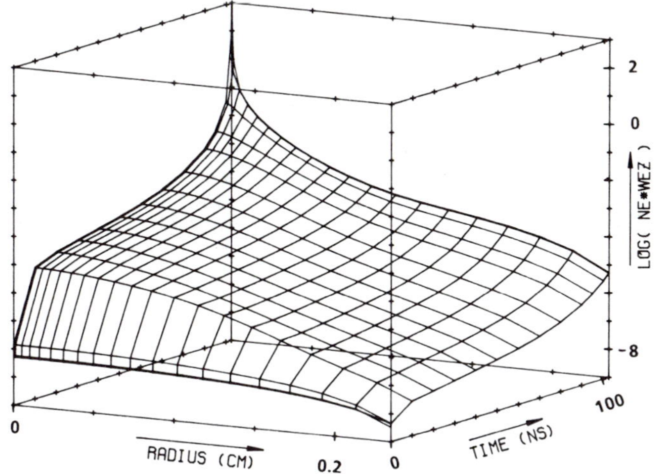

Fig. 2. Current density [Acm^2] at the hole, boundary model 1

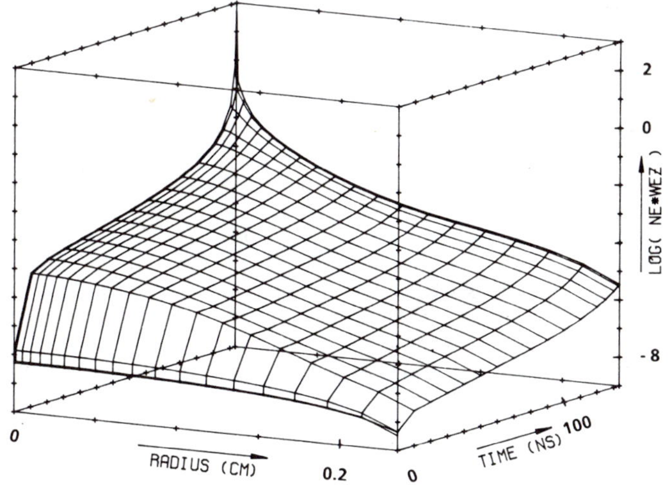

Fig. 3. Current density [Acm^2] at the hole, boundary model 2

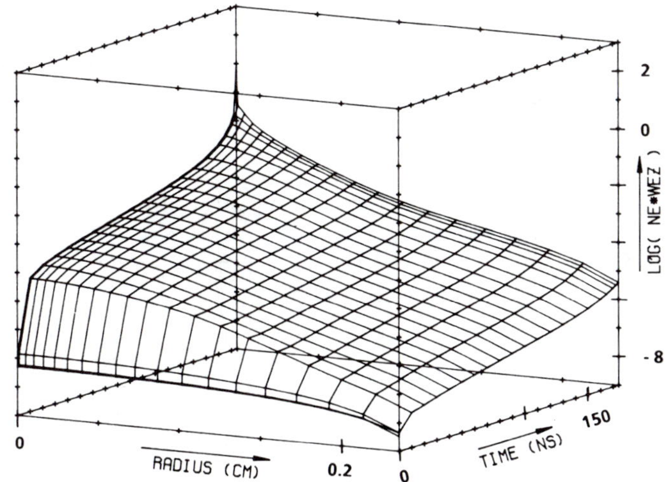

Fig. 4. Current density [Acm^2] at the hole, boundary model 3

the hole. It is well known that the method of characteristics tends to smear
out steep gradients, much more so than the more recently developed flux
corrected transport methods. In order to quantify this deficiency the following numerical experiment was set up.

Starting from a uniform electron density at the start of the simulation
the electron density at the wall opposite to the hole was increased instantaneously by a factor of 2 for radii less than 5 mm. The electron density
on the remainder walls was unchanged. This electron density distribution at
the walls was then kept constant in time, and the electron pulse leaving the
wall opposite to the hole was followed as it moved from the wall towards the
hole.

For this calculation the electric field was kept constant in time, hence
neglecting the space charge source term. Further the primary ionization by
electron collisions was set to zero, and the electron drift velocity was
approximated by a function exactly proportional to the electric field. From
these simplifications follows that the divergence of the electron drift
velocity is zero, the electrons move like an incompressible fluid, their
density remains unchanged as they drift from the walls towards the hole (the
increase in flux density reflects itself as an increase in the drift velocity
alone). This allows an easy way to calculate the exact time independent
solution of this test run.

On axis the time required for the electrons to traverse the hollow
cathode region can be calculated straight forward by numerical integration
to be 12.4 ns for a anode-cathode gap voltage of 20 kV and a background
nitrogen gas pressure of 133 Pa. At this time the front of the electron pulses
reaches the hollow cathode boundary on axis, whereas for larger radii this
front is still inside the hollow cathode region, because the drift velocity
is smaller off axis than on axis, also the drift path length is longer off
axis. The following plots are three-dimensional for the hollow cathode region
only. The origin of the r-z-coordinate system lies on the diode axis next
to the hole (see Fig. 1). In Fig. 5 the broadening of the pulse due to
deficiencies in the method of characteristics can clearly be seen. The pulse
maximum on axis at the hole is 22 % too small compared to the theoretical
solution.

The simulation was continued until a stationary solution was reached
(Fig. 6). On axis at the hole the numerical solution now agrees exactly with
the theoretical one, whereas off axis the pulse boadening is still appreciable. As is shown below, across the hole boundary the edge of the pulse
theoretically is at r=0.88 mm, where the pulse heigth is 28% too small. The
electron flux leaving the hole for radii less than r=0.88 mm is 19% less than
the corresponding flux leaving the opposite wall for radii less than 5 mm.
The electron flux up to a radius of 0.97 mm across the hole has to be included
in order that these source and sink fluxes are identical. This corresponds
to a pulse broadening in radius at the hole of 10%, and in area of 22%.
However, the total electron flux leaving the hole agrees to 0.6% with the
total flux leaving the metal walls inside the hollow cathode region.

In a similar manner the numerical solution was tested with respect to
the positive ion movement. For this case a step pulse of ions was entered
the hollow cathode region at the hole boundary for radii less than 0.88 mm.
The edge of this pulse should reach the opposing wall at a radius of 5 mm.
Fig. 7 shows the result after 1 ion passage on axis (1380 ns), when the pulse
is still 27 % to small on axis. Fig. 8 gives the answer at 2.9 ion passages
on axis (4000 ns). It is seen that instead of reaching the opposite wall some
ions are moving backwards towards the cathode back wall. They have diffused
out of the pulse to larger radii into comparitively higher electric field,
where they move faster than ions at the pulse edge. The pulse was not followed

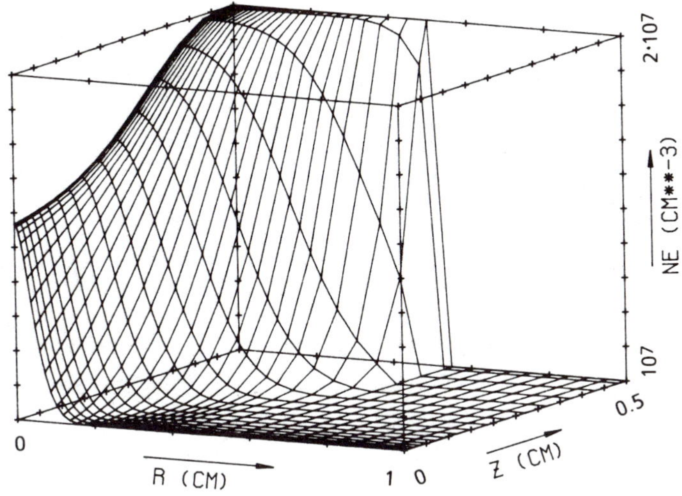

Fig. 5. Electron density after 1.0 electron passages on axis

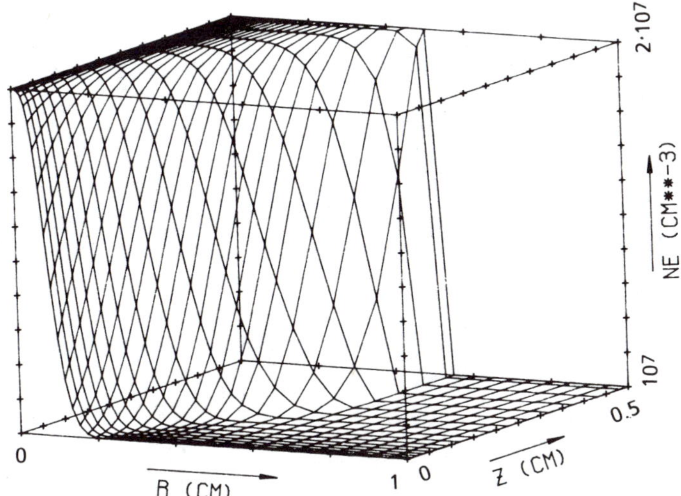

Fig. 6. Electron density, time independent solution

for longer times because of the exessive computation time required, and because the pseudospark behaviour develops much more rapidly.

Summing up this analysis, it is concluded that steep density gradients tend to be smeared out when the electrons or ions are traversing the hollow cathode region, the density at a sharp ridge is reduced typically by 20 to 30% compared to the theoretical solution. The flux diffusing out of the pulse amounts to about 20%, but the total flux is conserved perfectly to better than 1%. For the purpose of yielding a qualitative understanding of the prebreakdown phenomena in a pseudospark the accuracy is adequate, however. This the more so, because in the simulations to be reported an extremely localized high density plasma builds up in the vicinity of the hole, and any numerical errors related to the method of characteristics would just work in the opposite direction to smear out this localized density.

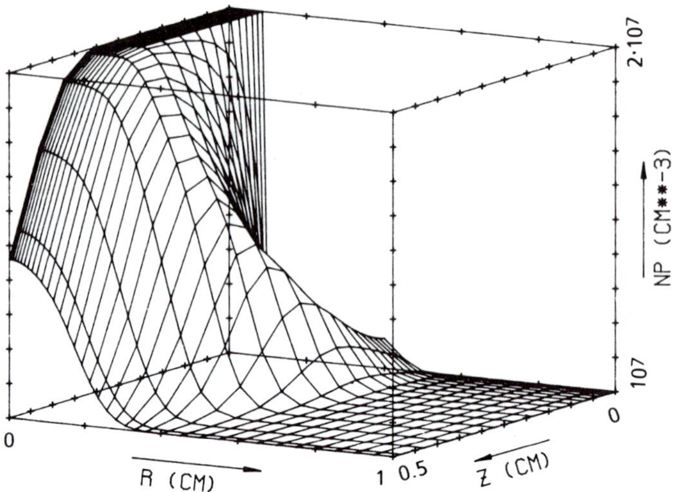

Fig. 7. Ion density after 1.0 ion passages on axis

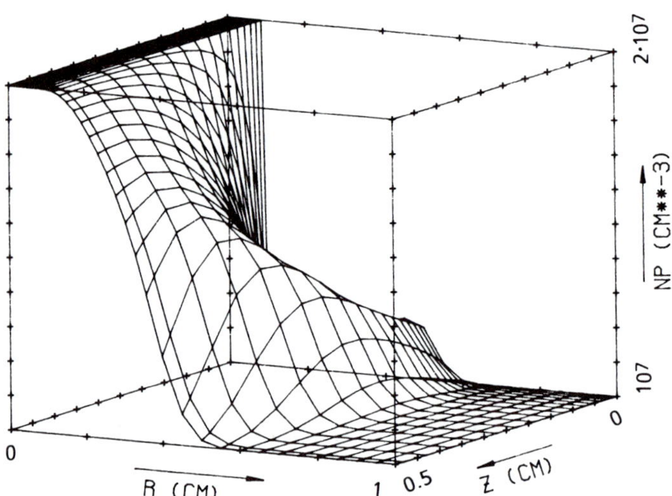

Fig. 8. Ion density after 2.9 ion passages on axis

A further test was to check on the exponential growth in electron density resulting from primary ionization when the electrons pass through the hollow cathode region. This can easily be checked only on axis, because there the electron motion is one-dimensional. For this numerical experiment the parameters were left as above, only turning on the primary ionization in addition. Evaluating the integral over the primary ionization coefficient along the axis, a growth factor of 5.98 is obtained. The method of characteristics yields a growth factor of 5.43, or 9% to small at 12.4 ns, when the pulse has just reached the hole. A growth factor of 7.00, or 17% too high is obtained for the stationary solution. The reason for this overestimate in growth lies in the four-point linear two-dimensional interpolation which is used in order to estimate the ionization coefficient between mesh points. A higher order interpolation formula would reduce this error, which seems to be tolerable for a qualitative analysis, however.

The scaling of the discharge behavior with voltage and pressure is another point of interest. The primary coefficients for ionization and photon exitation scale proportional to $N \cdot f(E/N)$. The drift velocities are proportional to E/N. Thus reducing E and N by the same factor while increasing these primary coefficients by the same factor should leave the discharge buildup unchanged. This was verified by comparing a run at 70 kV, 470 Pa with one at 20 kV and 133 Pa.

Another test of the computer code was done by comparing the results obtained after 100 time steps of 0.1 ns each with those obtained after 1000 time steps of 0.01 ns each. The densities agreed within 1 % for these two cases.

Also, it was verified that chosing a factor of 2 denser mesh did not change the results to more than 10% in the significant quantities.

RESULTS OF THE SIMULATIONS

<u>The Discharge Parameters</u>

The parameters of the simulations to be described are close to the experimental conditions under which pseudosparks are observed:

outer radius of the discharge chamber	= 10.0	mm
gap between anode and cathode	= 10.0	mm
disk thickness	= 3.0	mm
radius of the hole in the cathode	= 2.5	mm
length of hollow cathode region	= 5.0	mm
voltage between anode and cathode	= 20	kV
background gas pressure	= 133	Pa (1 Torr)
primary coefficients[10]		
secondary coefficient for photo effect	= 0.02	
secondary coefficient for ion impact	= 0	

The geometry, the numerical mesh chosen (in radial direction 38 lines, in axial direction 17 lines in the hollow cathode, 29 lines in the disk, and 27 lines in the gap) , and equipotential lines are shown in Fig. 1.

<u>The Dependence of the Current Density Growth Rate on the Primary Ionization Coefficient</u>

The current density on axis on the hollow cathode boundary is presented in Fig. 9. The parameters varied are the primary coefficients for ionization and photo excitation, which are multiplied by the constant factor ALC3 as compared to the fits of these coefficients to experimental data[10].

It is seen that if the primary coefficients are taken to be equal to the experimental data (ALC3 = 1) no pseudospark should occur according to the simulation , since the discharge dies out, the current density falls exponentially in time. This contradicts experimental results which show the occurence of the pseudospark at even lower pressure, corresponding to even smaller primary coefficients. The conclusion is that not all the effects relevant for ionization growth have been included in the model as yet.

Especially the influence of fast ions entering from the gap region and producing secondary electrons at the hollow cathode walls probably has to

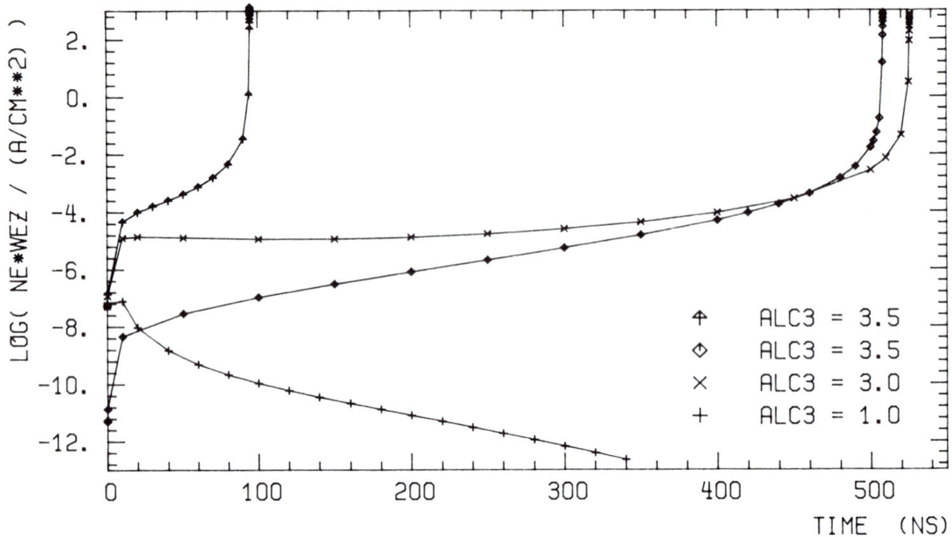

Fig. 9. Current density on axis on hollow cathode boundary

be taken into account. In order to get an idea how large the additional fast ion flux has to be to contribute significantly to to buildup of the discharge, the following model was tested. As a free parameter it was assumed that for every electron leaving the hole 0.1 electrons are created instantaneously by fast ion impact on the wall opposite the hole inside the hollow cathode at the same radial position. Thus the ion transit time and their radial movement were neglected. The magnitude of this free parameter can be justified by the results of the Monte Carlo simulations quoted above[9], and by a secondary coefficient for ion impact of 1, which is the typical order of magnitude for 20 kV nitrogen ions. Under these conditions a buildup of the discharge already was found for ALC3 = 1, and it developed similarly as in the simulations without fast ions.

Further, the ionization by the collisions of electrons with metastable molecules might be of importance in this context[10]. On the other hand, the contribution of slow ions produced by ionization in the discharge volume to the production of secondary electrons on the hollow cathode walls by ion impact will be small. The reason for this is the low ion velocity. During the predischarge time of typically several 100 ns the slow ions scarcely move, since it takes the fastest slow ions 1380 ns to transverse the hollow cathode region on axis, and most slow ions are born in the vicinity of the hole. So most slow ions will not have reached the walls before onset of the pseudospark.

As a first step to account for these other effects the primary ionization coefficients were multiplied by the free parameter ALC3 = 3.0 and 3.5 respectively. A nearly stable discharge is obtained in the first case, whereas a faster growth is seen in the latter one.

The effect of the magnitude of the initial ionization can be studied by comparing the two results given in Fig. 9 for ALC3 = 3.5. The initial density, uniform throughout the hollow cathode region, differs by a factor of 10^4 between these cases. After about 10 ns, which corresponds to the time which the electrons close to the axis need to pass the hollow cathode, the arbitrary initial electron distribution has been transformed into the physical solution. Thereafter the current density growths purely exponentially - as is predicted by the analytical solution of the electron continuity equation -, as long as space charge does not enhance the electric field appreciably. Once this occurs the current density growths overexponentially in time. For the case of the higher initial density this happens much earlier. Performing a transformation of the time coordinate on this curve it is seen that the overexponential growth behavior is identical for these two cases. Thus it is proven that the results are independent of the magnitude of the initial uniform distribution, which effects only the time required for the space charge to build up, and hence for the overexponential growth (pseudospark) to start.

<u>The Buildup of Charges in the Hollow Cathode Region</u>

For the parameters chosen the electrons and slow ions are in equilibrium with the electric field, they move along the field lines, and their local velocity is determined by the local electric field only, which is shown in Fig. 10 to 12 for zero space charge at the start (time T=0), and, for comparison, in Fig. 13 to 15 at the end of the simulation. The electric field magnitude grows from the walls towards the hole. A similar functional dependence holds for all field dependent functions, like drift velocities and primary ionization coefficient, therefore. The electric field extracts all electrons which are generated inside the hollow cathode through the hole into the gap region. Note that the axial electric field component changes sign inside the hollow cathode region.

On their way from the walls towards the hole the electrons are focused by the radial electric field to a considerable extent towards the axis. For zero space charge, and for the case that the electron drift velocity is exactly proportional to the electric field the motion of the electrons from the walls towards the hole can be traced by applying Gauss' theorem. As an example, it turns out that electrons leaving the wall opposite to the disk inside the hollow cathode region at a radius of 5 mm are leaving the hole at a radius of 0.88 mm. On axis the electron current density increases from the wall towards the hole boundary by a factor of 14 due to these purely geometrical effects alone. This increase manifests itself nearly completely as an increase in electron velocity. The electron density remains nearly constant, as the functional relation between drift velocity and electric field is close to linear, and hence the electrons - like the ions - nearly move as an incompressible fluid.

The electrons coming from the cylindric part of the wall move maximum 0.7 mm in 100 ns in radial direction. Hence they play a negligible role in the hollow cathode predischarge. Electrons which are leaving the walls for radii less than 5 mm need less than 160 ns to reach the hole, hence the secondary emission from the walls opposite the hole and from the backside of the cathode are the most important ones with respect to the buildup of the discharge. On axis at the hole the electrons have a velocity of 3.2 mm/ns, that of the ions is 2.8 mm/100ns; on axis at the hollow cathode wall these velocities are 2.2 mm/10ns and 1.7 mm/1000ns, respectively.

The buildup of electron and positive ion charges inside the hollow cathode is shown in Fig. 16 to 27 at various time steps for the case with ALC3 = 3.5 and an initial uniform density of 0.1 particle/cm^3 at the start of the simulation time T = 0. Contributions by ion impact on the walls were not taken into account. 100 ns after the start of the simulation (Fig. 16, 17) the arbitrary initial filling has transformed into the physical solution for the electrons. In comparison, the initial positive ions scarcely move, especially so far away from the axis. Their density gets enhanced by ions newly born by ionization. This ion density increase is least at the cylindric wall where the electric field is smallest, and gets larger towards the axis and towards the hole, having its maximum on axis at the hole. Next to the

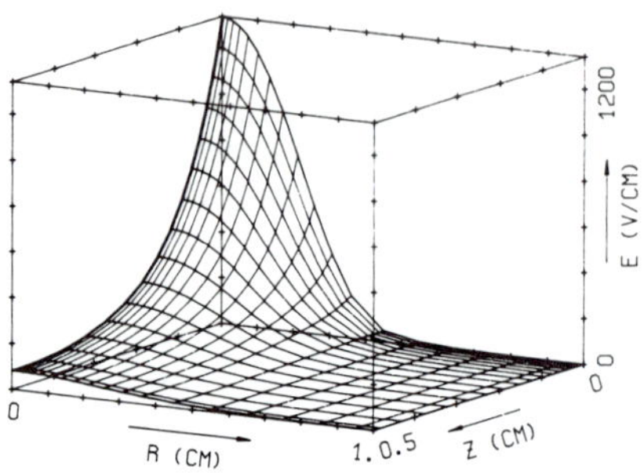

Fig. 10. Electric field at T=0

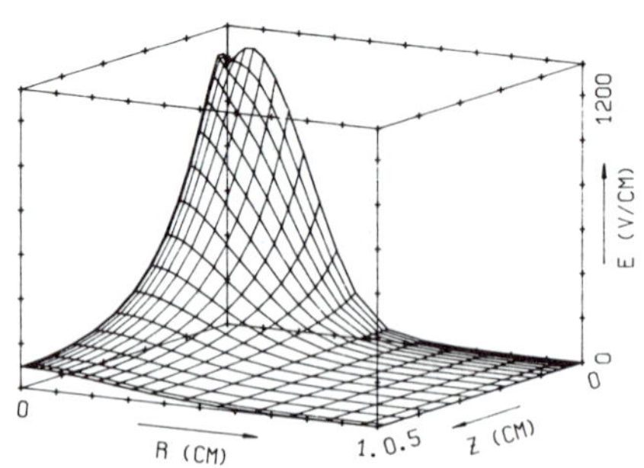

Fig. 13. Electric field at T=508 ns

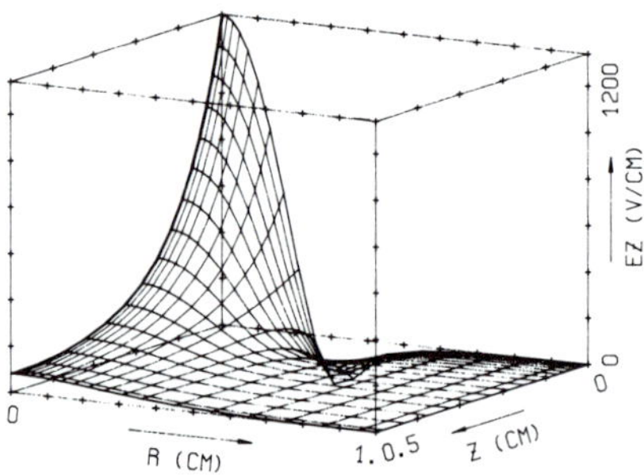

Fig. 11. Axial electric field at T=0

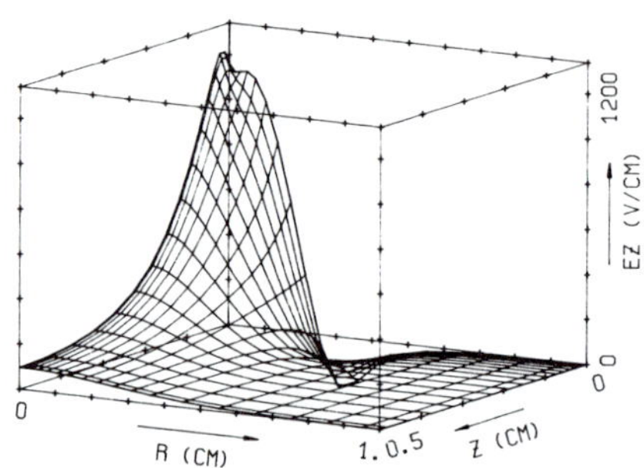

Fig. 14. Axial field at T=508 ns

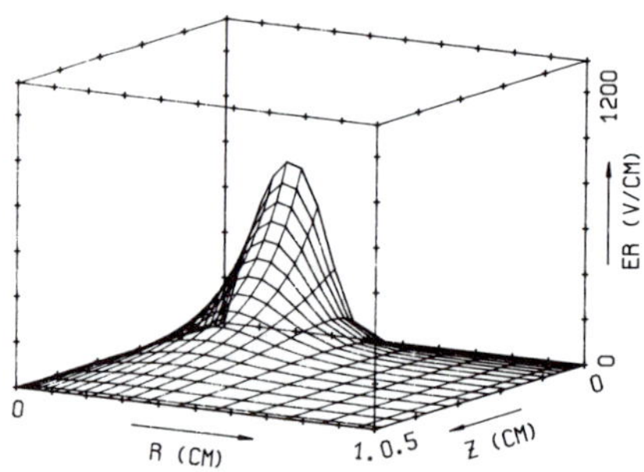

Fig. 12. Radial electric field at T=0

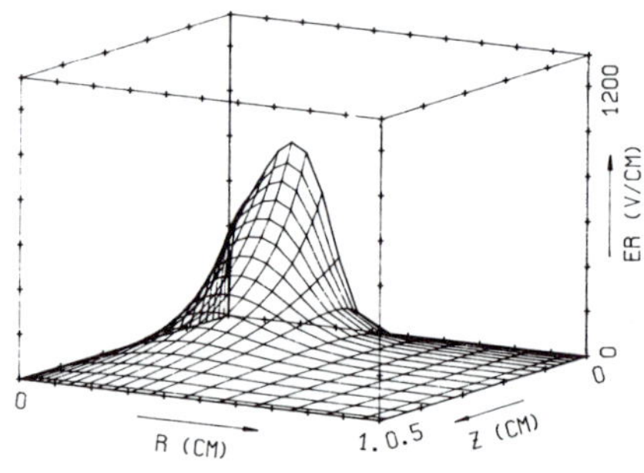

Fig. 15. Radial field at T=508 ns

cylindric wall both electrons and ions get more or less stuck during all of the simulation time, because of the extremely small drift velocities in this region.

From Fig. 9 it is seen that at 400 ns the current density at the hole still grows purely exponentially in time, since the space charge does not enhance the electric field as yet. At this time, the positive ion density at the hole is about three orders of magnitude larger than the electron density, because the electrons are drawn out of the hole rapidly, leaving the positive ions nearly stationary behind (Fig. 18, 19).

At later times (Fig. 20 to 27) the density peak appearing extremely rapidly during a few ns in the vicinity of the axis close to the hole is the most surprising feature of these simulations. Electron, positive ion and space charge density all grow continously as a function of time in the whole

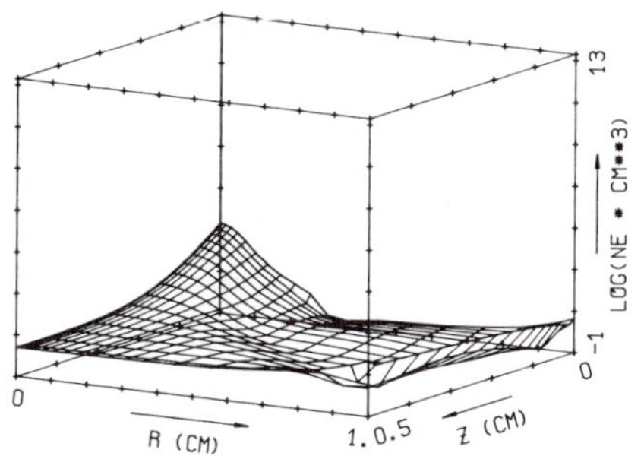
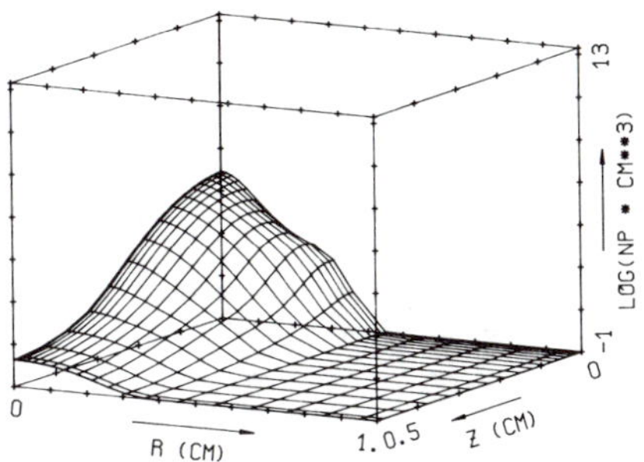

Fig. 16. Electron density at T=100 ns   Fig. 17. Ion density at T=100 ns

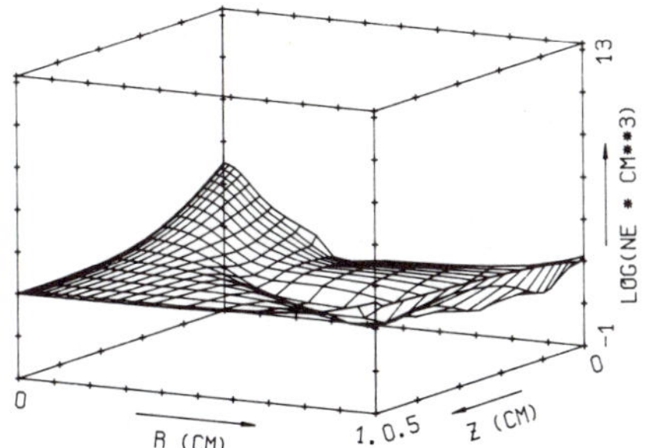
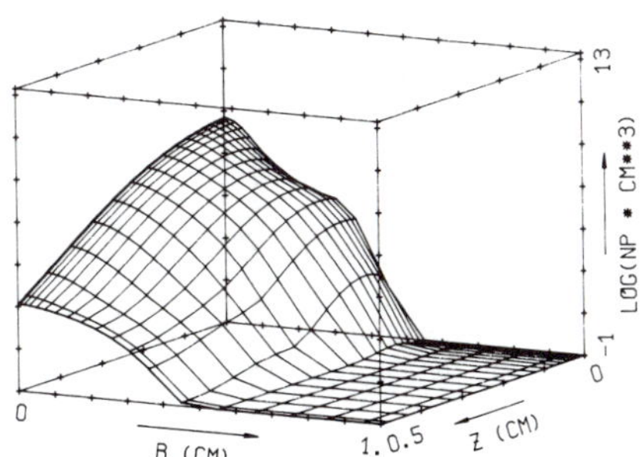

Fig. 18. Electron density at T=400 ns   Fig. 19. Ion density at T=400 ns

hollow cathode region during all phases of the simulation (apart from the very early stages mentioned above).

The effect of the positive space charge on the electric field at the end of the simulation is shown in Fig. 13 to 15, and for a magnified section at the hole it is given in Fig. 28 to 31. The radial electric field is increased drastically by up to a factor of 14 in the vicinity of the point-like positive charge 0.01 mm off axis, amounting already to 30% of its maximum value. At the radial electric field maximum, 1.9 mm off axis, this increase remains a less noticable 8%. A substantial increase in radial focussing of the electrons close to the axis at the hole is the consequence. Of course the electrons still follow the field lines, which have changed due to the space charge action.

The total field is reduced close to the hole and enhanced further inside the hollow cathode region as compared to the zero space charge field at T=0, as a result of the cumulative action of all the positive charges in the hollow cathode region. This field change amounts to a reduction of 14 % on axis at the hollow cathode boundary, and to a maximum increase of 30% in the interior. Its effect on the ionization a is less at the hole, where E/N is very high, but larger in the interior of the hollow cathode - resulting at the maximum to a factor of 2 increase at the hollow cathode wall opposite the hole -, because a varies with E as exp(-const/E).

The densities increase exponentially in time as long as the expression determining the growth due to ionization alone,

$$growth\ factor\ =\ \int a w_e\, dt\ =\ \int a\, ds$$

taken over the electron path, remains constant, and space charge influence

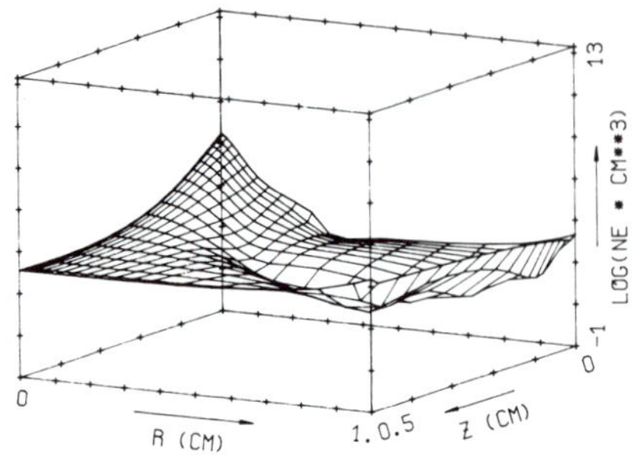

Fig. 20. Electron density at T=480 ns

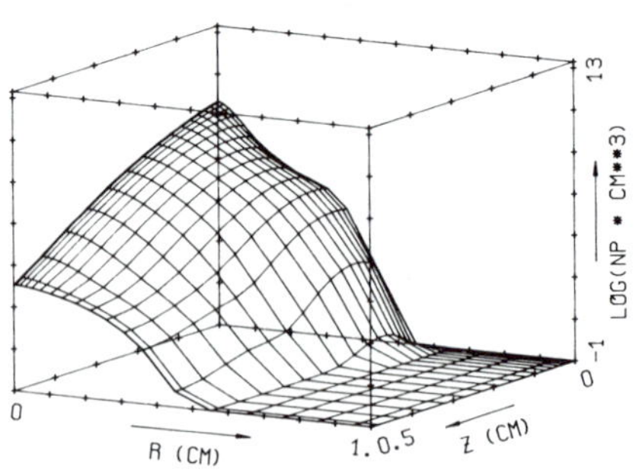

Fig. 21. Ion density at T=480 ns

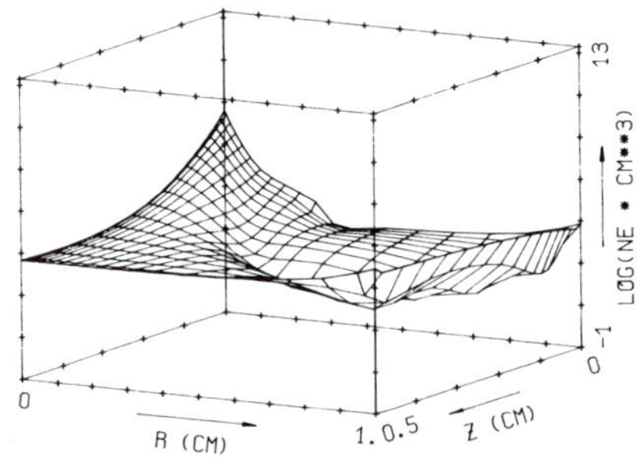

Fig. 22. Electron density at T=500 ns

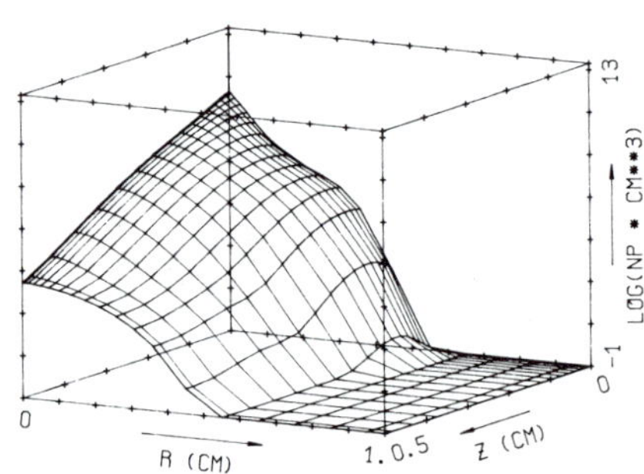

Fig. 23. Ion density at T=500 ns

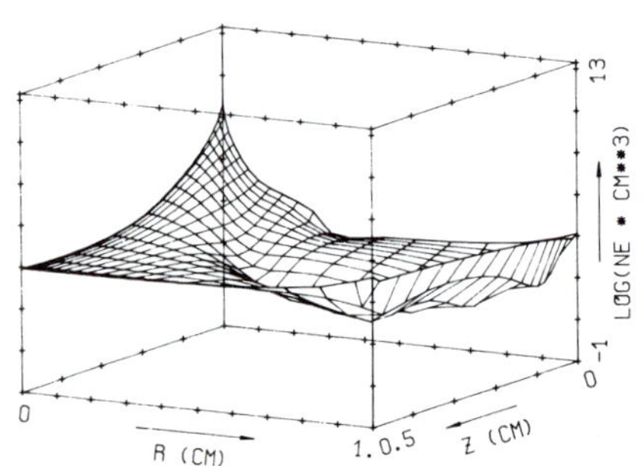

Fig. 24. Electron density at T=506 ns

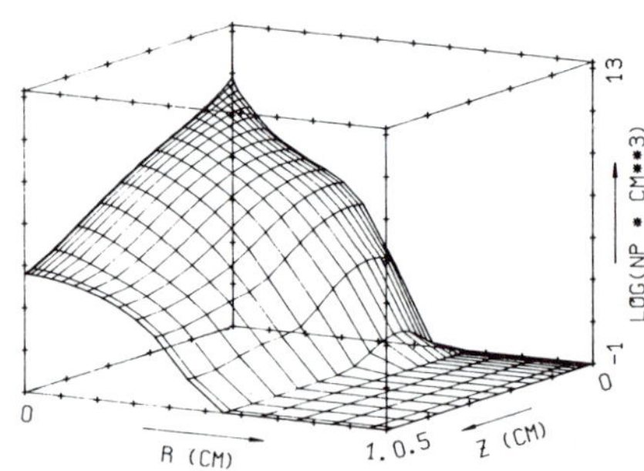

Fig. 25. Ion density at T=506 ns

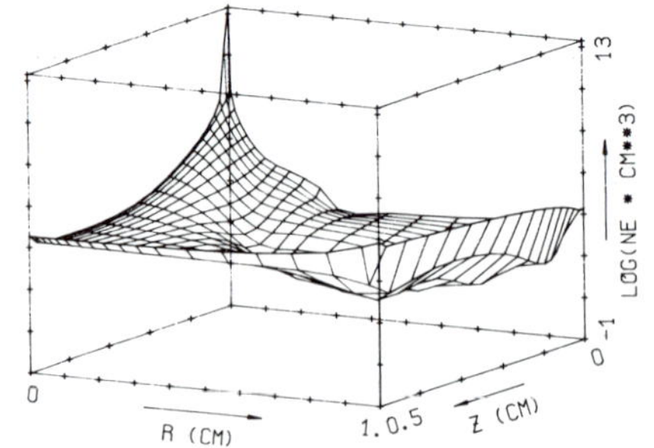

Fig. 26. Electron density at T=508 ns

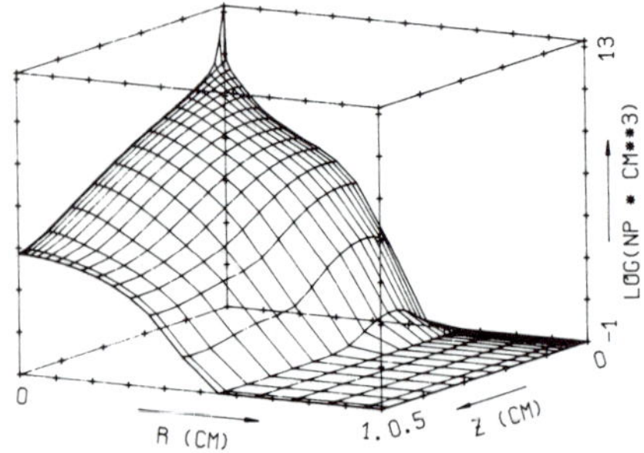

Fig. 27. Ion density at T=508 ns

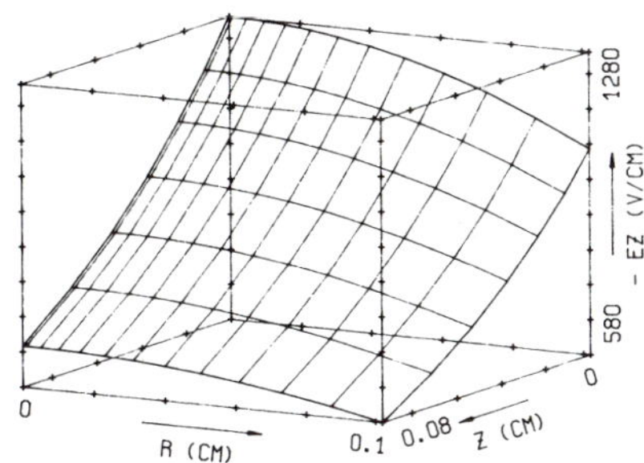

Fig. 28. Axial electric field at T=0

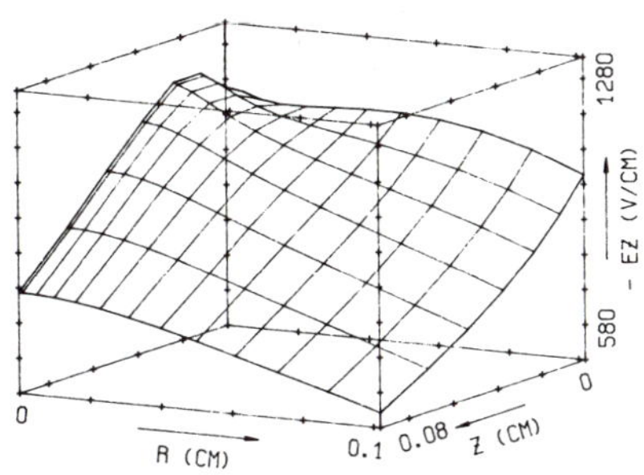

Fig. 30. Axial field at T=508 ns

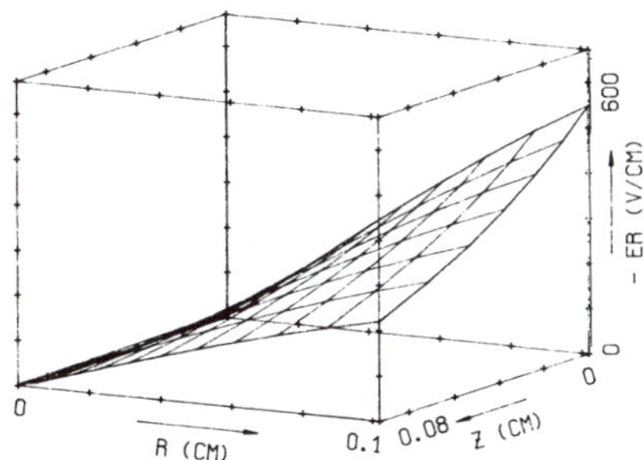

Fig. 29. Radial electric field at T=0

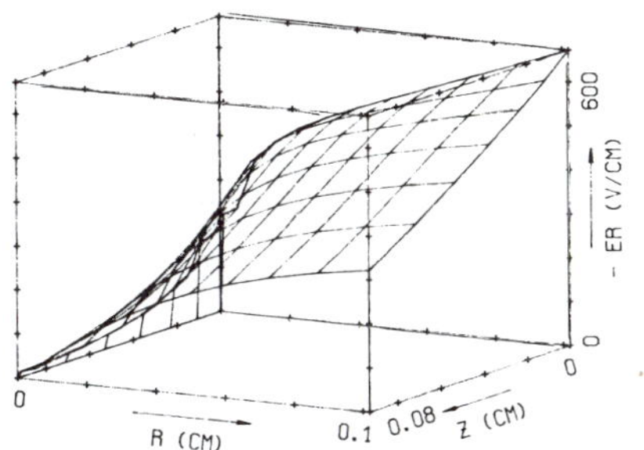

Fig. 31. Radial field at T=508 ns

can still be neglected. Once this happens, the growth factor increases in time, and the growth becomes overexponential. On axis, from the wall towards the hole the electron density growth caused by ionization amounts to a factor of 524 at T=0, but to 1750 at the late state of the simulation. The product a times electron drift velocity w_e which determines the relative change in electron density due to ionization in a given time interval is shown in Fig. 32 to 34, plotted over a magnified r-z-region close to the hole. It is seen that this function varies smoothly in both r and z, also at the final state of the simulation. Also the radial dependence of the growth factor is rather smooth, both with and without space charge. On axis it is only 1% larger than 0.1 mm off axis.

On the other hand, the electron densities differ by several orders of magnitude across this small radial distance at the end of the simulation. Analyzing the numerical results, it is found that the reason for the occurrence of the singularity on axis at the hollow cathode boundary is the action of the positive space charge, changing the electric field and thereby attracting electrons rapidly. (Note that the statements that the electrons are attracted by the positive charge or that they follow the field lines changed by this charge are both valid, since charge and field are equivalent in Maxwell's equations). During the last phase of the simulation the growth in electron density due to this effect is much larger than that caused by ionization, which can be seen by comparing Fig. 34 and 35. The drift term $\vec{\nabla} w_e$ is over a factor of 10 larger than the ionization term $a w_e$ at this location. The combined effect of ionization and drift caused by space charge attraction is shown in Fig. 36. The ionization term dominates inside the volume, whereas the divergence term produces the singularity on axis at the hole.

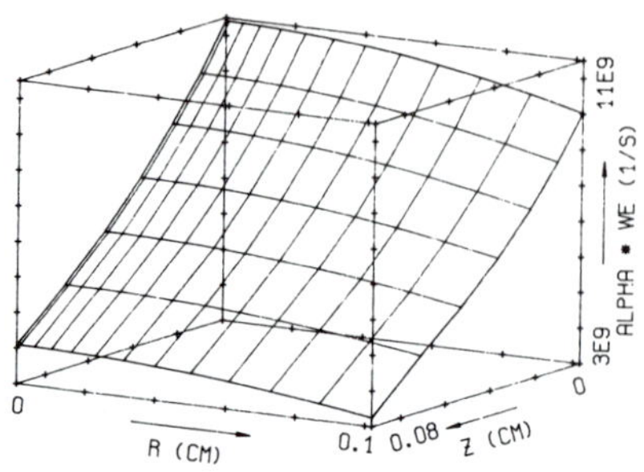

Fig. 32. Ionization coefficient times
electron drift velocity at T=0

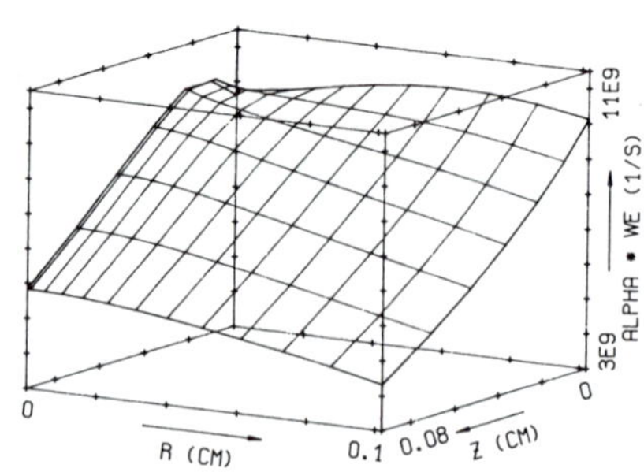

Fig. 33. Ionization coefficient
times velocity, T=508 ns

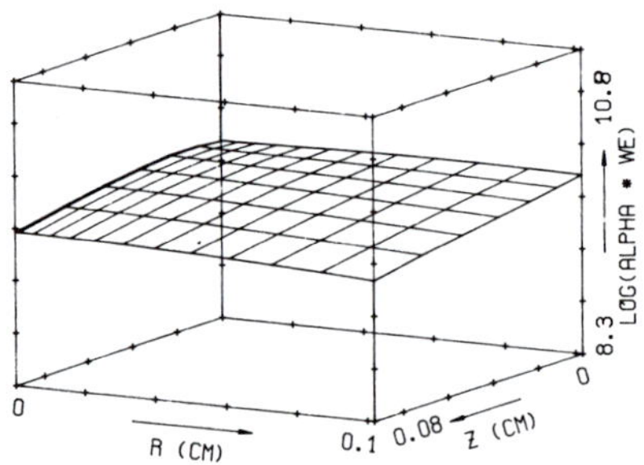

Fig. 34. $\log(\alpha w_e)$ at T=508 ns

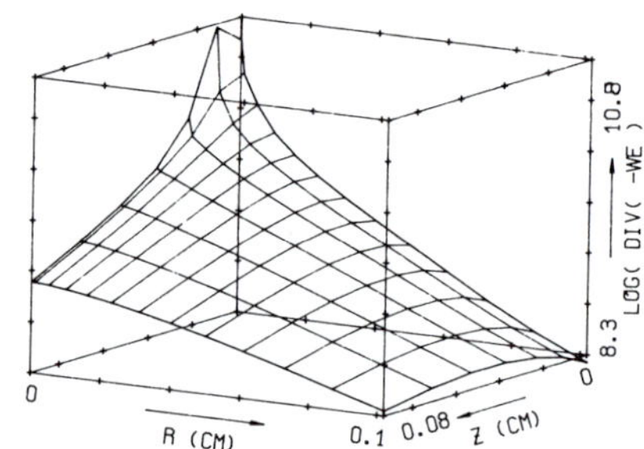

Fig. 35. $\log(- \nabla \vec{w}_e)$ at T=508 ns

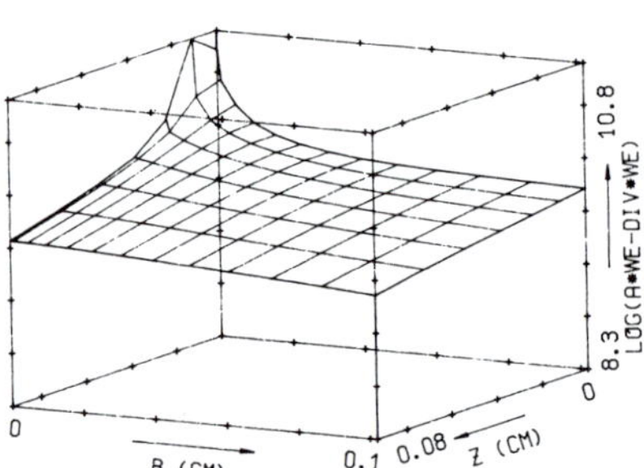

Fig. 36. $\log (\alpha w_e - \nabla \vec{w}_e)$ at T = 508 ns

Summarizing, the appearance of the singularity has its origin in the
fact that the electric field, the ionization coefficient and the drift
velocities all have their maximum on axis at the hole. Hence the rate at which
the charge carriers are multiplied is maximum here, too, and their exponential growth on axis is larger than off axis, which can be verified by analyzing Fig. 2 to 4. Consequently, the maxima of electron, ion and space
charge density are positioned also here. The space charge changes the electric field here most, and its attractive action on the electrons becomes here
noticable first, and also has its maximum here. Also the number of electrons
attracted by the space charge is largest here, since the electron density
maximum is located here, too. These attracted electrons also ionize, and
increase the positive space charge at the maximum even further, more than
elsewhere. The fact that the positive space charge itself was created by

electrons passing by and ionizing at previous times, causes a positive feedback mechanism, and an overexponential growth in electron density is the consequence.

This argument can be made more sound by an approximative solution of the electron continuity equation as follows. If in first approximation a linear relation between electron drift velocity and electric field is assumed, then

$$\nabla \vec{w}_e \approx const\ \nabla \vec{E} = const\ \rho = const\ (\rho_p - \rho_e)$$

$$\rho_e(t_2) = \rho_e(t_1)\ \exp\left(\int_{t_1}^{t_2} (a w_e - \nabla \vec{w}_e)\, dt \right)$$

$$\approx \rho_e(t_1)\ \exp\left(\int_{t_1}^{t_2} a w_e\, dt \right)\ \exp\left(\int_{t_1}^{t_2} const\ (\rho_p - \rho_e)\, dt \right)$$

(The integral is taken along a trajectory moving along with the electron fluid. ρ_e is the electron density at the start of the trajectory at time t_1). If the motion of the ions is neglected, the ion density at a given time t after the start of the discharge can be approximated by

$$\rho_p \approx \int_0^t a_e \rho_e w_e\, dt'$$

and it results

$$\rho_e(t_2) \propto \exp\left(const \int_{t_1}^{t_2} \int_0^t (a_e \rho_e w_e\, dt' - \rho_e)\, dt \right)$$

At the time when space charge starts to change the behavior of the discharge (Fig. 20 to 25) the positive charge is orders of magnitude larger than the negative one, in the vicinity of the hole close to the axis. It follows that the growth in electron density in this region depends in a complicated way exponentially on the time increasing electron density itself, which causes the overexponential growth. This happens first at the location where ρ_p has its maximum.

The positive space charge is not fully neutralized by the attracted electrons, because the latter are rapidly extracted out of the hollow cathode region. Because of their over a factor of 100 smaller mobility the ions are left behind, staying where they were created during the extremely rapid predischarge. At the time when the simulation was stopped, the positive ion density was $1.830 \times 10^{13}\ cm^{-3}$, and the electron density was $1.808 \times 10^{13}\ cm^{-3}$ on axis at the hole. It follows that the positive charge is compensated by the electrons to with about 1 %.

The electron drift velocity at this location corresponds to an electron energy of 22 eV. Setting this energy - for the sake of curiosity - equal to an electron temperature, the Debye length (which gives the distance over which a charge or electric field is shielded in a plasma) turns out to be 0.008 mm. In the code the smallest mesh size chosen at this point was 0.01mm in radial, and 0.12 mm in axial direction. It is surprising - and may be accidental - that the space charge shielding calculated by the code, and that calculated by a completely different method, based on the Debye length, turn out to be close together in value. The reason for this might be found in the fact that the electron drift velocity determines the speed at which the positive charge can be neutralized by the electrons, and that the thermal velocity plays a similar role in a plasma.

A temperature is neglected in our model - but could be introduced via a diffusion term, in principle. It is reminded that the numerical diffusion inherent in the method of characteristics plays this role in the simulation in a non-controllable manner.

CONCLUSIONS

The multi-fluid model to numerically simulate the prebreakdown of the pseudospark in the hollow cathode region describes the physics qualitatively correct. The method of characteristics chosen to solve the continuity equations tends to a pulse broadening of typically 20% for a single electron or ion passage through the hollow cathode region.

The model to estimate the flux of slow positive ions entering the hollow cathode region from the main discharge gap is based on extrapolating the positive ion density from the hollow cathode towards the hole. This model is adequate for a qualitative understanding of the underlying physics, since the buildup of a positive space charge in the hollow cathode is mainly due to ionization by electron impact and not by an incoming positive ion flux from the hole region.

Thus the occurence of a well focused high intensity electron beam on axis originating in the hollow cathode at the hole - interpreted as the onset of the pseudospark - certainly is not related to numerical errors or instabilities but corresponds to similar observations in real experiments[11][12].

The physical picture of the prebreakdown phase of the pseudospark based on the results of the numerical simulations is summarized as follows:

Due to the special hollow cathode geometry with a hole in the cathode the electric field is extremely inhomogenous in the hollow cathode region. The field increases from the walls towards the hole with a radial maximum on axis. A similar space dependence follows for all field dependent quantities, like electron and ion drift velocities, as well as primary ionization and photon emission due to collisions of electrons with the background gas molecules. The electrons are focused to a considerable extent towards the axis by the radial electric field.

As long as the space charge does not change the electric field the initial charge densities grow or fall exponentially in time, depending on whether the electrons being extracted out of the hollow cathode through the hole towards the anode can be replenished by electrons newly created in primary or secondary collisions, like ionization by electrons and photo effect at the walls. This exponential growth follows directly from the electron continuity equation, because the change of electron density in time is proportional to the electron density itself. During these initial stages of the discharge all quantities vary smoothly both in axial and radial directions. As a consequence of the electric field maximum on axis the electron, the ion, and the space charge densities also have a maximum on axis, and increase towards the hole.

Since the electrons move over a factor of 100 faster than the ions, they are extracted out of the hollow cathode region rapidly. The positive ions are left nearly stationary behind. Their density grows predominantly due to ions newly born by ionization, and at first gets several orders of magnitude larger than the electron density in the course of the predischarge. At some time this positive space charge starts to shield off the electric field entering through the hole from the anode. Because of this, the electric field is decreased close to the hole and is increased in the interior of the hollow cathode region as compared to the zero space charge case. The influence of the space charge on the spatial dependence of the primary coefficients and the drift velocities is similar, as these quantities all increase with the electric field, all of them still vary smoothly in space. The growth factor for the electrons due to ionization increases because of this change in the electric field, and consequently the particle densities grow overexponentially in time.

The most important effect of the positive space charge in changing the electric field is the attraction of electrons by the Coulomb force, however. First, this causes an even stronger focussing of electrons towards the axis. They are extracted through the hole, however, so that they do not succeed in fully neutralize the positive charge.

Second, the pseudospark singularity with an overexponential density growth in time on axis at the hole is triggered by the following positive physical feedback mechanism. The relative rate at which the electron density increases close to the hole is mainly determined by the attraction of electrons due to space charge, which is largest on axis at the hole where the positive space charge has its maximum. Also the number of attracted electrons is largest here, since the electron density maximum is located here, too. These attracted electrons also ionize, and increase the positive space charge at the maximum even further, more than elsewhere. As this positive space charge itself was created by electrons passing by and ionizing at previous times, an overexponential growth in electron density, and consequently also in ion density, occurs.

It results an extremely well focused high intensity, but still low current electron beam on axis - interpreted as the onset of the pseudospark - which is extracted out of the hollow cathode region towards the anode.

These predictions agree qualitatively with corresponding experimental results obtained with a streak camera[11], or with more sophisticated optical spectroscopy[12]. In order to achieve a more quantitative agreement between simulations and experiments the contribution of fast ions entering the hollow cathode from the anode - cathode region, as well as the ionization of metastable background molecules[10] probably have to be incorporated into the numerical model.

ACKNOWLEDGEMENT

Stimulating discussions with W. Bauer, P.Choi, A. J. Davies, W. Hartmann, Y. Kaufman, W. Niessen, and C. Schultheiss were valuable and encouraging for this work, which was initiated by W. Schmidt. Especially the constructive criticism by P. Choi is appreciated. The parallel plate version of the code to solve the continuity equations was supplied together with considerable introductary help by A. J. Davies. The Poisson solver was contributed by E. Halter, who also was very helpful in incorporating it into the other code. Y. Kaufman updated the coefficients for ionization and photon emission after a thorough literature search. The plots were produced using computer routines by K. Thurnay who was never tired of answering my questions.

REFERENCES

1. K. Frank, J. Christiansen, O. Almen, E. Boggasch, A. Görtler,
   W. Hartmann, C. Kozlik, A. Tinschmann, G. F. Kirkman,
   A 40 kV/20 kA Pseudospark Switch for Laser Applications,
   Proc. of the Int. Soc. for Optical Engineering,
   Innovative Science and Technology Symposium, Los Angeles, 173 (1988)

2. W. Bauer, H. Ehrler, A. Rogner, C. Schultheiss, Review of Effects of the High Power Pseudospark, Proc. of the IX Int. Conf. on Gas Discharges and their Applications, Venice, Italy, 677 (1988)

3. P. Billaut, H. Riege, M. van Gulik, E. Boggasch, K. Frank, R. Seeböck, Pseudospark Switches, CERN-Report 87-13 (1987)

4. K. Mittag, Numerical Simulation of Prebreakdown in Pseudospark Discharges with a Multi-Fluid Model, Proc. of the IX Int. Conf. on Gas Discharges and their Applications, Venice, Italy, 673 (1988)

5. K. Mittag, W.Niessen, Numerical Simulation of a Pseudospark Gas Discharge, Kerntechnik 52, 188 (1988)

6. A. J. Davies, C. J. Evans, P. M. Woodison, Simulation of the Growth of Axially Symmetric Discharges Between Plane Parallel Electrodes, Computer Physics Communications 14, 287 (1978)

7. A. J. Davies, Discharge Simulation, IEEE Proceedings, Vol. 133. Pt. A, No. 4, 217 (1986)

8. E. Halter, Die Berechnung elektrostatischer Felder in Pulsleitungsanlagen, Kernforschungszentrum Karlsruhe, KfK-Report 4072 (1986)

9. W. Niessen, Monte-Carlo-Simulation of the Prebreakdown Phase for a high E/N-Discharge, Proc. of the IX Int. Conf. on Gas Discharges and their Applications, Venice, Italy, 670 (1988)

10. Y. Kaufman, P.Choi, private communication (1989)

11. P. Choi, H. Chuaqui, J. Lunney, R. Reichle, A. J. Davies, K. Mittag, Plasma Formation in a Pseudospark Discharge, IEEE Transaction for Plasma Science, to be published

12. D. Dietrich, C. Schultheiss, K. Mittag, to be published

SELF-CONSISTENT MODELS OF DC AND TRANSIENT GLOW DISCHARGES

J.P. Boeuf

Centre de Physique Atomique de Toulouse
CNRS Unité de Recherche Associée 277
Université Paul Sabatier, 118, route de Narbonne
31062 Toulouse, CEDEX, FRANCE

INTRODUCTION

In this paper we report some recent progress in the modeling of DC
and transient glow discharges. The cases and examples which are discussed correspond to typical low current glow discharges (current densities of the order of 1 mA/cm^2) at intermediate pressure (around one
torr) and for simple electrode configurations. However, the approach
and the methods which are described here could be generalized and extended to more complex conditions such as those occurring in high
power hollow cathode glow switches.

In the discharge models presented below, the coupling between
charged particle transport and the electric field is taken into account in a self-consistent way. These models are based on the coupled
solutions of transport equations describing the motion of charged particles and Poisson's equation for the electric field. The background
gas properties are supposed to be unaffected by the discharge. This is
only a first order assumption which is valid for low current densities. For higher current densities, the gas temperature and composition can change (creation of metastable molecules and dissociation
products), and the charged particle transport should be coupled to the
neutral and metastable molecule kinetics.

The models described below could therefore be qualified in a re-
strictive way as self-consistent "electrical" models (the physico-
chemistry of the discharge is not considered). Even in this restric-
tive sense, developing a self-consistent discharge model is a difficult task, for the following reasons:

- 1) one must find a simple and accurate way to describe the
transport of charged particle transport

- 2) the basic data characterizing the system (gas and electrodes) must be available (cross-sections, transport coefficients, secondary emission coefficients etc...)

- 3) a numerical technique able to deal with very steep density gradients and a large plasma density in a reasonable amount of computation time must be developed

These points are discussed in the first part of this paper. Results concerning one- and two-dimensional DC glow discharges are presented in the second part. The third part contains some comments on the basic properties of hollow cathode discharges, and the last section illustrates the complementarity of numerical models and laser diagnostic techniques to study the steady state and transient behavior of a glow discharge.

PHYSICAL AND NUMERICAL BASIS OF THE MODELS

A self-consistent electrical model of a discharge consists in solving the charged particle transport equations coupled with Poisson's equation with proper boundary conditions (which can depend on the external circuit). The solution of these equations yields the space and time variations of the electric field, charged particle densities and current densities, and excitation and ionization rates.

Physical basis

The ideal way to describe the charged particle transport in the conditions of a glow discharge is to use the electron and ion Boltzmann equation. Although important progress has been achieved in the past few years in the numerical techniques for solving Boltzmann equation for a given electric field distribution, the coupling between Boltzmann equations and Poisson's equation cannot be handled in a reasonable amount of computational time in practical situations, and one has to seek a simpler description of the electron and ion kinetics.

The classical simplification of the transport problem is to replace the Boltzmann equation by a finite number of moment equations. The Boltzmann equation being equivalent to an infinite number of moment equations, this simplification implies some further assumptions in order to close the system of moment equations. A classical approach consists in considering only the first three moments, i.e. continuity, momentum transfer and energy equations. The unknown variables of this system are the density, mean velocity and mean energy. The need for some further assumptions comes from the fact that some of the terms contained in these equations, such as the ionization, momentum transfer and energy loss rates cannot be expressed in a simple way as a function of these three variables but depend on the velocity distribution function which is unknown. It is therefore necessary to make some hypothesis concerning the shape of the distribution function. Different possible solutions are listed below:

- A rough approximation consists in assuming that the electron distribution function at a given location depends only on the value of the electric field at this location (local equilibrium between electron kinetics and electric field) (Ward, 1962, Boeuf, 1987, 1988); in that case the unknown rates can be deduced from measurements (or calculations) under uniform electric field conditions (drift tube experiments) giving the hydrodynamics transport coefficient.

- A commonly used approximation consists in assuming that the distribution function is Maxwellian; the rates can then be simply expressed in term of the electron density and mean energy (temperature) and the system of moment equations is closed by this assumption (Graves and Jensen, 1986). Although this assumption is reasonable in uniform plasmas such as the positive column of a glow discharge, it cannot be used to describe the electron distribution function in the cathode region. Figure 1 shows an example of the shape of the EEDF (electron energy distribution function) in the cathode region of a glow discharge in helium. The electrons emitted by the cathode gain energy in the sheath where they undergo only a few inelastic collisions. Some of these electrons enter the glow with an energy as large as the cathode fall voltage. The field being almost zero in the glow region, the high energy electrons release their energy in the form of inelastic collisions, contributing to the formation of a low energy, high density group of electrons which forms the bulk of the distribution in the glow. This situation can obviously not be described by a Maxwellian EEDF.

- A third approach consists in coupling the fluid model with a microscopic treatment of the high energy electron (Surendra et al., 1989): the fluid equations are still used with Poisson's equation to calculate the self-consistent electric field while a Monte Carlo simulation is used (iteratively) to calculate the ionization rates

- A simpler but realistic solution is to split the EEDF in two parts: the first one represents the electrons emitted by the cathode and accelerated in the sheath, the second part corresponds to low energy electrons created in the glow. This approach has been used in the examples presented in this paper and is described in more details below.

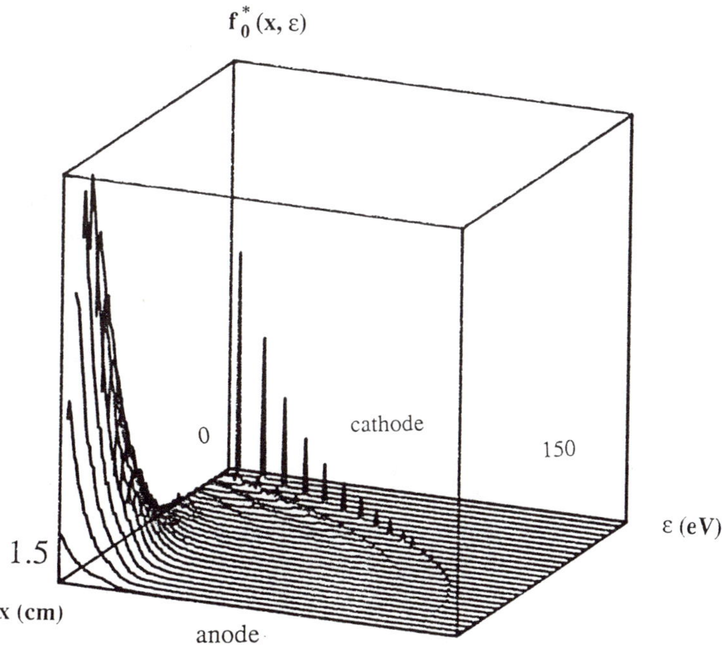

Figure 1. Spatial variations of the electron energy distribution function $f_0^*(\varepsilon, x)$ (in this representation, the electron density is given by $n_e(x) = \int f_0^*(x, \varepsilon)\, d\varepsilon$) in helium (V=150 V, sheath length d_c=1.3 cm, d=1.5 cm, p=1 torr)

<u>Two electron group model</u>

As suggested by the EEDF shown in Fig. 1, the sheath acts as an electron gun accelerating the cathode emitted electrons into the glow. These electrons provide an ionization source for the negative glow plasma which can be compared to a plasma sustained by an external source (contrary to the case of the positive column plasma).

In our two electron group model, the kinetic properties of each electron group are defined as follows (a similar model has been used for the modeling of RF glow discharges - see Boeuf and Belenguer, 1989):

1) The fast electrons (or beam electrons) are emitted by the electrodes under ion bombardment. They are assumed to form a monoenergetic beam (Ingold, 1978, Phelps et al., 1987) whose velocity is directed toward the anode. Secondary electrons created by ionization by the fast electrons are assumed to belong to the beam if they are created in the sheath; when the ionization takes place in the glow, the secondary electrons join the low energy group which is described below. The transport of the fast electrons is described by a continuity equation and an energy equation (see below). The beam is supposed to release instantaneously its energy in the discharge (the continuity and energy equations for the beam are steady state equations).

2) The low energy electrons (bulk electrons) are created in the glow by ionization due to fast electrons; when the beam energy falls below the ionization threshold, fast electrons join the low energy group. The kinetics of the low energy group is described by a continuity equation and a momentum transfer equation (see below). The momentum transfer equation for the bulk electrons (as well as for the ions) is simplified by omitting the time derivative term and neglecting the drift energy with respect to the thermal energy, as in Boeuf (1987); this neglect of the inertia terms is reasonable only in a collisional regime and leads to the representation of the particle flux by the sum of a drift term and a diffusion term.

The transport equations describing the two electron groups are coupled with continuity and momentum transfer equations for ions and with Poisson's equation. This leads to the following system of equations:

$$\frac{\partial (n_b v_b)}{\partial x} = A \; n_b v_b \; Q_i(\varepsilon_b), \quad \text{if } \varepsilon_b > \varepsilon_i; \quad v_b = \left(\frac{2}{m} \varepsilon_b \right)^{1/2} \quad (1a)$$

$$\frac{\partial (n_b v_b)}{\partial x} = - \frac{n_b \; v_b}{\lambda_b}, \quad \text{if } \varepsilon_b < \varepsilon_i \quad (1b)$$

$$\frac{\partial \varepsilon_b}{\partial x} = eE - \Sigma \varepsilon_k Q_k(\varepsilon_b) - A \; \varepsilon_b \; Q_i(\varepsilon_b) \quad (2)$$

$$\frac{\partial n_e \langle v_e \rangle}{\partial x} = (1-A) \; n_b \; v_b \; Q_i(\varepsilon_b) - r \; n_e \; n_p, \quad \text{if } \varepsilon_b > \varepsilon_i \quad (3a)$$

$$\frac{\partial n_e <v_e>}{\partial x} = \frac{n_b \; v_b}{\lambda_b} - r \; n_e \; n_p, \qquad\qquad \text{if } \varepsilon_b < \varepsilon_i \qquad\qquad (3b)$$

$$n_e <v_e> = n_e W_e - \frac{\partial (n_e D_e)}{\partial x} \qquad\qquad (4)$$

$$\frac{\partial n_p <v_p>}{\partial x} = n_b \; v_b \; Q_i(\varepsilon_b) - r \; n_e \; n_p \qquad\qquad (5)$$

$$n_p <v_p> = n_p W_p - \frac{\partial (n_p D_p)}{\partial x} \qquad\qquad (6)$$

$$\frac{dE}{dx} = \frac{|e|}{\varepsilon_0} (n_p - n_e - n_b) \; ; \qquad\qquad \int E(x) \; dx = -V \qquad\qquad (7)$$

The coefficient A is set to 1 in the sheath and to 0 anywhere else (the sheath glow boundary is supposed to be situated at the point where the electric field is less than a given arbitrary value - 10 V/cm/torr). The subscripts b, e and p refer respectively to the beam electrons, bulk electrons and positive ions; n_s, v_s, W_s, D_s are respectively the number density, mean velocity, drift velocity and diffusion coefficient of species s ($W_s = \mu_s E$ where $\mu_s(E)$ is the mobility); ε_b is the energy of the beam, ε_i the ionization threshold and ε_k the threshold for the inelastic process number k. $Q_k(\varepsilon) = N \; \sigma_k(\varepsilon)$ where N is the gas density and σ_k is the electron-molecule cross-section for the inelastic process number k (or for ionization if $k \equiv i$). The summ in eq. (2) includes all the inelastic processes. λ_b is the distance characterizing the rate at which the beam electrons join the bulk electron group when the beam energy is less than the ionization threshold (this parameter is chosen as small as possible but is finite in order to avoid discontinuities in the numerical treatment). r is the electron-ion recombination coefficient. e and m are the electron charge and mass, E is the electric field, V is the applied potential and d the gap length.

The boundary conditions for the charged species are as follows. The beam electrons are created by ion bombardment of the cathode according to

$$\varphi_b(0) = - \gamma \, \varphi_p(0) \qquad\qquad (8)$$

where $\varphi_b(0)$ and $\varphi_p(0)$ are the beam flux and the positive ion flux on the cathode respectively; γ is the secondary electron emission coefficient due to ion bombardment. The charged particle densities are set to zero on the electrodes; these boundary conditions are different from those used in Boeuf (1987, 1988) which did not account for the possibility of the existence of a net positive electron flux on the anode in the case of a field reversal (in that case, the diffusion term and conduction term of the flux are opposite and the diffusion becomes larger than the conduction term in order to satisfy current continuity).

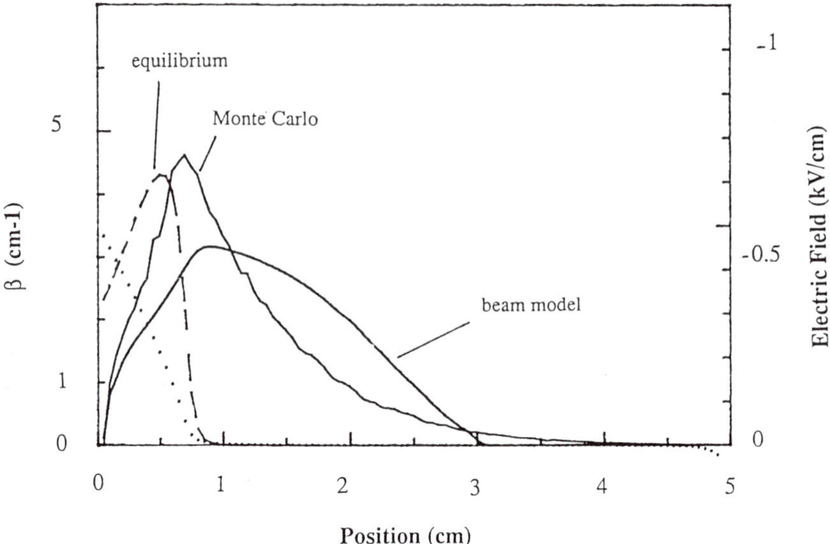

Figure 2.   Spatial variations of the reduced ionization rate β (see text) calculated by three different ways: beam model, Monte Carlo simulation, equilibrium model; the electric field is also represented (dotted line) and corresponds to the conditions of Fig. 3: helium, V=220 V, d=5 cm, p=1 torr, γ=0.2

An idea of the accuracy of the representation of the electrons accelerated in the sheath by a monoenergetic beam is shown in Fig. 2 which presents a comparison of the reduced ionization rate calculated by three different ways: 1) with the beam model described above, 2) using a Monte Carlo simulation and 3) assuming local equilibrium between electron kinetics and electric field. The reduced ionization rate β is defined as:

$$\beta(x) = \alpha(x) \; \varphi_e(x)/\varphi_e(0) = \alpha(x) \; \exp\left(\int_0^x \alpha(y) \; dy\right) \qquad (9)$$

where $\alpha(x)$ is the local ionization coefficient and $\varphi_e(x)$ is the electron flux at position x ($\alpha(x)\varphi_e(x)$ is the ionization rate). The electric field distribution is also plotted in Fig. 2 and corresponds to the self-consistent field shown in Fig. 3; the conditions are: helium, V=220 V, d=5 cm, p=1 torr. As expected we see in Fig. 2 that the equilibrium model does not predict any ionization in the glow. The penetration depth of the fast electrons predicted by the beam model is in reasonable agreement with the Monte Carlo simulation although the beam model tends to exaggerate ionization in the glow and to underestimate ionization in the sheath.

Data

The results presented in this paper correspond to helium. The transport coefficients (mobility and ionization coefficient for the equilibrium model) for electrons and ions have been taken from Ward (1962). The electron-atom cross-sections for the beam electrons and for the Monte-Carlo simulations are the same as those used in Boeuf and Marode (1982). The diffusion coefficients are supposed to be constant and equal to 10^6 and 10^3 $cm^2/V/s$ at one torr, for electrons and ions respectively. Electron-ion recombination is not considered (r=0).

Numerical methods

A few comments concerning the numerical method which has been used in this work to solve the fluid equations coupled with Poisson's equation are given in this section. This method is similar to the methods widely used in the numerical modeling of semiconductor devices such as MOSFET. An important effort has been focused on the modeling of semiconductor devices in the past years, due to the development of the VLSI technology. Since the equations describing the transport of electron and holes in semiconductors are very similar to those describing the transport of electrons and ions in gases, we can take advantage of the work which has been dedicated to semiconductor device modeling.

The method uses a finite difference technique with a very efficient and stable implicit discretization scheme for the continuity and momentum transfer equations; this scheme is similar to the Scharfetter and Gummel scheme used in the modeling of semiconductor devices (Gummel 1964, Scharfetter and Gummel 1969, Selberherr and Ringhofer 1984).

Continuity equations and Poisson's equation can be solved successively until steady state is reached, as in Boeuf (1987, 1988). However this (explicit) method does account for the variations of the electric field during the motion of the charged particles so that the time step must be limited (the time step must be proportional to the inverse of the electron density; typically: a few 10^{-10} s for a plasma density of 10^9 cm^{-3} at 1 torr). It is therefore difficult to follow the time evolution of a high density plasma during a long time using this explicit method.

In implicit methods one tries to solve transport equations and Poisson's equation together instead of successively. Since the system is strongly non-linear, it is first necessary to linearize it with respect to the three variables (electron density, ion density, electric potential): this leads to a bloc-tridiagonal system which can be solved by standard methods. This method can be easily implemented in a one-dimensional situation, and the 1D results presented in the next section and in the last section have been obtained with this method. Since there is almost no limitation on the time step, this implicit method is considerably faster than the explicit one. In 2D situations the method cannot be fully implicit because of the prohibitively large size of the matrix to be inverted. The 2D results presented in the next section have been obtained with an explicit method; a semi- implicit method which will allow to study plasma with much higher densities in a reasonable amount of computation time is being developed in Toulouse.

Details on these numerical methods can be found in Kurata (1982) in the context of semiconductor device modeling.

1D AND 2D MODELS OF DC GLOW DISCHARGES

1D model

Figure 3 shows the space variations of the electric field and charged particle densities in the gap, in a DC discharge in helium,

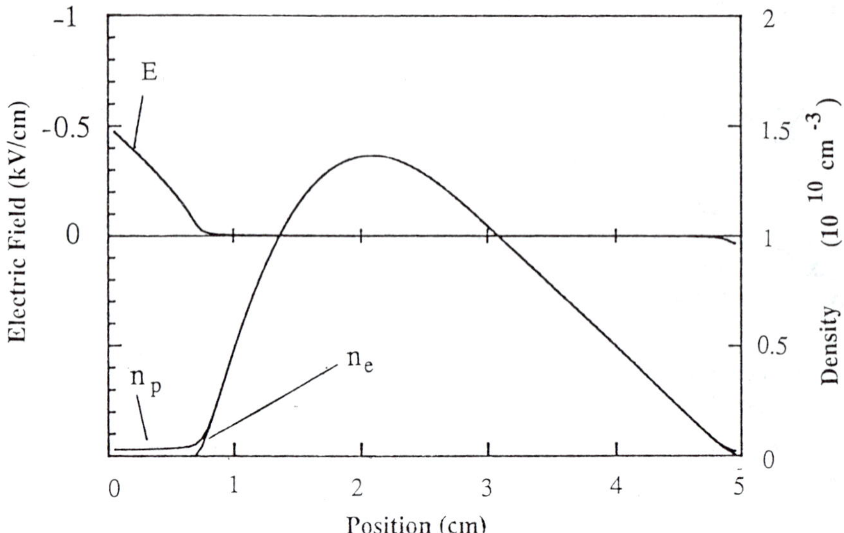

Figure 3. Spatial distribution of the electric field and charged particle densities in a helium DC discharge (two electron group model): V=220 V, d=5 cm, p=1 torr, γ=0.2

discharge voltage V=220 V, gap length d=5 cm, pressure p=1 torr, secondary electron emission (ion impact) γ=0.2.

The sheath field decreases roughly linearly, the sheath length being around 6 mm. The ion density increases abruptly from a few 10^8 cm^{-3} in the sheath to $\sim 1.5 \ 10^{10}$ cm^{-3} in the plasma glow. An important point which appears in Fig. 3 is the existence of a field reversal in the negative glow (the location of the field reversal corresponds to the maximum plasma density): the direction of the electric field near the anode is such that the field tends to confine electrons in the plasma. Since the electrons are trapped in the plasma by the electric field, the plasma density must increase until the diffusion term of the electron flux overcomes the conduction term, allowing some electrons to reach the anode. Due to the field reversal the plasma potential is larger than the anode potential, the potential barrier being of the order of the electron temperature. Note that the total current density (Fig. 4a) is about the same as in Boeuf and Ségur 1988 (Figs 4.6 and 4.7), but the gap voltage is lower; this is due to the fact that in the results presented in Boeuf and Ségur (1988), the discharge had not completely reached steady state (the numerical method was explicit).

Figure 4a shows the space variations of the electron, ion and total current densities in the gap. As expected, the ion current in the glow becomes negative at the point corresponding to the field reversal and a small ion current is collected by the anode. Consequently, the electron current is larger than the total current in the part of the glow close to the anode. The possible existence of a field reversal in the negative glow has been mentioned for a long time and is discussed by Chapman (1980), but numerical models have been able to predict field reversals only recently (Graves and Jensen, 1986, Boeuf and Ségur, 1988). The field reversal in the glow has been demonstrated experimentally recently by Gottscho et al. (1989) by measuring the direction of the ion velocity throughout the gap; the location of the

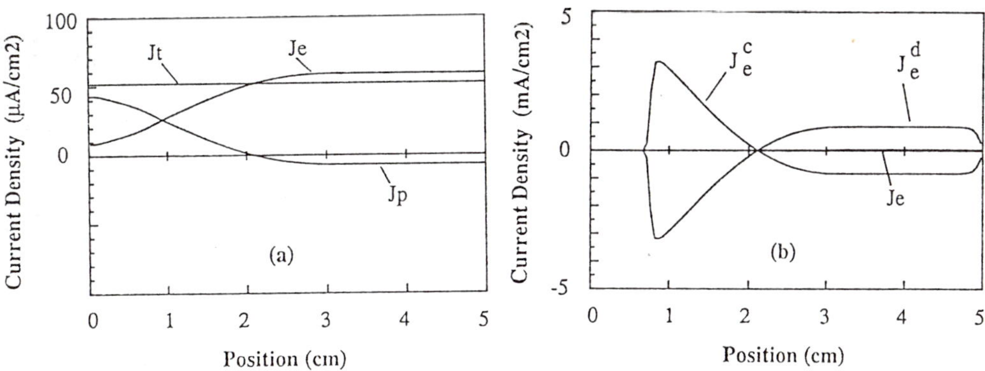

Figure 4.   Spatial distribution of the current densities in the conditions of Fig. 3; a) electron (Je), positive ion (Jp) and
total (Jt) current densities; b) electron current density
(Je) and its conductive ($J_e^c = en_eW_e$) and diffusive ($J_e^d = -$
$e\,\partial(n_eD_e)/\partial x$) components

field reversal has also been obtained from these measurements. Note
that in the above discussion, it is assumed that the anode is in the
glow region and that there is no positive column.

As mentioned above, the diffusive part of the electron flux must
be necessarily larger than the conductive part in the region close to
the anode. Figure 4b shows the space variations of the two components
of the electron current density. It appears that 1) the absolute value
of each term (conduction and diffusion) of the electron current is
much larger than the total electron current in the glow, and 2) that
the conduction term is larger than the diffusion term at the sheath
plasma boundary but smaller than the diffusion term near the anode;
these properties are characteristic of ambipolar diffusion.

Figure 5 shows the space variations of the electric field and
charged particle densities in the same conditions as in Fig. 3 but
from calculations in which the ionization coefficient is assumed to
depend only on the local electric field (ionization coefficient from
Ward, 1962, as in Boeuf 1988); this model is termed as equilibrium
model below. We see that although this model seems to give a reasonable picture of the discharge in term of sheath length and ion density
in the sheath, the plasma density is smaller than in the two group
model by about two orders of magnitude. The important features of the
glow (field reversal, ambipolar diffusion) described above cannot be
predicted by the equilibrium model: this is due to the fact that the
equilibrium model assumes that the negative glow plasma is sustained
by the local electric field as in a positive column plasma. The negative glow plasma is actually sustained by the electrons emitted by the
cathode and accelerated in the sheath, and there is no direct relation
between the ionization rate at a given location in the glow and the
electric field at this location (as in an externally sustained
plasma). The equilibrium assumption is reasonable only close to the
normal regime (lower voltage and larger sheath length) where most of
the ionization by cathode emitted electrons occurs in the sheath.

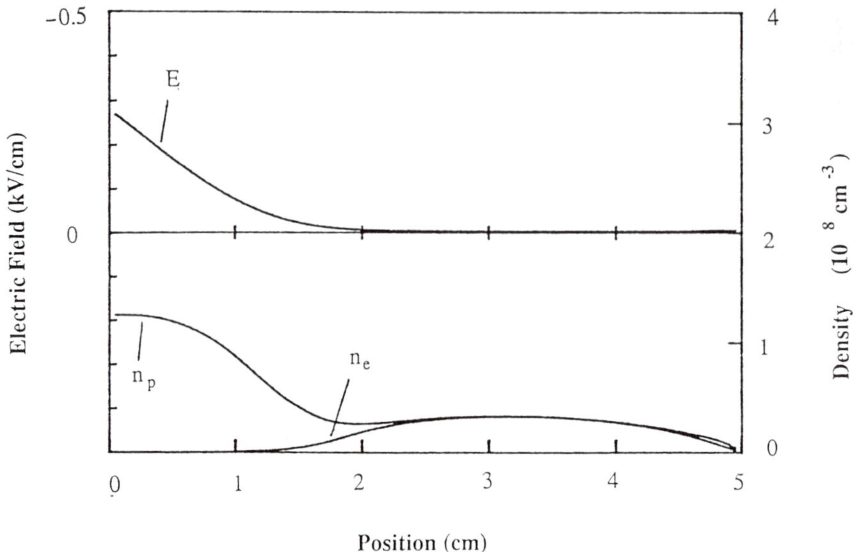

Figure 5.   Spatial distribution of the electric field and charged particle densities obtained with the equilibrium model for the
same conditions as in Fig. 3

2D model

The results presented below correspond to a 2D cylindrical geometry as shown in Fig. 6: OC corresponds to the cathode, AB to the anode, OA is the discharge axis and CB the wall (cylinder). Electron
emission due to ion bombardment is supposed to occur only on the part
of the cathode corresponding to OD (i.e. γ is set to zero outside the
disk of radius OD). The external circuit consists in a variable resistor and a DC power supply. The potential is assumed to increase linearly on the wall AB. This 2D model has been used to study the transition between normal and abnormal regimes in a DC glow discharge. Similar calculations have been published in Boeuf (1988) in the case of a
Cartesian geometry.

The transition between abnormal and normal regime occurs when,
while the discharge current is decreased (e.g. by increasing the external resistance), the "discharge area" becomes smaller than the
cathode area. During the transition, the cathode fall reaches its minimum voltage (normal cathode fall). In the normal regime, the
longitudinal properties of the discharge (cathode fall voltage, sheath
length, total current density) stay roughly constant when the current
is decreased, while the radial dimensions of the discharge decreases
in order to maintain a constant current density. The ability of the
model to predict this transition has been checked in helium, for the
geometry of Fig. 6 with OC=4 cm, OA=8 cm, OD=2 cm, p=1 torr, γ=0.2 on
OD. For a constant voltage of the generator V_g, the resistance of the
external circuit has been varied in order to study the region of the
discharge characteristic around the minimum gap voltage.

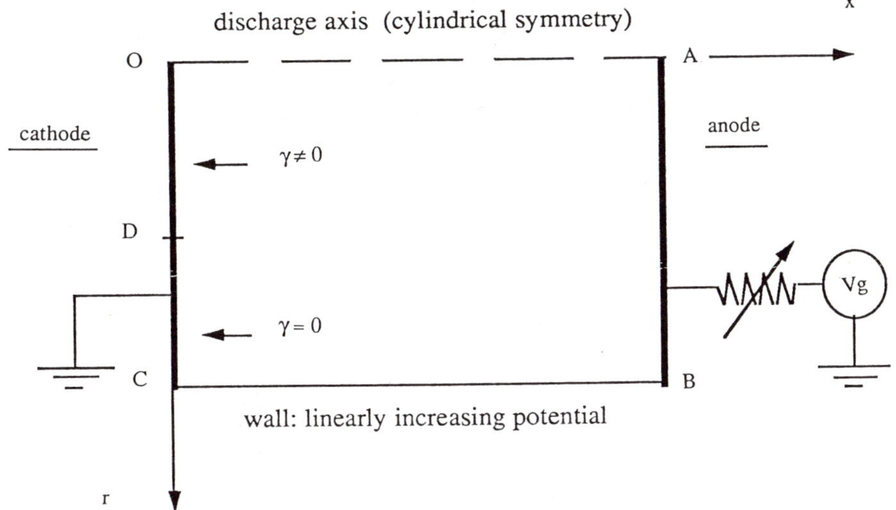

Figure 6.   Geometry considered in the 2D results shown in Figs. 7, 8

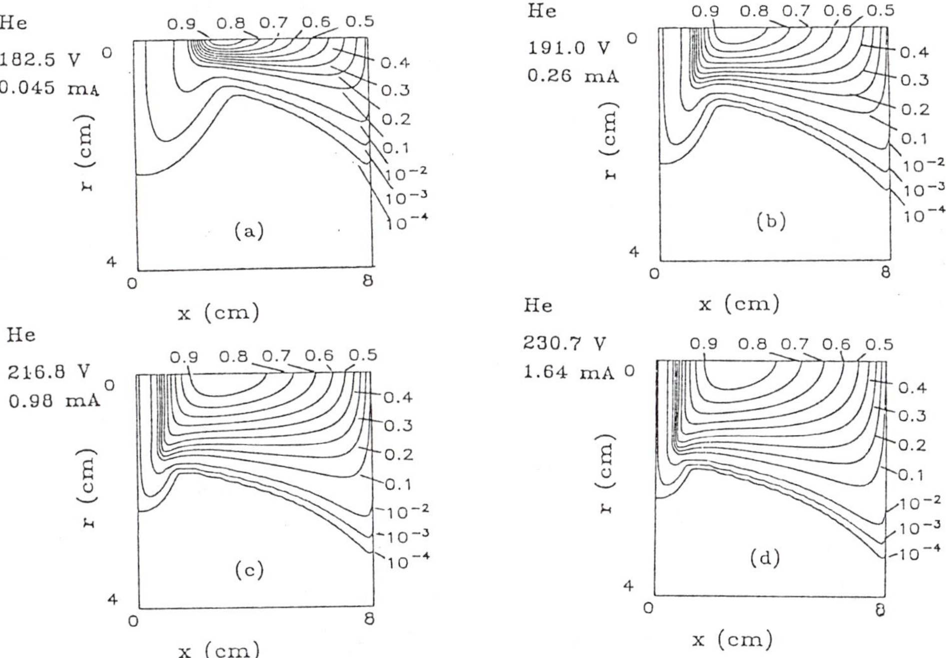

Figure 7.   Contours of constant electron number density for 4 different cases showing the transition between normal and abnormal regimes. Helium, geometry of Fig. 6 with OC=4 cm, OA=8 cm, OD=2 cm, p=1 torr, γ=0.2 on OD.   Units of contours: (a) 8.3 10^{-7} cm^{-3}, (b) 1.8 10^{8} cm^{-3}, (c) 4.3 10^{8} cm^{-3}, (d) 5.45 10^{8} cm^{-3}

Figure 7 shows the contour plots of the electron number density for four different points of the current-voltage curve. Assuming that the discharge area which can be observed experimentally is actually the area of the negative glow, and that the luminosity of the negative glow is proportional to the electron density, then the variations of the discharge area with current and voltage can be deduced from Fig.7.

It appears that the discharge area is roughly equal to the cathode area (emissive part only) in Figs. 7c and 7d and smaller than the cathode area in Figs. 7a and 7b. On the other hand, the discharge current increases by a factor of 40 from Fig. 7a to Fig. 7d. It is therefore clear that Figs. 7a to 7d show the transition from the normal regime to the abnormal regime.

The above results have been obtained with an equilibrium model, as in Boeuf (1988); the results corresponding to the largest gap voltages are therefore quantitatively doubtful since ionization by fast electrons in the glow is not considered.

Figure 8 presents some results obtained with a 2D two electron group fluid model (the beam electrons are assumed to be directed forward and do not follow the field lines) in the case of helium in the geometry of Fig. 6. The discharge corresponding to the results shown in Fig. 8 has not yet reached its steady state regime. The equipotential curves of Fig. 8a show that, as in the 1D case, there is a field reversal in the glow and that the plasma potential is larger than the anode potential. The ion density shown in Fig. 8b presents similar features as in the 1D case.

As illustrated in this section, 2D models represent a very useful and powerful tool for studying gas discharge devices. These models can be extended very easily to more complex geometries (Boeuf, unpublished results). However, as indicated above, the computational time of the

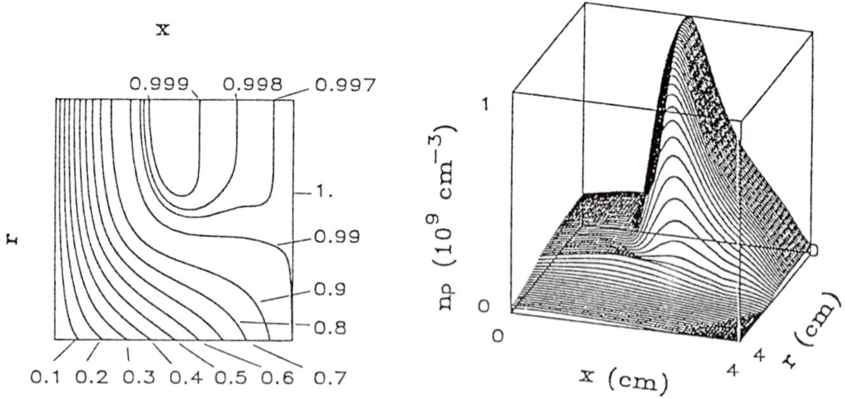

Figure 8. a) equipotential curves and b) space variations of the ion density in a helium discharge. Geometry of Fig. 6, with OC=4 cm, OA=4 cm, OD=2 cm, γ=0.1, V=291 V (I=0.48 mA). The values of the equipotential curves are given in units of the anode voltage (291 V)

explicit method used in the 2D model increases with the plasma density, and it is still difficult to model steady state DC discharges with current densities larger than a few 100 $\mu A/cm^2$ in a reasonable amount of computational time (the steady state regime may be reached, under some conditions in times as large as tens of ms at one torr, so that it is very important to avoid any limitation on the time step). Work is continuing to extend the implicit method to 2D models.

PENDULUM EFFECT IN HOLLOW CATHODE DISCHARGES

In this section we present some comments on the pendulum effect in a hollow cathode discharge, based on Monte Carlo simulations of the electron kinetics.

We consider the simplest geometry of a hollow cathode, where the cathode consists in two parallel disks separated by a distance d, the common anode being a circular ring whose radius is sufficiently larger than the cathode radius and located in the mid-plane between the two cathodes (see Fig. 9). When the distance d between cathodes is large enough, two independent discharges develop between each cathode and the common anode: in that case the distance between the cathode and the common anode is larger than the length of the negative glow (which is defined as the relaxation distance of the high energy electrons coming from the sheaths). When the cathode distance is decreased, the applied voltage being kept constant, the discharge current stays roughly constant until the negative glows corresponding to each cathode overlap. At this point the discharge current starts to increase as d decreases, and can reach values several order of magnitudes larger than the initial current; the intensity of the light emission from the glow increases in the same way. This effect is known as the hollow cathode effect and has been described and analyzed in a number of papers. Recent reviews on hollow cathode discharges includes papers by Selvin and Harrison (1975), Rosza (1980), Pillow (1981) and Mavrodineau (1984).

Although the physical phenomena which might be responsible for the hollow cathode effect have been known for a long time, there is still

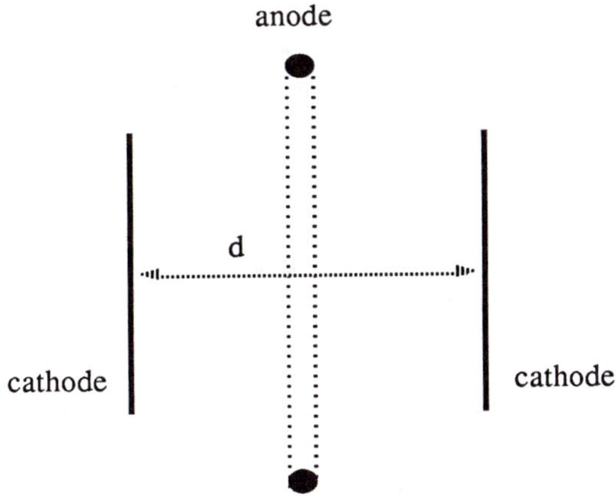

Figure 9. Geometry of the hollow cathode discharge considered

some doubt and discussion on the relative importance of the contribution of each of these phenomena to the hollow cathode effect. Some of these phenomena are listed below (see Sturges and Oskam (1964) or Selvin and Harrison (1975) for a more detailed discussion):

- Due to the confining geometry of the cathode, the secondary emission due to the cathode bombardment by photons might be substantially enhanced (Little and von Engel, 1954)
- The larger concentration of metastables in the discharge can induce an increase in the secondary emission due to the bombardment of the cathode by these excited species; also ionization might be enhanced due to superelastic collisions or stepwise ionization
- the presence of sputtered particles in the gap can enhance the total ionization
- Fast electrons are trapped in the cavity formed by the cathode and can oscillate (pendulum effect) in the gap, being reflected by the sheaths (von Engel and Steenbeck 1934)

Sturges and Oskam (1966) showed that the role of metastables in the hollow cathode effect does not seem to be dominant.

In this section we try to estimate quantitatively the contribution of pendulum electrons to the high efficiency of the hollow cathode discharge. This analysis is based on the use of Monte Carlo simulation to describe the electron kinetics. The goal of this study is to determine whether or not the pendulum effect can by itself explain the increase in the discharge current observed experimentally (Guntherschulze 1930, von Engel 1965) when the distance between cathode is decreased. In other words, the question is: is it possible to reproduce the experimental curves of Guntherschulze (1930) (see Fig. 10) if one consider only the specific properties of the electron kinetics in a hollow cathode, all other effects (photons, metastable molecules etc...) being neglected ?

The Monte Carlo simulation is used to calculate the electron energy distribution function (EEDF) and the ionization rates throughout the discharge for a given constant discharge voltage and for different values of the distance between cathodes. The discharge current density J is calculated for each value of d, and the curve $J(d)/J_0$ is compared to the experimental curves of Guntherschulze (J_0 is the current density when the distance between cathode is large enough so that the two glows do not overlap).

The Monte Carlo code is similar to the one described in Boeuf and Marode (1982); for symmetry reasons (see Fig. 9), only the part of the gap between one cathode and the common anode is considered in the calculations: electrons are supposed to be reflected by the mid plane. As soon as their energy becomes too small to allow them to undergo any more inelastic collision, electrons are no longer followed by the Monte Carlo simulation. The results presented here have been published in Boeuf and Ségur (1988); similar calculations have been performed by Hashiguchi and Hasikuni (1988). The electric field distribution in the gap is obtained in a "semi" self-consistent way, as follows:

- the field is supposed to be linearly decreasing in the sheaths and zero in the glow
- the sheath length is iteratively adjusted in such a way that the self-sustaining condition is satisfied, i.e. ($M = 1 + 1/\gamma$) where M is the electron multiplication and γ the secondary emission coefficient due to ion bombardment.

– the current density is deduced from Poisson's equation in the sheath
and by assuming that the sheath is collisional for ions; in that case,
for a constant applied voltage V, $J \propto (d_c)^{-5/2}$ where d_c is the sheath
length. Therefore:

$$\frac{J}{J_0} = \left(\frac{d_{c0}}{d_c}\right)^{5/2}$$

It is important to keep in mind that the calculations are performed with the following assumptions:

– secondary electron emission due to photons or metastable atoms is
neglected
– all the ions created in the glow (and in the sheath) are collected
by the cathodes
– the secondary electron emission due to ion impact on the cathode
does not change with ion energy
– only ionization by electron impact of the ground state atoms is considered
– the sheath is collisional for ions
– Coulomb collisions are not considered

The calculated $(J(d)/J_0)$ curve is plotted in Fig. 10 in the case
of helium, discharge voltage V=400 V and is compared with the experi-

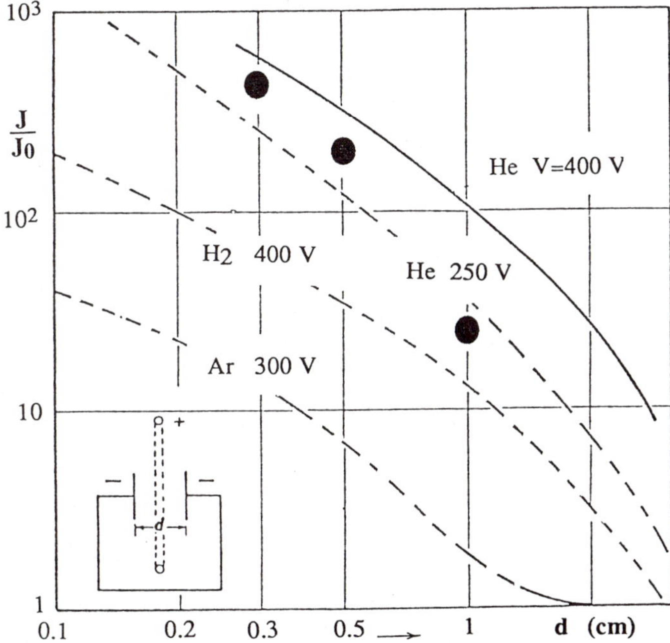

Figure 10. Variations of the hollow cathode efficiency J/J_0 (see text)
as a function of the distance between cathodes. Full lines
and dashed lines: experimental results from Gunterschulze
(1930) (after von Engel, 1965). (•): numerical results
(Monte Carlo simulations) for V_c=400 V in helium, γ=0.1.
After Boeuf and Ségur (1988)

mental results of Gunterschulze. We see that in spite of the assumptions mentioned above, the agreement between calculated and experimental results is reasonable. The calculated results are very sensitive to the choice of the value of the secondary emission coefficient which was supposed to be $\gamma=0.1$, but the slope of the curve $(J(d)/J_0)$ is less sensitive to γ. The most important point of this comparison is that the model is able to reproduce well the important increase in the current density when the distance is decreased. We can therefore conclude that the pendulum effect by itself can explain the increase in the efficiency observed in a hollow cathode discharge.

Let us now comment on the mechanisms of the pendulum effect and on the reasons of the enhancement of the ionization efficiency when high energy electrons oscillate between the sheaths. Let us consider a particular hollow cathode discharge regime such as the regime corresponding to the point (see Fig. 10) V=400 V, d=1 cm (1 torr). For this regime the calculated sheath length was 0.3 cm and the electron multiplication was $M = 1 + 1/\gamma = 11$. If we now consider a "non-hollow cathode" regime with the same voltage and sheath length the electron multiplication will be less than M because in that case, the electrons created by ionization outside the sheath can only loose energy in the zero field of the negative glow. In the hollow cathode case, due to the oscillation of fast electrons, the probability is much larger that ionization takes place in one of the sheaths; electrons created by ionization in the sheaths can in turn gain some energy from the sheath field. In summary, the ionization efficiency of the pendulum effect is due to the fact that the oscillations of high energy electrons lead to the enhancement of the *ionization in the <u>sheaths,</u>* allowing the electrons created by ionization (secondary electrons) to gain energy from the sheath field and to create further ionization. Note that some hollow cathode models (Kagan, 1985) assume that ionization by secondary electrons is negligible; as mentioned by Valentini (1987), such models as well as models assuming that the sheath is non collisional for electrons cannot explain the hollow cathode effect.

$f_0(x,\varepsilon)$ (log. scale, a.u.)

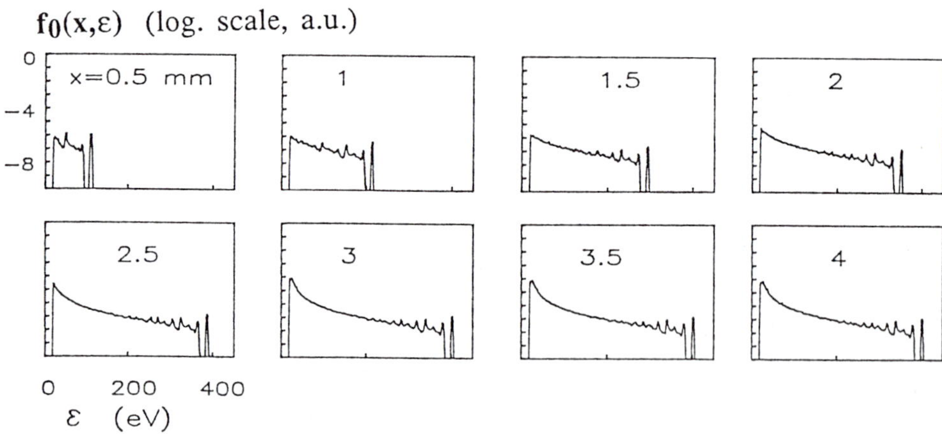

Figure 11. Spatial variations of the electron energy distribution function in a hollow cathode discharge in helium: V=400 V, distance between cathode d=1 cm, sheath length d_c=0.3 cm, p=1 torr. In this representation the electron density is related to f_0 by $n_e(x) \propto \int f_0(x, \varepsilon) \, \varepsilon^{1/2} \, d\varepsilon$ (a Maxwellian distribution is a straigth line). Only the distribution above ε = 20 eV is represented. After Boeuf and Ségur (1988)

Finally, Fig. 11 illustrates the properties of the electron kinetics in a hollow cathode discharge. The variations of the electron energy distribution function is represented in this figure in the case of helium, V=400 V, d=1 cm, d_c=0.3 cm, p=1 torr. Only the EEDF between one cathode (x=0) and the mid-gap (anode, x=0.5 cm) is plotted, and the low energy group (ε < 20 eV) is not represented. Note that in the glow (x > 0.3 cm) the high energy part of the EEDF does not change very much with position. On the other hand, the slope of the EEDF in the high energy part is extremely small; a Mawellian distribution with the same slope would have a temperature of hundreds of eV.

TRANSIENT BEHAVIOR OF A GLOW DISCHARGE

In this section we report some experimental and numerical studies of the transient behavior of a glow discharge following a perturbation from DC steady state. More details on the diagnostic techniques and numerical models can be found in Debontride et al. (1989) (see also Derouard et al., 1989). Similar experiments and calculations have been performed by Mitchell et al. (1989).

The DC discharge is perturbed by a laser pulse illuminating the cathode and creating significant photoelectron emission from the cathode. Diagnostic techniques based on laser induced fluorescence are used to measure the space and time variations of the sheath electric field after the laser pulse. The experimental results are compared with numerical results obtained with the two electron group fluid model described above.

The laser diagnostic of the electric field uses the NaK molecule as a spectroscopic probe; for this reason, the gas mixture used in the discharge was Ar (98%) and K(2%) (see Derouard et al., 1986, 1987 and Debontride et al., 1989 for more details on the diagnostic technique).

The laser used to induce photoelectron emission on the cathode was a pulsed Cu vapor laser; the pulse duration was around 60 ns and the repetition rate 6.5 kHz.

The conditions of the DC discharge considered were: pressure 0.4 torr, gap length 3.5 cm, gas temperature 210° C, current 100–500 μA (14–70 μA/cm^2), discharge voltage 154–224 V. The estimated current density of the photoelectron emission was larger than 500 μA/cm^2 (much larger than the steady state DC current density of the discharge) so that the laser pulse creates a very strong perturbation of the discharge.

Figure 12 shows the experimental results corresponding to the following conditions: discharge current 100 μA (14 μA/cm^2), discharge voltage 154 V, sheath length 10 mm. The space and time variations of the measured electric field and emission of the Ar I (419 nm) line in the sheath region are represented in this figure. At time t=0, the discharge is in its steady state DC regime and the electric field decreases roughly linearly in the sheath. The laser pulse then hits the cathode for 60 ns. As can be seen in this figure, the photoelectron avalanches initiated by the laser pulse have induced an important contraction of the sheath. This contraction is due to the increase of the ion space charge in the sheath and shows that ionization in the sheath is important in this low current regime.

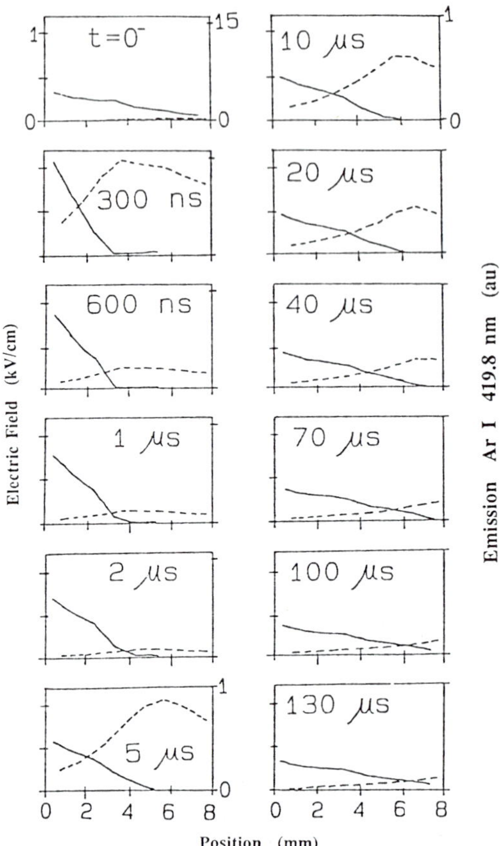

Figure 12. Space and time variations of the electric field (continuous line) and Ar atom emission at 419.8 nm ($5p[1/2]_0 - 4s[1/2]_1$ transition) profiles as a function of time. Discharge conditions: 100 μA (14 μA/cm^2), V=154 V, Ar(98%)+K(2%), 0.4 torr, 210°C. Note the change in the scale of the Ar emission signal at t = 5 μs. After Debontride et al. (1989)

The intensity of the light emission in the glow has increased by a factor of ten at time t=300 ns. The discharge then relaxes towards its steady state DC regime, and the sheath expands and reaches its steady state value (around 1 cm) after about 100 μs.

Numerical calculations have been performed for the same conditions in pure argon, using the two electron group fluid model (the influence of the presence of potassium atoms in the discharge has been studied — see Debontride et al., 1989, but will not be discussed in this paper). The secondary electron emission coefficient due to ion bombardment of the cathode is first adjusted (the gap voltage being equal to the experimental one: 154 V) so that the calculated sheath length is in

agreement with the measurement corresponding to the DC regime; the
value obtained was $\gamma=0.07$; the calculated current density for these
conditions was 8 $\mu A/cm^2$, i.e. about 50 % less than the experimental
value. The laser pulse is then simulated by adding an electron current
at the cathode equal to 500 $\mu A/cm^2$ during 60 ns. Figure 13 shows the
space and time variations of the calculated electric field, electron
and ion density after the laser pulse. The laser is fired at t=10 ns
(the first curve, t=10ns corresponds to the unperturbed DC discharge).
We see that the model reproduces well the experimental measurements of
the electric field. The sheath contraction corresponds to about one
half of its steady state DC length in agreement with the experiment.
The discharge then relaxes towards the DC regime. The time constant of
the relaxation (around 100 μs) is close to the experimental value and
is related to ambipolar diffusion. Note the important increase in the
ion density in the sheath immediately after the laser pulse; the
plasma density perturbation is also important, showing that for this
low current regime, the number of supplementary charges created in the
volume by the photoelectron avalanches is not negligible with respect
to the total number of charges present in the plasma volume in the unperturbed state. This is no longer true for higher DC discharge currents: the calculations (see Debontride et al. 1989) show that for a
discharge current of 55 $\mu A/cm^2$ corresponding to a gap voltage of 224 V,
the plasma density in the DC state is more than one order magnitude
larger than in the case of Fig. 13; the perturbation induced by the
laser pulse in the plasma glow is not so important in that case.

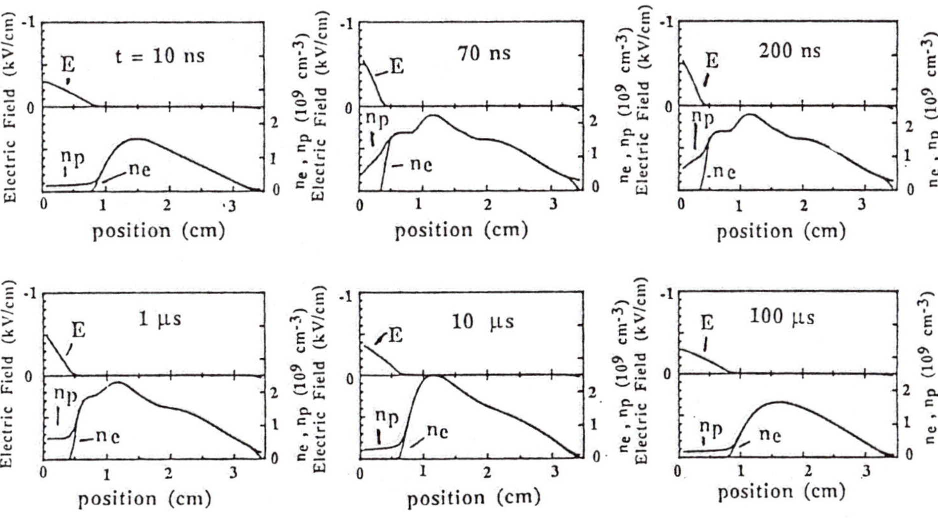

Figure 13. Electric field and charged particle density profiles as a
function of position, at different times, as predicted by
the two electron group fluid model. Discharge conditions
corresponding roughly to Figure 12 (8 $\mu A/cm^2$, 154 V, but
pure Ar, 0.25 torr at 300 K). The photoemission current is
supposed to be constant (500 $\mu A/cm^2$) during the time interval 10 ns < t < 60 ns. After Debontride et al. (1989).

The results described above show that the numerical models can reproduce satisfactorily the experimental results, and that the powerful laser diagnostic techniques developed recently can be very helpful to check the validity of these models.

CONCLUSION

From the results presented in this paper, we can draw the following conclusions:

- Efficient self-consistent electrical models of glow discharges have been developed recently. These models can reproduce well the experimental observations. The efficiency of the numerical methods in the 2D case can still be improved.

- In such models, the use of a realistic description of the ionization rate throughout the gap is of paramount importance (equilibrium models cannot predict the specific properties of the plasma glow).

- The combination of such calculations with the recently developed laser diagnostic techniques is a very efficient way to improve our understanding of glow discharge devices.

REFERENCES

Boeuf J.P. and Marode E., 1982, _J. Phys. D: Appl. Phys._ 15, 2169
Boeuf J.P., 1987, _Phys. Rev._ A 36, 2782
Boeuf J.P., 1988, _J. Appl. Phys._ 63, 1342
Boeuf J.P. and Ségur P.S., 1988, in "Interactions Plasma Froids Materiaux", ed. by C. Lejeune, GRECO 57 CNRS, Les Editions de Physique, p. 113.
Boeuf J.P. and Belenguer Ph., 1989, in "Non equilibrium processes in partially ionized gases", NATO ASI series, Eds M. Capitelli and J.N. Bardsley, to be published (Plenum Press, 1989)
Chapman B., 1980, 'Glow Discharge Processes', (John Wiley and Sons: New York)
Debontride H., Derouard J., Edel P., Romestain R., Sadeghi N. and Boeuf J.P., 1989, "Transient current and sheath motion following the photoelectron initiated avalanche in DC glow discharges", to appear in _Phys. Rev._ A
Derouard J. and Sadeghi N., 1986, _Opt. Comm._ 57, 239
Derouard J., Debontride H. and Sadeghi N., 1987, . _Physique_ 48, Colloque C7, 725
Derouard J., Debontride H., Alberta M.P. and Sadeghi N., 1989, in "Non equilibrium processes in partially ionized gases", NATO ASI series, Eds M. Capitelli and J.N. Bardsley, to be published (Plenum Press, 1989)
von Engel A., 1965, 'Ionized Gases', (Clarendon Press: Oxford)
von Engel A. and Steenbeck M., 1934, Elektrische Gasentladungen (Springer Verlag: Berlin), Vol 2, P 115
Graves D.B. and Jensen K.F., 1986, _IEEE trans. Plasma Science_ PS-14 78
Gottscho R.A. ,Mitchell A.,Scheller G.R. ,Chan Y.Y. and Graves D.B., "Electric Field Reversal in DC Negative Glow Discharges", to be published
Guntherschulze A., 1930, _Z. Phys._ 11, 49; 1930, _Z. Phys._ 59, 433
Gummel H.K., 1964, _IEEE Trans. Electron Devices_ ED-30, 1097
Hashiguchi S. and Hasikuni M., 1988, _Japan. J. Appl. Phys._ 27, 1010

Ingold J.H., 1978, in "Gaseous Electronics", Eds Hirsh M.N. and Oskam H.J. (Academic Press, 1978), pp. 19-64

Kagan Yu. M., 1985, _J. Phys. D_ 18, 1113

Kurata M., 1982, "Numerical Analysis for Semiconductor Devices", D.C. Heath and Company, (Lexington Mass, 1982)

Little P.F and von Engel A., 1954, _Proc. Roy. Soc._ A224, 209

Mavrodineau R., 1984, _J. Res. of the National Bureau of Standards_ 89, 143

Mitchell A., Scheller G.R., Gottscho R.A and Graves D.B., 1989, to appear in _Phys Rev._ A

Pillow M.E., 1981, _Spectrochimica Acta_ 36B, 821

Phelps A.V. ,Jelenkovic B.M. and Pitchford L.C., 1987, _Phys. Rev._ A 36, 537

Rozsa K., 1980, _Z. Naturforsch_ 35a, 649

Selberherr S. and Ringhofer C.A., 1984, _IEEE Trans. Computer-Aided Design_ CAD-3, 52

Selvin P.J.and Harrison W.W., 1975, _Applied Spectroscopy Reviews_ 10, 201

Sharfetter D.L.and Gummel H.K., 1969, _IEEE Trans. Electron. Dev._ ED-16, 64

Sturges D.J. and Oskam H.J., 1964, _J. Appl. Phys._ 35 2887

Sturges D.J. and Oskam H.J., 1966, _J. Appl. Phys._ 37, 2405

Surendra M., Graves D.B. and Jellum G.M., 1989, "Self-consistent models of DC glow discharges: treatment of fast electrons", to be published

Valentini H.B., 1987, _Contrib. Plasma Phys._ 27, 331

Ward A.L., 1962, _J. Appl. Phys._ 33, 2789

WEAK COLLISIONS IN STRONG DOUBLE LAYERS

H. Schamel[*]
P. Hatjiimanolaki and P. Nicoletopoulos[**]
[*]Institut für Physik, Univ. Bayreuth, 858 Bayreuth
FRG
[**]Faculté de Sciences, Université Libre de Bruxelles
1050 Bruxelles, Belgium

ABSTRACT

The theoretical basis of strong double layers with cold drifting particles is extended by including collisional friction of the primary electron beam. It is found that finite electric fields of different size exist in the regions adjacent to the double layer in accordance with stress balance. A comparison is made with a recent experiment on electron transmission spectroscopy.

I. INTRODUCTION

Double layers (DLs) are one-dimensional, electrostatic structures in the volume of a plasma consisting of two space-charge layers in close proximity and connecting monotonically two different levels of the electrostatic potential (Fig. 1). The conditions for their excitation which can vary from experiment to experiment are to a large extent unknown. However, when they are excited they represent an effective means of controlling the plasma's dynamical evolution, e.g. via current disruptions and the associateed potential relaxation oscillations (e.g. Iizuka et al., 1985). Only in few cases it was possible to resolve temporarily the steps towards the formation of DLs (e.g. SATO & OKUDA, 1981; BARNES et al., 1985). Generally speaking, DLs seem to be preceded

Physics and Applications of Pseudosparks
Edited by M. A. Gundersen and G. Schaefer
Plenum Press, New York, 1990

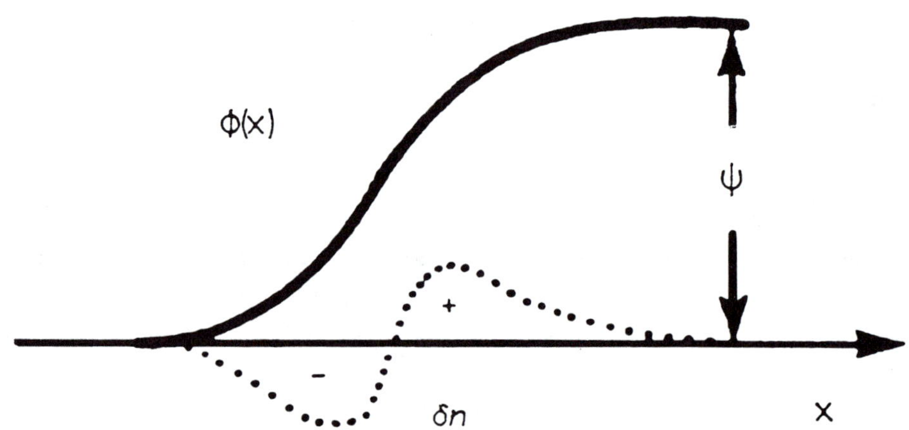

Fig. 1. A qualitative plot of a double layer potential. ψ denotes its amplitude and δn the surplus on space charge.

by solitary like potential pulses, such as ion and electron holes, negative potential dips etc. which serve as a triggering mechanism (see, for example, Chan's and Pécseli's article in Williams and Moorehead (eds.), 1986; or Schamel 1986).

Experimentally, DLs are found in current- and/or voltage-driven plasmas (see e.g. reviews of TORVÉN 1980 and N. SATO in Michelsen and Rasmussen, 1982) in triple plasma devices (Quon and Wong, 1976; Coakley and Hershkowitz, 1979), in Q-machines etc., especially under conditions in which collisional effects are weak or negligible. Theoretical work, on the other hand, to our knowledge has focussed exclusively on the collisionless approximation, i.e. on time-independent DL solutions of the Vlasov-Poisson system. A kinetic approach is usually necessary to cope with distributions deviating strongly from Maxwellian.

In the present paper, a first step is done in incorporating collisional effects in theory and a comparison is made with a recent FRANCK-HERTZ experiment which is seemingly masked by space charge effects. In section II a brief summary is given of collisionless DLs and a special, analytically easily tractable class of strong DLs is presented.

This class of DLs is then extended in section III by introducing collisional effects and section IV deals with the FRANCK-HERTZ experiment and its possible explanation in terms of DLs. A short summary concludes the paper.

II. SURVEY OF COLLISIONLESS DLs AND A SIMPLIFIED VERSION OF STRONG DLs

DL solutions of the 1-D, time-independent Vlasov-Poisson system

$$v\partial_x f_j - \frac{q_j}{m_j}\phi'(x)\, \partial_v f_j = 0 \qquad j = i,e \qquad (1)$$

$$\phi''(x) = \frac{e}{\varepsilon_0}[\int dv f_e - \int dv f_i] \qquad (2)$$

where the notation is standard, can be obtained in three steps (Schamel 1972, 1975, Schamel & Bujarbarua 1983).

(i)   first, one looks for invariants of the two Vlasov equations which are the energy $E_j = \frac{m_j}{2} v^2 + q\phi(x)$ and the sign of the velocity

σ = sgn v for particles which are not reflected by the potential structure, the so-called free particles or beam particles (drifting particles).

(ii) second, one specifies the distributions as functions of the invariants.

In regions, where reflected (or trapped) particles are absent, a shifted Maxwellian is in many cases a good representation of the experimental situation. For the trapped species a Maxwellian with a different temperature (which may be negative) joining continuously with the free particle distribution at the separatrix seems to be most natural.

(iii) third, one inserts these distributions which are functions of ϕ into (2) and solves the latter according to the DL boundary conditions: $E(x = -\infty) = 0 = E(x = +\infty)$ and $E'(x = -\infty) = 0 = E'(x = +\infty)$, where $E = -\phi'(x)$ is the electric field.

Three classes of DL solutions are found by this minimum set of distributions (a richer structure of the distributions can easily be incorporated) as shown schematically in Fig. 2 (Schamel 1983). Details of these solutions which are found in the original papers, are dropped here. Only the strong DL, represented by the first column, is discussed in more detail. The space pattern for ions (second row) and for electrons (third row), respectively, reveals that two well separated groups of particles, the trapped ones within the separatrix (dashed line) and the free ones, are needed. The free particles have to enter the DL region with finite drift velocity (called Bohm criterion) and in the region where both groups are present, the distributions are double-humped.

In cases where the thermal width of the free species is negligible (e.g. when the accelerating potential beyond the DL region is large) the following simplified analysis for strong DLs holds.

Integrating the distributions over the velocity one gets the following normalized density expressions:

$$n_e = a \, \varepsilon(\phi) + (1 + 2\phi v_0^{-2})^{-\frac{1}{2}} \qquad (3)$$

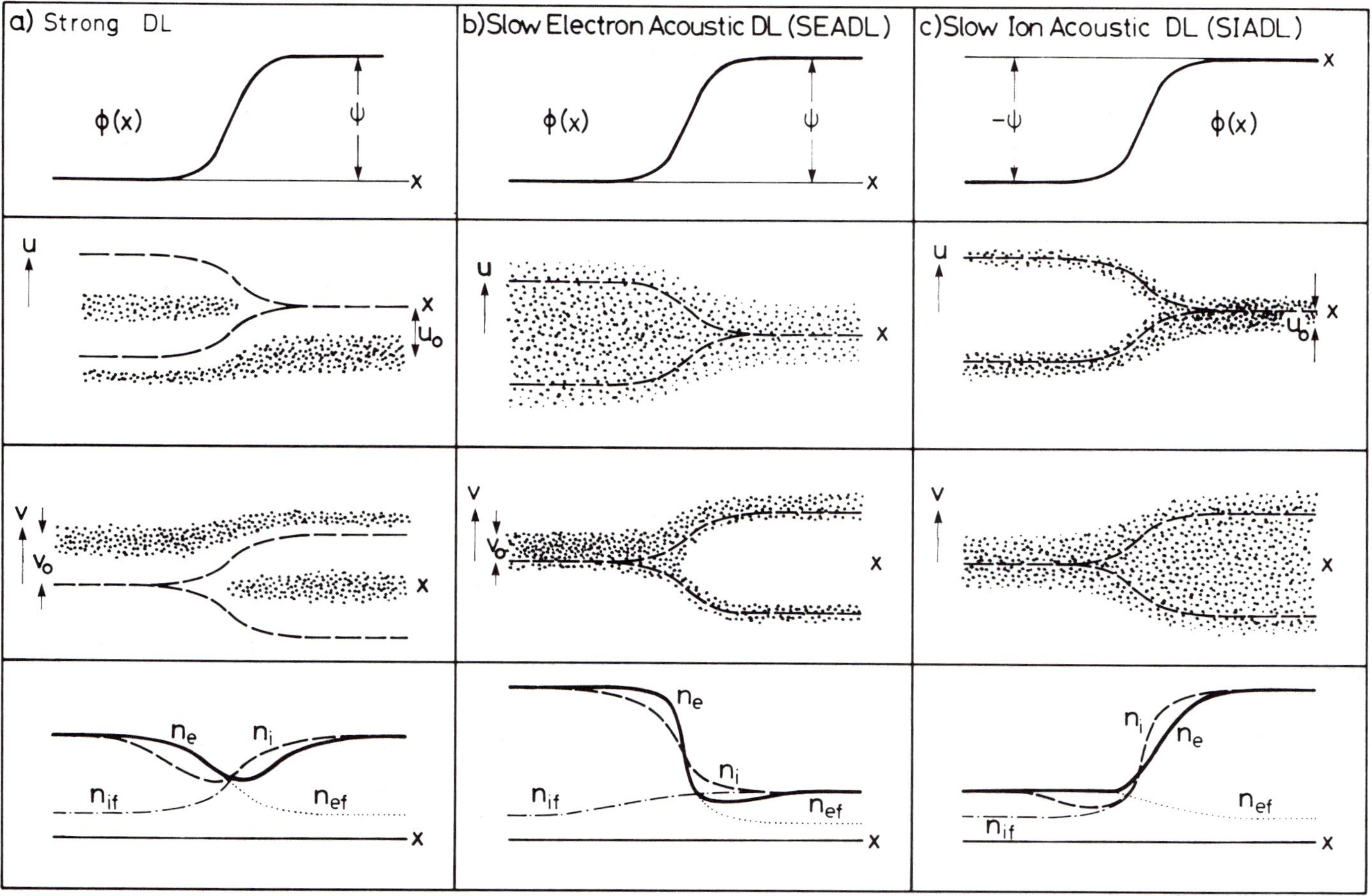

Fig. 2. Schematic plot of three types of double layers. The second (third) row represents the ion (electron) phase space, whereas the various densities are plotted in the fourth row.

$$n_i = N\left[b\varepsilon\ (\tau(\psi-\phi))\ +\ (1+2(\psi-\phi)\ u_0^{-2})^{-\frac{1}{2}}\right] \qquad (4)$$

This densities are normalized by the plasma density n_0 at $\phi = 0$. The electric potential is normalized by $T_{et}e^{-1}$ where T_{et} is the thermal energy of the trapped electrons. The maximum value of ϕ is denoted by ψ (Fig. 1). The first expression in (3) and (4) represents the contribution of trapped particles, the strength of which is measured by a and b, respectively. The function $\varepsilon(x)$ is defined by

$$\varepsilon(x) = \exp x\ \mathrm{erf}\ x^{\frac{1}{2}} \qquad (5)$$

It is obtained in the electron case by integrating the (normalized) distribution $f_e \sim \exp[-\frac{1}{2}(v^2-2\phi)]$ over the trapped region $|v| \le (2\phi)^{\frac{1}{2}} \equiv v_{tr}$, where v_{tr} is the trapping velocity. Note that $\varepsilon(x)$ vanishes for $x = 0$. There are, hence, no trapped electrons in the low potential ($\phi = 0$) and no trapped ions in the high potential ($\phi = \psi$) region.

The second expression is due to the free particles; v_0 and u_0 which can have both signs represent the normalized drift velocity of free electrons at $\phi = 0$ and that of the free ions at $\phi = \psi$, respectively (Fig. 2); they are normalized by $(T_{et}m_e^{-1})$ and $(T_{it}m_i^{-1})^{\frac{1}{2}}$, respectively. The normalized density at $\phi = \psi$ is given by N. $\tau \equiv T_{et}/T_{it}$ represents the temperature ratio of the trapped particles. At this stage one has therefore introduced seven parameters a, b, τ, v_0, u_0, ψ and N, which are, however, not entirely independent, as shown next.

Integrating Poisson's equation(2), one gets

$$\phi'(x)^2/2\ +\ V(\phi)\ =\ 0 \qquad (6)$$

where the "classical potential" $V(\phi) \equiv \int_0^{\phi}[n_i(\phi')- n_e(\phi')]\ d\phi'$, which vanishes together with the electric field at $\phi = 0$, becomes

$$V(\phi) =$$
$$ah(\phi)+v_0^2[1-g(\phi,v_0)]-Nb\tau^{-1}[h(\tau\psi)-h(\tau(\psi-\phi))]+Nu_0^2[g(\psi,u_0)-g(\psi-\phi,u_0)] \qquad (7)$$

where

$$h(x)\equiv 2(x/\pi)^{\frac{1}{2}}-\varepsilon(x)$$

$$g(x,y) \equiv (1 + 2xy^{-2})^{\frac{1}{2}} \qquad (8)$$

Eq.(6) describes the stress balance within the DL and can be viewed as the analogon to Bernoulli's law in hydrodynamics.

Three further conditions (conform with (iii) in section II) have to be satisfied to get DL solutions:

a) quasineutrality at $\phi = 0$

$$1 = N[b\varepsilon(\tau\psi) + g(\psi,u_0)^{-1}] \tag{9}$$

b) quasineutrality at $\phi = \psi$

$$a\varepsilon(\psi)+g(\psi,v_0)^{-1}=N \tag{10}$$

c) no electric field at $\phi = \psi$ $(V(\psi) = 0)$

$$ah(\psi) + v_0^2[1-g(\psi,v_0)] = Nb\tau^{-1}h(\tau\psi) + Nu_0^2[1-g(\psi,u_0)] \tag{11}$$

A DL solution is finally obtained if in addition holds

$$V(\phi) < 0 \qquad \text{in } 0 < \phi < \psi \tag{12}$$

A sketch of $V(\phi)$ is given in Fig.3.

That the system (9)-(12) admits DL solutions can easily be seen by setting $N = 1$, $u_0 = v_0$, $\tau = 1$, $b = a$. In this symmetric case (11) is fulfilled identically and (9), (10) coincide and become

$$a\varepsilon(\psi)+g(\psi,v_0)^{-1}=1 \tag{13}$$

The classical potential, on the other hand, behaves near $\phi \approx 0$, like

$$V(\phi) = \frac{-4a}{3\sqrt{\pi}}\phi^{3/2} - \frac{\phi^2}{2} (g-1)\left[\frac{1}{g} - \frac{g+1}{2\psi}(1+g^{-3})\right] + 0 \, (\phi^{5/2}) \tag{14}$$

where g stands for $g(\psi,v_0)$ and an identical expression is obtained for V near $\phi = \psi$ whereby ϕ in (14) is replaced by $\psi-\phi$. Since g>1, and since a is generally small (see later), conditions (12) requires that the bracket in (14) is positive. This yields

$$\psi > \frac{g(g+1)}{2}(1+g^{-3}) \tag{15}$$

In view of g>1 the DL amplitude must be finite and exceed unity. Small amplitude strong DLs do not exist in agreement with Schamel & Bujarbarua (1983). This is the reason why the DL is called strong.

Using g as the classifying parameter one gets from (13)

$$a\varepsilon(\psi) = \frac{g-1}{g} \tag{16}$$

and from the definition of g

$$v_0{}^2 = \frac{2\psi}{g^2-1} \tag{17}$$

If g only slightly deviates from unity $v_0{}^2 = \frac{\psi}{g-1} \gg 1$. v_0 decreases if g increases. In the case of large g, (15) demands $\psi > g^2/2$, hence ψ must be large. From (17) one gets $v_0{}^2 > 1$. Hence $v_0 \approx 1$ is the lowest possible value for v_0. (g $\gg$ 1 is equivalent with $1 \approx a\varepsilon(\psi) \approx ae^{\psi}$ and hence $a \approx e^{-\psi}$ which is small).

One thus arrives at the BOHM criterion for strong DLs which says that for its existence the free species have to be preaccelerated and to form a beam before they enter into the DL region.

This concludes the analytic proof that strong DLs exist. Of course, a larger variety of solutions including different densities (N≠1), different trapped temperatures (τ≠1), different drift velocities ($u_0≠v_0$) and different contributions of trapped particles (b≠a) can be obtained.

III. STRONG DOUBLE LAYERS IN WEAKLY COLLISIONAL PLASMAS

We now turn our attention to the influence of collisons. Since it is primarily the electron beam which experiences collisions we concentrate on this dissipation mechanism.

First, we note that the free electron contribution to the density in (1) can be obtained from the normalized continuity and momentum equation

$$\partial_t n + \partial_x (nu) = 0 \tag{18}$$

$$\partial_t u + u\partial_x u = \partial_x \phi \tag{19}$$

Indeed, looking for stationary solutions and using the boundary conditions: n = 1, u = v_0 at ϕ = 0 we get nu = v_0 and $\frac{u^2}{2} - \phi = \frac{v_0{}^2}{2}$ from which we get n = $(1+2\phi/v_0{}^2)^{-\frac{1}{2}}$, the desired result. It is then clear how to incorporate collisional friction. In the simplest possible fashion we add on the rhs of (19) a friction term $-\nu u$, which models the momentum loosing collisions between electrons and neutrals in a weakly ionized plasma or the electron-ion collisions in a fully ionized plasma. Dropping the time derivative we replace (19) by

$$u\partial_x u = \partial_x \phi - \nu u \tag{20}$$

The collision frequency ν may depend on x, or ϕ via $x = x(\phi)$, the inverse of the DL expression $\phi = \phi(x)$.

If the x-dependence of u is through $\phi(x)$, (20) yields

$$[uu'(\phi) - 1]\,\phi'(x) = -\nu u \qquad (21)$$

This expression has the interesting consequence that in regions where ν and u are both nonzero there must exist a <u>finite</u> electric field $E(x) = -\phi'(x)$, since in general the bracket in (21) does not vanish. Hence, in a collisional plasma the asymptotic region adjacent to DLs can no longer be field free. The DL is now embedded in a non-vanishing electric field region. This field will be generally quasi-neutral ($\phi'' \approx 0$) and its relations to the electron current is given by classical resistivity.

To see how ν changes the DL solution we modify the DL boundary conditions as follows

$$\phi = 0,\ \phi' = -E_1 \qquad \text{at } x = x_1$$
$$\phi = \psi,\ \phi' = -E_2 \qquad \text{at } x = x_2 \qquad (22)$$

where $x_1 \leq x \leq x_2$ is the region where the electric potential changes fast, i.e. on the Debye scale, the short scale length. This is our modified DL region (Fig.4). As before, we demand

$$u = v_0,\quad n = 1 \qquad \text{at } \phi = 0 \ (x = x_1) \qquad (23)$$

Spatial integration of (20) gives

$$\frac{u^2}{2} = \frac{v_0^2}{2} + \phi - \int_{x_1}^{x} dx'\,\nu(x')u(x') \qquad (24)$$

an equation which can be solved iteratively in the case finite ν or perturbatively for small ν. Assuming weak collisions we follow the second approach and assume $\nu \ll 1$.

If the zeroth order ($\nu \equiv 0$) solution is denoted by $u_0(\phi) \equiv \sqrt{v_0^2 + 2\phi}\,$. sgn v_0 the following solution, first order in ν, holds

$$u(\phi) = u_0(\phi) - \frac{1}{u_0(\phi)}\int_0^{\phi} d\tilde{\phi}\,\nu(\tilde{\phi})u_0(\tilde{\phi})\Big/\frac{d\phi}{dx}\Big|_{x=x(\tilde{\phi})} \qquad (25)$$

where we substituted $\tilde{\phi} = \phi(x')$ resp. its inverse and set $\nu(\tilde{\phi}) \equiv$

$\nu(x'(\tilde{\phi}))$. To lowest order, the denominator under the integral can be replaced by $\sqrt{-2V_0(\tilde{\phi})}$, where we used (6) and the subscript zero refers to the unperturbed classical potential (7).

Introducing the abbreviation

$$u_1(\phi) \equiv u_0^{-2}(\phi) \int_0^{\phi} d\tilde{\phi}\,\nu(\tilde{\phi})u_0(\tilde{\phi})[-2V_0(\tilde{\phi})]^{-\frac{1}{2}} \tag{26}$$

(25) can be written as

$$u(\phi) = u_0(\phi)\,[1-u_1(\phi)] \tag{27}$$

Substituting this expression into the integrated continuity equation $nu = v_0$, where we used (23), we get

$$n = (1+2\phi/v_0^2)^{-\frac{1}{2}}[1+u_1(\phi)] \tag{28}$$

This is the corrected version of the free electron density and replaces the corresponding expression in (3).

The correction to the classical potential is then easily found from the integral in the text which follows (6) and reads

$$V(\phi) = V_0(\phi) + V_1(\phi) \tag{29}$$

with

$$V_1(\phi) = -v_0\int_0^{\phi} d\hat{\phi}\frac{\hat{u}_1(\hat{\phi})}{u_0(\hat{\phi})} - \frac{E_1^2}{2} \tag{30}$$

where (22) has been used. A quasi DL solution is finally obtained if $E_2^2/2 + V_1(\psi) = 0$ holds which follows from the second boundary condition in (22). We then have

$$v_0\int_0^{\psi} d\hat{\phi}\frac{\hat{u}_1(\hat{\phi})}{u_0(\hat{\phi})} = \tfrac{1}{2}(E_2^2 - E_1^2) \tag{31}$$

This is our final result. A finite value of the integral in (31) implies <u>different</u> electric fields in the adjacent quasineutral regions. Since both u_0 and u_1 carry the sign of v_0, the quasineutral electric field $|E_2|$ at the high potential side exceeds that of the low potential side $|E_1|$ if v_0 is positive, i.e. if the electron beam is injected from the low potential side. If v_0 is negative and hence injected from the high potential side the reverse holds $|E_1| > |E_2|$. Both situations are sketched in Figs. 4a,b.

286

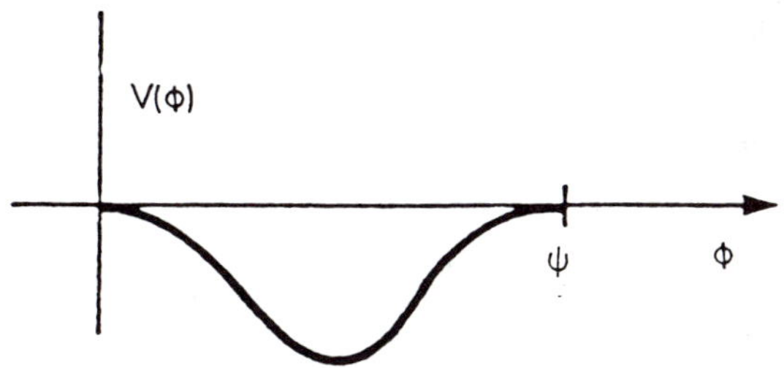

Fig. 3. The classical potential as a function of the electric potential, qualitatively.

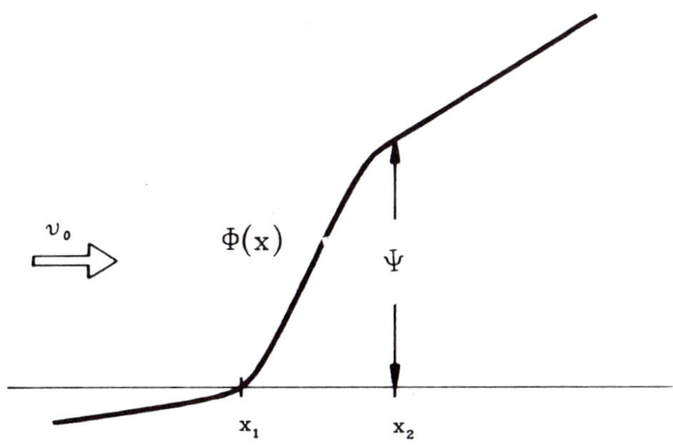

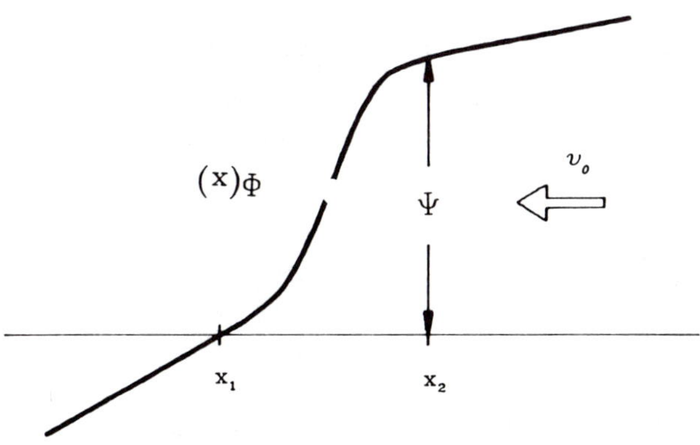

Fig. 4. A qualitative plot of the double layer transition modified by collisional friction with primary electrons streaming in a) from the low potential side and b) from the high potential side.

The symmetry with respect to $v_0 \leftrightarrow -v_0$ is now broken. This result can be interpreted in terms of the stress or pressure balance. The left hand side of (31) represents the frictional contribution to the dynamical pressure. It is positive (negative) if $v_0 > 0$ ($v_0 < 0$) and adds (subtracts) to the electric field stress term $E_1^2/2$. It is a kind of ram pressure which must be balanced by a higher electric field stress term on that side which is opposite to the injection region. This result has therefore a simple physical explanation.

Note that the double integral appearing in (31) can be reduced to a simple one by exchanging the order of integration. The stress balance equation (31) then becomes

$$v_0 \int_0^\psi d\tilde{\phi} \frac{\nu(\tilde{\phi})}{[-2V_0(\tilde{\phi})]^{\frac{1}{2}}} \left[1 - \frac{u_0(\tilde{\phi})}{u_0(\psi)}\right] = \tfrac{1}{2}(E_2^2 - E_1^2) \tag{32}$$

The double layer region $\Delta \equiv x_2 - x_1$ is affected as follows. Following Schamel & Bujarbarua (1983), Δ is given by

$$\Delta \equiv \psi\, d^{-\frac{1}{2}} \tag{33}$$

where d is the depth of the classical potential . If subscript 0 denotes unperturbed quantities we find

$$\Delta = \Delta_0 \left[1 + \frac{V_1(\phi_{min})}{2d_0}\right] \tag{34}$$

where ϕ_{min} is the location in ϕ where V_0 becomes minimum. Since V_1 is strictly negative, friction narrows the transition region, as expected from the ram pressure argument. Furthermore, since $V_1(\phi)$ for $v_0 > 0$ is generally less than $V_1(\phi)$ for $v_0 < 0$ the situation sketched in fig.4a has the smaller transition region, as expected.

We conclude that DLs in collisional plasmas are accompanied by pre- and post sheath neutral regions, the strength of the sheath electric fields depending on the direction of injection of the electron beam.

IV. INDIRECT EVIDENCE OF STRONG DOUBLE LAYERS IN A FRANCK-HERTZ EXPERIMENT

As a possible application we briefly describe the results of a recent electron transmission spectroscopy experiment (Nicoletopoulos 1988). The tube used in this experiment is a commercially available Franck-Hertz tetrode, a schematic view of which is shown in Fig.5. Whereas grid G2 was essentially at the anode potential, the voltage at grid G1 was varied and the anode current was recorded as a function of the potential difference U_{KG1} between cathode and grid G1. If there was no potential difference between anode and G1 ($\Delta U = 0$) the current-voltage characteristic is given in Fig.6 by the solid line. It exhibits the various excitation levels of the mercury atoms. If, however, an effective potential drop ($\Delta U \neq 0$) between G1 and G2 was applied, the current-voltage characteristic was characteristically modified as seen by the dashed line in Fig.6. We see two different kinds of structures, a "standing" spectrum denoted by characters and a "moving" spectrum denoted by numbers. The moving spectrum is systematically shifted by an amount ΔU on the energy scale in comparison with the $\Delta U = 0$ case.

A possible explanation is indicated in Fig.7 which shows schematically a possible potential distribution in the tube without (solid line) and with (dashed line) a potential drop. In the latter case a standing DL is assumed which divides the drift region in two parts denoted by I and II. Whereas region I is claimed to be responsible for the standing spectrum, in this region the primary electron beam is the same as for $\Delta U = 0$, region II may account for the moving spectrum. In the latter case, due to the deceleration of the primary electron beam by the DL, the grid potential has to be increased by ΔU to reach the same excitation conditions as for $\Delta U = 0$. The well defined shift of the moving spectrum suggests a narrow internal potential drop, the suggested DL. This DL is probably a strong one of the type sketched in Fig.4b with a reversed x-direction. Due to collisional friction a presheath on the high potential side exists which accelerates the ions, created by impact ionization, to perform an ion beam satisfying the Bohm criterion. These free streaming particles together with the trapped ones, low energy ions on the low and low energy electrons on the high potential side, thus may provide the space charge conditions necessary for the maintenance of a strong double layer. More details, especially a more quantitative analysis, are presented in a forthcoming publication (Schamel, Hatjimanolaki and Nicoletopoulos, 1989). We conclude that there is some indirect evidence of a strong DL in a Franck-Hertz experiment being due to space charge effects.

Fig. 5.   Sketch of the Franck-Hertz tetrode.

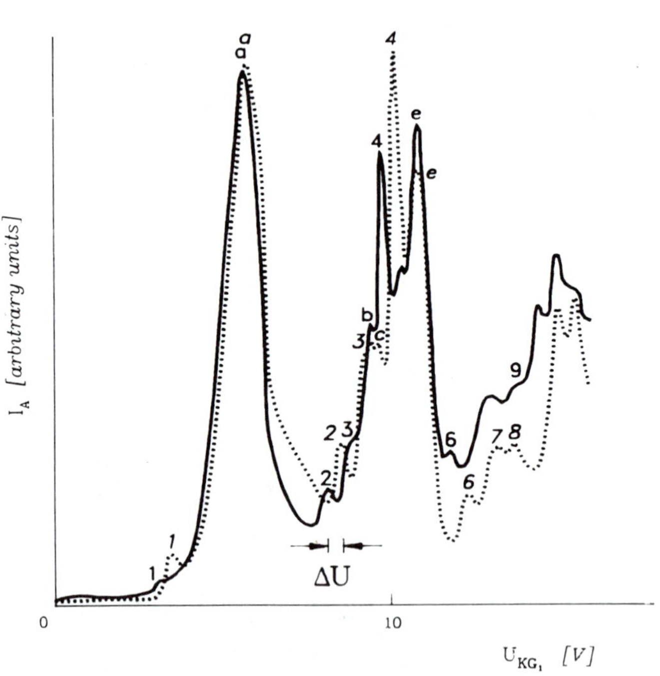

Fig. 6.   The anode current I_A as a function of the cathode-grid1 potential U_{KG1} without an applied potential drop (ΔU = 0) between the two grids (solid line) and with ΔU ≠ 0 (dashed line). Characters denote the standing and numbers the moving spectrum.

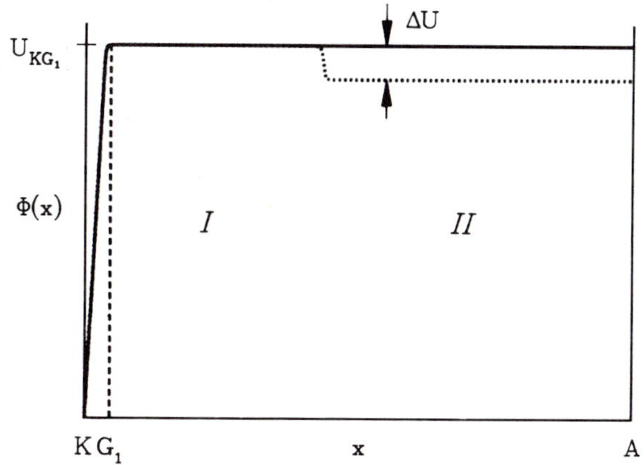

Fig. 7.   The suggested potential distribution for the two situations of
         Fig. 6, showing the excitation of a double layer in the case
         ΔU ≠ 0.

V. SUMMARY

In summary, we have included the effect of weak collisions on the class
of strong double layers consisting of monoenergetic beam electrons (and
ions). It is found that there is a unique relation between the
collisional friction and the existence of a pre- and a postsheath with a
finite quasineutral electric field. The symmetry with respect to the
direction of injection is broken and a narrowing of the double layer
width is observed. Finally, comparison with a recent Franck-Hertz tube
experiment shows evidence of a stong double layer caused by space
charge.

REFERENCES

Barnes C., Hudson M.K., and Lotko W. Phys. Fluids $\underline{28}$, 1055 (1985)

Coakley P., and Hershkowitz N., Phys. Fluids $\underline{22}$, 1171 (1971)

Iizuka S., Michelsen P., Rasmussen J.J., Schrittwieser R., Hatakeyama
R., Saeki K., and Sato N., J.Phys.Soc. Japan $\underline{54}$, 2516 (1985)

Nicoletopoulos P. (1988), to be published

Sato T., and Okuda H., J.Geophys.Res. $\underline{86}$, 3357 (1981)

Quon B.H., and Wong, A.Y., Phys.Rev.Lett. $\underline{37}$, 1393 (1976)

Schamel H., Plasma Phys. $\underline{14}$, 905 (1972)

Schamel H., J.Plasma Phys. $\underline{13}$, 139 (1975)

Schamel H., and Bujarbarua S., Phys. Fluids $\underline{26}$, 190 (1983)

Schamel H., Z. Naturforsch. $\underline{38a}$, 1170 (1983)

Schamel H., Phys. Reports $\underline{140}$, 161 (1986)

Schamel H., Hatjimanolaki P., and Nicoletopoulos P. (1989), to be published

Torvén S., in Proc.Int.Conf. Plasma Physics, Nagoya, Japan, Vol.II, Fusion Research Association of Japan, p. 89 (1980)

Williams A.C., and Morehead T.W. (eds.): Proceeding of the Workshop on Double Layers in Astrophysics, G.C. Marshall Space Flight Center, Huntsville, Alabama, USA, NASA Conf.Publ. No. 2469 (1986).

THE EFFECT OF PENDEL ELECTRONS ON BREAKDOWN AND SUSTAINMENT OF A HOLLOW CATHODE DISCHARGE

K.H. Schoenbach, L.L.Vahala, G.A. Gerdin, N. Homayoun, F. Loke

Old Dominion University
Department of Electrical & Computer Engineering
Norfolk, Virginia 23508-0246

and G. Schaefer

Weber Research Institute, Polytechnic University
Farmingdale, N.Y. 11735

Although the basic properties of hollow cathode discharges have been known for over 70 years [1], the applications of these discharges have been rather limited. They have been used mainly as a spectroscopic light source of high emission efficiency, low power consumption and small doppler broadened line width [2]. Recently, there has been considerable renewed interest in hollow cathode discharges because of their possible application as high power switches. They utilize a cold cathode and under various operating conditions have switched over 100 kA, demonstrated over $2\text{x}10^{12}$ A/s dI/dt, and achieved subnanosecond jitter [3].

Two types of pulsed hollow cathode discharges have been observed in experiments [4] with He as fillgas: a low current discharge ($\leq$ 1 A) at low pressures and a high current discharge ($\geq$ 100 A) at higher pressures. It is suggested, from these experimental results that the two types of discharges represent two temporal stages of the hollow cathode discharge at a given pressure. In modeling the breakdown of the hollow cathode discharge, it is assumed that a predischarge is initiated by the Paschen breakdown mechanism. In the Paschen curve of the breakdown potential versus pxd (where p is the pressure and d the electrode spacing), hollow cathode discharges operate to the left of the Paschen minimum. Thus the breakdown into the low current mode, the predischarge, occurs over the longest possible separation distance between the anode and the cathode. In the geometry of Fig. 1, this maximum axial distance occurs between the anode and the bottom of the cylindrical hollow cathode. The breakdown voltage for the predischarge was measured as a function of pressure and is shown in Fig. 2 for He as filling gas (lower curve). It resembles a Paschen curve with a minimum in the 1 Torr range. The glow discharge, which develops along the axis in our device, is confined radially and carries currents in the range of several 100 mA at high sustaining electric fields E. In this discharge, a large percentage of the electrons do not collide with the gas atoms and these collisionless electrons will constitute an electron-beam of energy determined by the applied voltage [5].

Before a radially confined plasma column is formed on the discharge axis, the

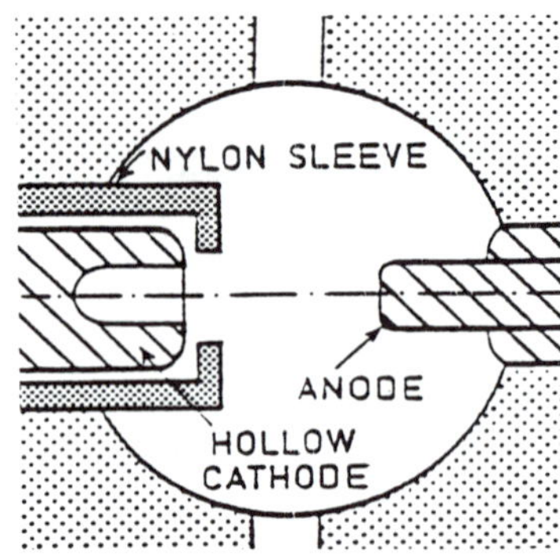

Fig.1. Cross-sectional view of the discharge chamber [4].

potential V(0,h) on axis at the top of the cathode hole will be approximately equal to the cathode potential. However, after the thin plasma column is formed, V(0,h) will approach the anode potential. The electric field distribution in the hollow cathode region will change in such a way that there is a substantial increase in the radial component of the electric field at the orifice. For sufficiently high fields in this region, a second breakdown can be expected from the walls of the cathode wall to the center of the axial plasma column. This will rapidly lead to the establishment of a high current hollow cathode discharge, the main discharge. The voltage required for the transition into the high current discharge is shown in Fig. 2 (upper curve). Although seemingly quite different in shape from the Paschen curve, it may be transformed into a similar curve by dividing the pressure by a constant factor. This transformation corresponds to a reduction in distance by the same factor, since the Paschen curve is dependent on the product pxd, rather than on p alone. This indicates that the second breakdown occurs over a smaller distance than the first one. The fact that this breakdown path is a radial one is proven by experimenting with axial dc-magnetic fields. It could be shown that the application of magnetic fields in the direction of the first axial discharge, prevents the breakdown into the second mode in a certain pressure range; in our opinion,

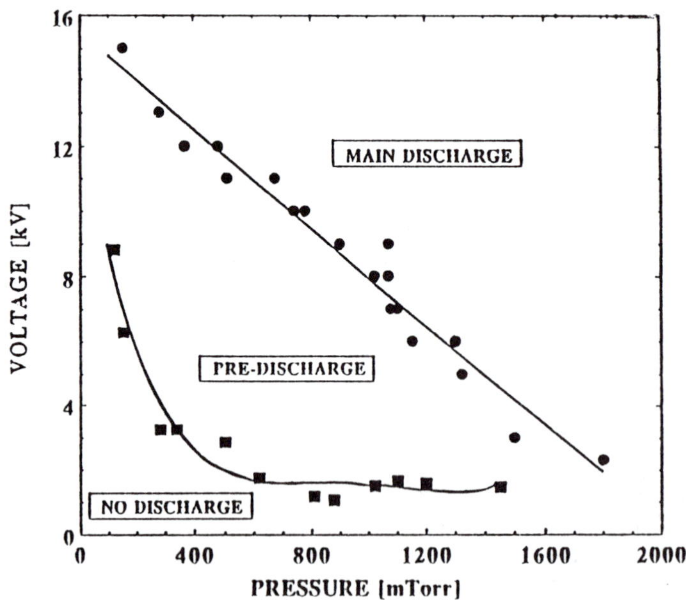

Fig.2.Breakdown voltage for the predischarge (lower curve) and the main discharge (upper curve) versus pressure [4].

294

an indication of a radial breakdown. This radial breakdown will most likely occur between the edges of the cathode hole and the center plasma column, because of the maximum of radial electric field intensity in this plane. The distance, the radius of the cathode hole is 0.43 cm compared to a distance of 2.3 to 3.6 cm from the anode to the bottom of the cathode hole.

The stages of the discharge development are schematically shown in Fig. 3. A possible third stage, which was observed in experiments at higher current levels is a discharge determined by a hot cathode operation, the superdense glow discharge [6]. This transition into the superdense mode might be caused by accumulative heating of the region about the hollow cathode edge through ion impact during the second phase, the main discharge phase, observed in our experiments [4]. The current gain of the main hollow cathode discharges over standard glow discharges between plane parallel electrodes is assumed to be caused by electrons trapped in the potential well of the axial plasma column. Electrons emitted from the walls of the cylindrical cathode cavity are accelerated towards the cylindrical axis of the cavity. These electrons contribute mainly to the ionization in the negative glow which is on the axis of the hollow cathode and serves to transport the thermalized electrons in the axial direction towards the anode. Some electrons, however, will cross the negative glow region without significant energy loss. In a linear discharge, these electrons would play no further role in the negative glow region. However, in the hollow cathode geometry these electrons will be radially reflected back into the negative glow region from the cathode fall in front of the opposite cathode surface and traverse it again. This pendulum motion will result in an enhanced ionization rate within the negative glow.

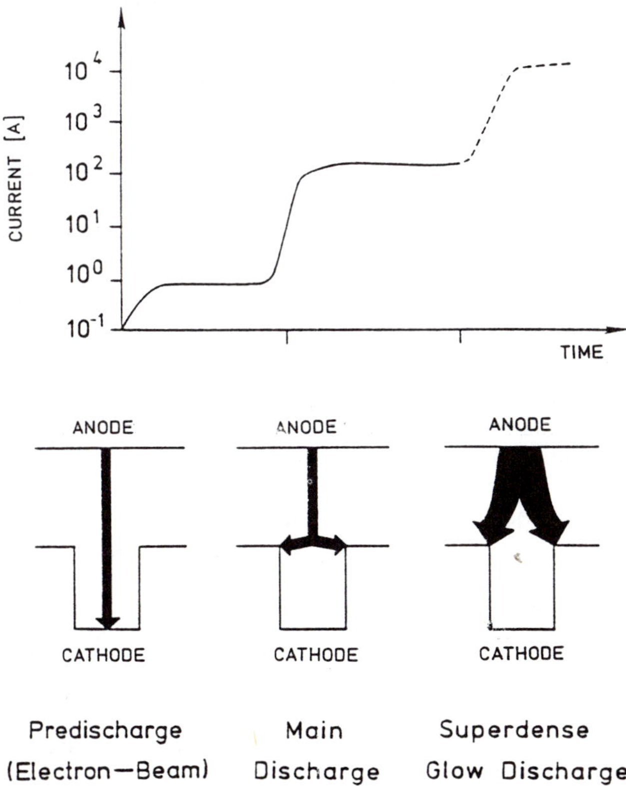

Fig.3. Temporal development of the current in a hollow cathode discharge. The different related stages of the discharge are depicted graphically in the lower diagram.

The "pendel-electron" effect [7], may also explain the second breakdown in the radial discharge between the cathode orifice and the filamentary plasma on the axis. In order to obtain the breakdown condition it is generally assumed that the electrons released from the cathode by ion impact generate on their way to the anode enough ions to provide for current continuation at the cathode. The equations which lead to the so-called Paschen curve are:

$$n_c = \gamma n_i$$

and

$$n_i = n_c[\exp(\int_0^R \alpha(v(x))dx) - 1]$$

where n_c is the electron density at the cathode, n_i is the ion density at the cathode, α is the first Townsend coefficient with $\alpha(v) = n\sigma_1(v)$, n being the gas density and $\sigma_i(v)$ the velocity dependent cross-section for ionization; γ is the coefficient for electron generation at a metal surface by ion impact. The distance from the wall to the center of the hollow cathode is R. For the Paschen breakdown it is generally assumed that α can be expressed as a function of E and p

$$\alpha = Ap \exp\left(\frac{-B}{(E/p)}\right)$$

with A, B being gas constants and E and p, the electric field intensity and the pressure, respectively. The voltage between anode and cathode is $V = ER$.

In the radial hollow cathode discharge the number of ions reaching the cathode and providing for secondary electrons is not only determined by the ionization coefficient α in the space between cathode and anode but also by ion generation in the filamentary plasma which serves as anode. Electrons, which for an ordinary Townsend discharge reach the anode with a certain energy, deposit this energy in the form of thermal energy in the anode. In a radial hollow cathode discharge, however, the electrons which reach the center plasma column will oscillate through it and gradually lose their energy through collisions, part of which are ionizing collisions.

In order to take the pendel effect into account for the modeling of the breakdown, a modified ionization coefficient can be introduced [8], to describe the electron behavior in hollow cathode discharges. An increase in effective ionization will cause a lowering of the breakdown voltage compared to the ordinary situation where breakdown between plane-parallel electrodes is described. A breakdown model which uses an equilibrium approach, however, can give us only qualitative results. A quantitative approach requires the study of nonequilibrium electrons kinetics in the hollow cathode geometry by means of microscopic techniques. For the description of the pendel electrons especially, for which the velocity is certainly not in equilibrium with the local electric field, this approach is imperative.

In order to study the effect of these pendel electrons on the sustainment of a hollow cathode discharge, we have performed one dimensional steady state Monte Carlo calculations for the electrons in the hollow linear cathode region. As a first step, we have ignored all axial z-dependences and performed Monte-Carlo simulations along a line between the two cathodes perpendiculum to the z-axis. The initial electric field E_x is chosen consistent with an uniform ion density in the hollow cathode region $0 < x < R$: $E_x(x) \approx x$. $E_x(0) = 0$ and at the cathode wall $E_x(R) = 2$ kV/cm with R being

0.5 cm. For this E_x-field, a Monte-Carlo code is used to determine the electron kinetics perpendicular to the main axis of the discharge system [electron velocity distribution, average electron velocity, rate coefficients and the electron collision frequency], propagating away from the cathode wall, x = -R. The gas which was modeled is He. The cross-sections which we have used were taken from a report by Hayashi [9]. The data consists of the total cross section; the cross sections for momentum transfer and ionization. Scattering was assumed to be isotropic and it was also assumed that the energy between outgoing electrons after an ionization was shared equally between the electrons. The initial electron energies at the cathode surface are assumed to be uniformly distributed between zero and some appropriate upper energy. The electrons moving through the center part of the hollow cathode discharge will (eventually) encounter a potential well. Some of the electrons will be trapped in this well for a sufficient time that, in the full two-dimensional problem, would then be accelerated axially to the anode.

Phase diagrams for randomly selected electrons show the possible motion in phase space: in Fig. 4a the electron traverses through the center without collisions and reaches the opposite wall. In Fig. 4b the electron "pendels" once before it is trapped. In Fig. 4c the electron is trapped after several collisions. Since our one-dimensional model cannot take into account the axial electric field E_z explicitly, we mimic this effect by stipulating in the Monte-Carlo simulation that any electron localized around the z-axis after a time Δt is removed from the calculation. This time is estimated on an average electron speed in axial direction and is typically chosen to be 80 ns.

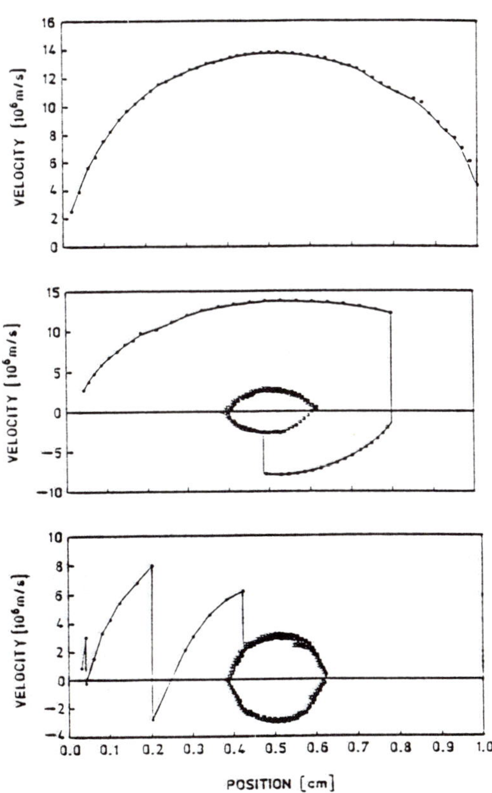

Fig.4. Phase diagrams for electrons in a hollow cathode geometry with two plane parallel cathodes separated by a distance of 1 cm. The electric field is given as E(x) = [2|x|/R]kV/cm .

The Monte-Carlo technique was used to compute the electron velocity distribution for two different filling pressures : p = 150 mTorr, and p = 500 mTorr. The velocity distribution as a function of position along the x-axis are shown in Fig. 5 for p = 500 mTorr. The electrons emitted from the wall at x = -R form a beam with an increasing electron speed towards the center of the hollow cathode system. Electrons which are scattered out of the beam or electrons which have "pendeled" through the center form the distribution with negative velocities. Their number is small compared to the forward moving electrons at positions close tot he cathode. Closer to the middle of the

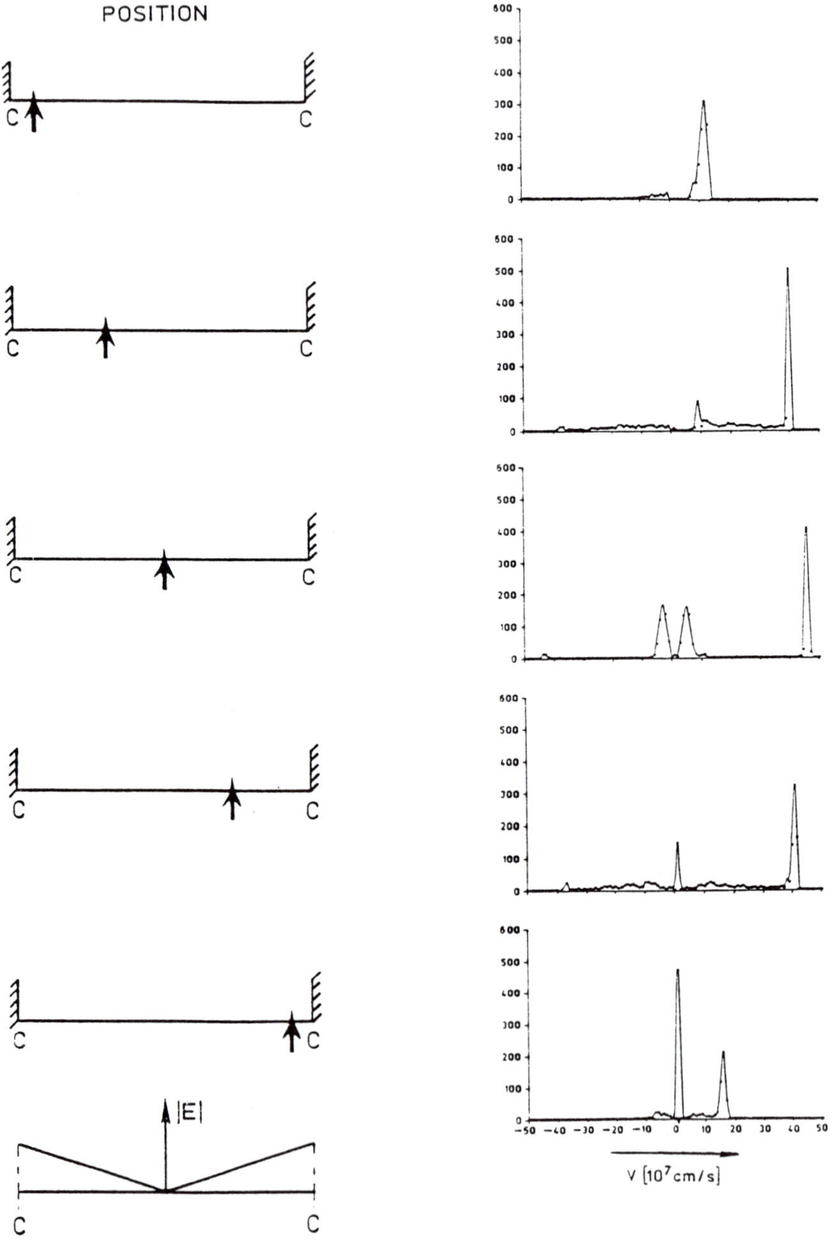

Fig.5.Velocity distributions of electrons, emitted from the left cathode, at different positions along the gap between the two cathodes for p = 500 mTorr.

system we find an increasing number of trapped electrons with relatively small velocities; the drop in the center of the distribution function (v 0) arises since we follow the electrons for a limited time only (80 ns) which is too short to allow relaxation of the electron velocity in the x-direction to a value close to zero. Following the distribution across the center shows a slowdown of the beam electrons and an increasing number of electrons with small energies. The peak close to $v = 0$ corresponds to electrons which turn at this position and pass through the velocity $v = 0$. A similar behavior appears in the case of $p = 150$ mTorr (Fig.6). The main difference is the

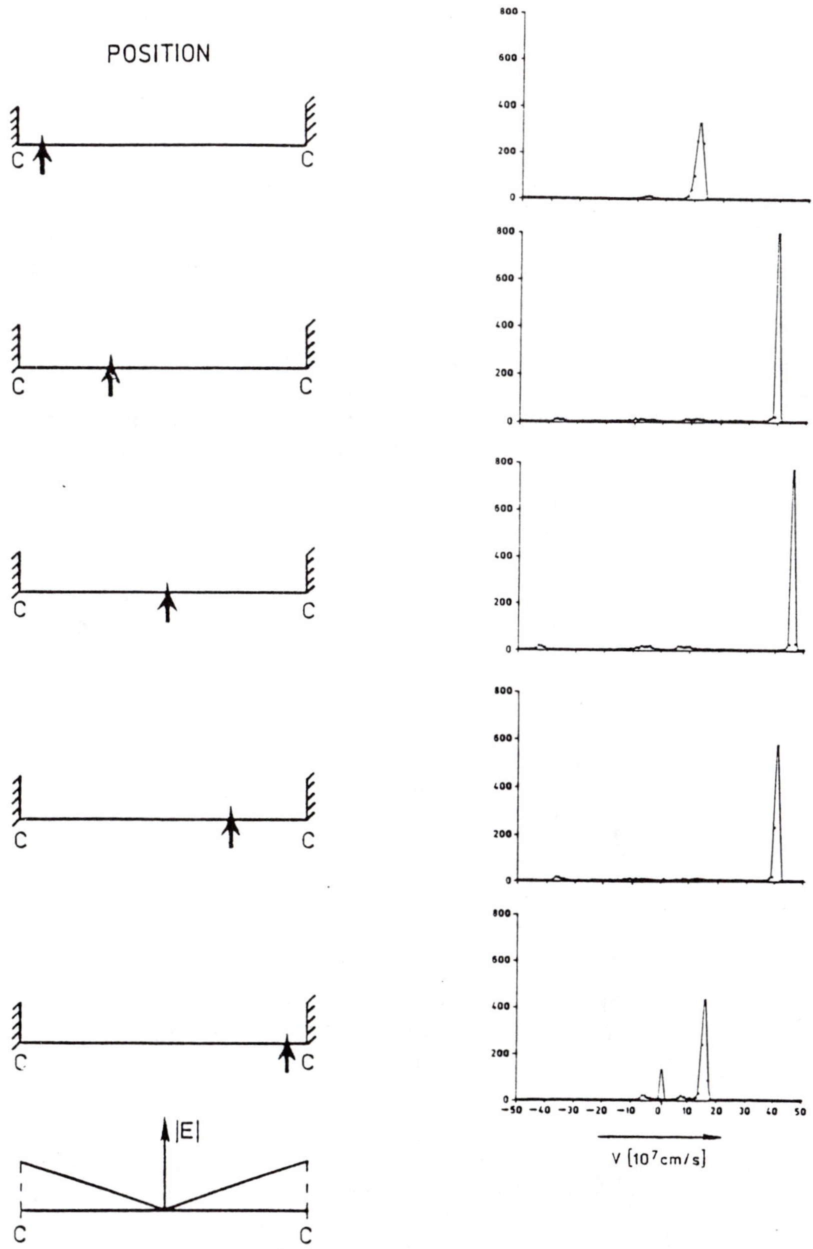

Fig.6. Velocity distributions of electrons, emitted from the left cathode, at different positions along the gap between the two cathodes for $p = 150$ mTorr.

smaller concentration of electrons in the velocity range near to the center of the system; a fact which can be explained by the reduced number of collisions over the observation time. It should be noted that the velocity distributions represent the electrons emitted from one cathode only. In the symmetric hollow cathode discharge, the real velocity distribution is given by the superposition of the distribution of electrons from both cathodes.

Remarkable in both cases is that a large fraction of the electrons hit the opposite wall. In the case of p = 150 mTorr it is 75% of the electrons emitted from the cathode, for p = 500 mTorr is still 45%. The electron reaching the opposite cathode have generally such a small energy that they are not able to generate secondary electrons, they are lost for the discharge. This electron loss which was experimentally studied by Helms [10], is assumed to cause the extinction of the hollow cathode discharge at lower pressure, where the percentage of collisionless electrons reaching the opposite electrode approaches the value of 100%.

With increasing pressure this loss process becomes less important and the pendeling electrons trapped in the center between the cathodes provide for a high ionization rate coefficient in this region. This is shown in Fig. 7 for p = 150 mTorr and p = 500 mTorr. The plasma sheet which is formed in the center between the cathodes forms the discharge channel in axial direction. In a geometry where the hollow cathode has a cylindrical shape, an even more pronounced ionizations peak along the axis can be expected. This pendel electron effect seems to define the shape of the hollow cathode discharge, at least before the superemissive cathode effect sets in and it explains the high currents observed in the main discharge mode. The plasma generated at the axis by Pendel electrons, serves as a virtual cathode with a high electron emission rate for the axial discharge.

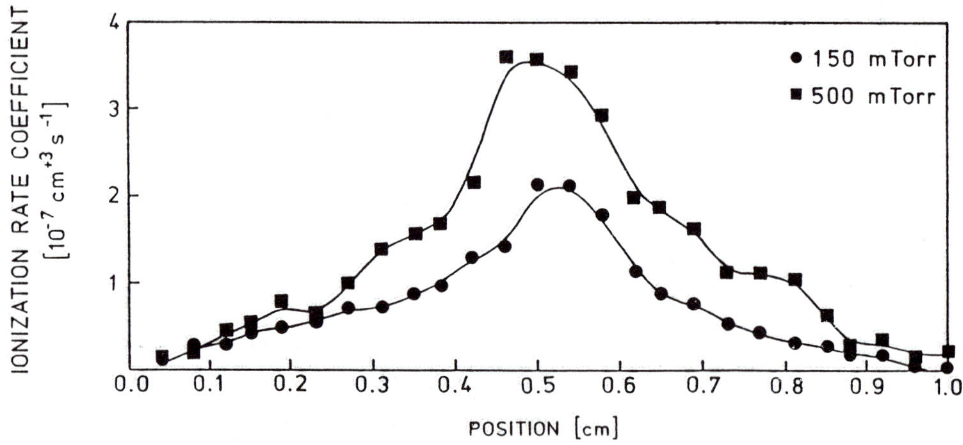

Fig.7.The distribution of the ionization rate coefficients in the gap between the two cathodes for p = 150 mTorr and p = 500 mTorr. The distribution was calculated by considering electrons emitted from one cathode only.

The discussed model gives us an understanding of the physics of the transition from the predischarge into the main discharge and the sustainment of the hollow cathode discharge. It is by no means complete. Steps to make the electric field in the hollow cathode region self consistent are under way [11]. The expansion into a two-dimensional model, which also takes axial charge transport into account will follow.

ACKNOWLEDGEMENT

This work was supported by the Strategic Defense Organization/Innovative Science and Technology Office and managed by the office of Naval Research under Contract No. N00014-85-K-0602. The program monitor for this project is Gabriel Roy.

REFERENCES

[1] F. Paschen, Ann. Physik **50**, 901 (1916).

[2] S. Caroli, *'Improved Hollow Cathode Lamps for Atomic Spectroscopy'*; [Ellis Horwood Ltd., Chichester, UK, 1985].

[3] K. Frank, E. Boggasch, J. Christiansen, A. Goertler, W. Hartmann, C. Kozlik, G. Kirkman, C. G. Braun, V. Dominic, M. A. Gundersen, H. Riege, and G. Mechtersheimer, IEEE Trans. Plasma Sci. **16**, 317 (1988).

[4] M. T. Ngo, K. H. Schoenbach, G. A. Gerdin, and J. Lee, submitted to IEEE Trans. Plasma Science.

[5] B. Wernsman, H. F. Ranea-Sandoval, J. J. Rocca, and H. Mancini, IEEE Trans. Plasma Sci. **14**, 518 (1986).

[6] W. Hartmann, V. Dominic, G. F. Kirkman, and M. A. Gundersen, Appl. Phys. Lett. <u>53</u>, 1699 (1988).

[7] A. Guentherschulze, Z. Physik **19**, 313 (1923).

[8] H. Helm, F. Howorka, and M. Pahl, Z. Naturforsch, <u>27a</u>, 1417 (1972).

[9] M. Hayashi, "Recommended values of transport cross sections for elastic collisions and total cross section for electrons in atomic and molecular gases," Inst. Plasma Phys., Nagoya University, Report IPPJ-AM-19 (Nov. 1981).

[10] H. Helm, Z. Naturforsch. **27a**, 1812 (1972).

[11] K. H. Schoenbach, H. Chen, and G. Schaefer to appear in J. Appl. Phys., January 1990.

A TWO-ELECTRON-GROUP MODEL FOR A HIGH CURRENT

PSEUDOSPARK OR BACK-LIGHTED THYRATRON PLASMA

H. Bauer, G. Kirkman, and M. A. Gundersen

Department of Electrical Engineering-Electrophysics
University of Southern California
Los Angeles, California 90089-0484

ABSTRACT

A two-electron-group model is applied to a hydrogen pseudospark and/or back-lighted thyratron switch plasma with peak electron density of $1\text{-}5\text{x}10^{15}$ cm^{-3} and peak current density of 10^4 A/cm^2. In addition to a Maxwellian "bulk" plasma a second group of monoenergetic non-thermal electrons (energy ≥ 100 eV) is assumed to be produced in the high electric field region of the cathode fall. Collisional and radiative processes between electrons, ions and atomic hydrogen are modeled by a set of rate equations and line intensity ratios are compared with measurements. Under these high current conditions, for an initial density $n_{H2} \approx 10^{16}$ cm^{-3} the evaluated "bulk" plasma parameters are electron density of $1\text{-}3\text{x}10^{15}$ cm^{-3} and electron temperature of $1\text{-}1.5$ eV, the estimated "beam" density is $\approx 10^{12} - 10^{13}$ cm^{-3}. Results obtained from a Fokker-Planck and a transport momentum equation suggest the possibility of producing in a simple way a very high density electron beam.

I. INTRODUCTION

The intention of this work is to provide quantitative information about electron temperature, energy, and density of two electron components, i.e. an isotropic "bulk" plasma and anisotropic electron beam, in either a pseudospark or a back-lighted thyratron (BLT) switch plasma during the conduction phase.

The pseudospark and the BLT are new thyratron type glow discharge switches [1] characterized by hollow electrode geometries, small high voltage gap separation of 2-3 mm, center electrode holes of 2-5 mm diameter, a self heated cathode, high current densities of about 10^4 A/cm^2, and peak currents of $10^4 - 10^5$ A -- which are higher than for a typical high power thyratron. A detailed knowledge of the distribution functions of the plasma particles and a quantitative knowledge of the plasma parameters such as temperature and density are necessary to understand the transport properties of the device plasma and the inherent limitations of the switch, and to determine new applications that include the production of particle beams, plasma based particle accelerators [2], and plasma lens devices [3].

In a high voltage pseudospark operation at 10-50 kV with hydrogen gas at an initial density of $n_{H2} \approx 10^{16}$ cm^{-3} the peak electron density was measured as $1\text{-}5\text{x}10^{15}$ cm^{-3}. In earlier work [4]-[6] it was found that for the same initial density a typical hydrogen thyratron "bulk" plasma is Maxwellian with electron temperature of 1eV, electron density of about 5×10^{14} cm^{-3} and dissociation degree of 0.2. In this paper calculations are performed for a single gap geometry of a typical BLT or pseudospark and for external circuits with about 0.5-1 μs time to maximum current. The high electron densities and an estimated electron temperature of about 1 eV result in the formation of a Maxwellian "bulk" plasma. Consideration of the current density and electric field strength lead to the assumption of non-thermal electrons which are produced in the cathode fall region characterized by a small width of several μm and a high electric field of about 10^6 V/cm [7]-[14]. Hence, because the device geometry is confined and the cathode fall is close to the "bulk" conducting region, it is necessary to consider a strong anisotropic electron component, e.g. an electron beam.

The distribution function of initially monoenergetic electrons which are injected into a Maxwellian plasma was evaluated from a Fokker-Planck equation [15] and the penetration was analyzed with a transport momentum equation [8],[16]. It was found that injected electrons will not become thermalized during the gap penetration of 3mm length, i.e., their distribution function becomes not too broadened and is therefore assumed to be a delta function in energy [17].

The model thus consists of two electron groups, a monoenergetic electron beam and a Maxwellian "bulk" plasma, hydrogen ions and atomic hydrogen. The production of Maxwellian electrons and excited atomic states is evaluated from a set of rate equations which take into account impact excitation, de-excitation, ionization, recombination and radiative transitions [4], [18]-[26]. Re-absorbtion of spectral lines is modeled by the radiation escape factors [27], [28]. Calculations are performed for different initial densities of atomic hydrogen and for optically thick Lyman series. Ionization and excitation due non-thermal electron impact is investigated for a beam pulse duration of 100-300 ns. A steady state solution for a purely Maxwellian plasma is used to obtain a first estimate of the population of atomic levels and the "bulk" electron density by comparison of calculated and measured line intensity ratios of H$_\alpha$ and H$_\beta$ lines.

The assumption of a steady state may be in appropriate for the short current pulses of 0.5-1 μs and preliminary results obtained from a time dependent solution are presented. Since the rate coefficients depend on temperature an energy balance equation was solved in addition. Calculations of the temporal behaviour of the population of atomic levels, electron temperature, and line intensity ratio of H$_\alpha$ and H$_\beta$ lines are in agreement with experiment, but also indicate that energy losses of the electrons have to include diffusion and wall processes which can be important for small spatial dimensions of about 3 mm.

II. TWO-ELECTRON-GROUP PLASMA MODEL

Several considerations support the assumption of two electron groups, i. e. particles of the same species but different distribution functions, during the conductive phase. In the case of an almost fully ionized plasma where electron-neutral collisions can be neglected the evaluation of the distribution function requires the solution the Fokker-Planck equation [13],[14]. A further simplification is the expansion of the distribution function in an isotropic and an anisotropic part which can be applied when the electric field is small compared to some critical or "runaway" electric field [9] defined as

$$E_c = 2\pi \left(\frac{e^2}{4\pi\varepsilon_0}\right)^2 \ln\Lambda \, \frac{n_e}{kT_e} \tag{1}$$

where $\ln\Lambda$ is the Coulomb logarithm, T_e the electron temperature and n_e the density of the "bulk" electrons. For a typical BLT plasma with $T_e=1$ eV and $n_e=1\text{-}5\times10^{15}$ cm^{-3} the value of E_c is $1\text{-}6\times10^3$ V/cm. It will be shown that the electric field strength inside the cathode fall region of the BLT is higher and therefore an appropriate electron distribution function has to be defined and justified ad hoc.

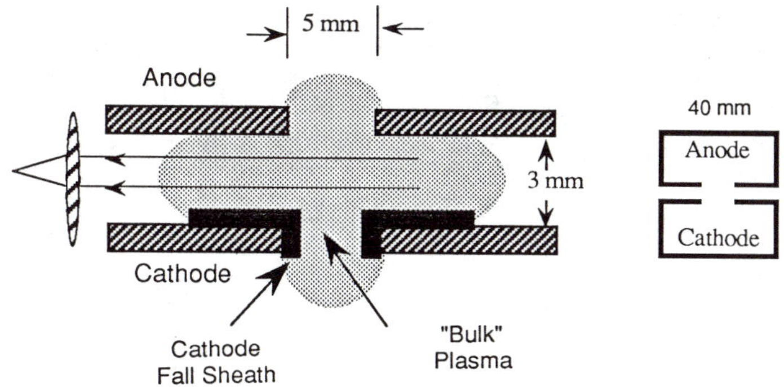

Fig. 1) Schematic drawing of the discharge geometry. The "bulk" plasma extends from the gap space into the hollow electrodes of anode and cathode. The thickness of the cathode fall plasma sheath is several μm.

In the following the analysis is presented for the time during the conduction phase around current maximum, illustrated in Fig. 1, and two electron components will be considered.

A. The "Bulk" Plasma

High densities lead to high electron-ion collision frequencies ($\approx 10^{10}$ s^{-1}) and short mean free paths ($\approx$ several μm) which are much smaller than the characteristic length of the plasma ($\approx$ 3mm). Thus the times when electrons with energies of several eV become thermalized are much shorter than the total current pulse length (0.5-1 μs) and one can assume that electrons in the "bulk" space which have been produced by ionization of neutrals develop a Maxwellian distribution function.

The electric field E in the "bulk" region which is required to conduct the high current densities is estimated from an analysis [10]-[12] based on the expansion of the distribution function, described above. The assumption of a steady state condition, neglect of spatial gradients, and a Maxwellian isotropic distribution results in

$$E = \eta\, j, \quad \eta = (\frac{e}{4\pi\varepsilon_0})^2 (\frac{\pi}{2kT_e})^{1.5} \frac{\ln\Lambda \sqrt{m}}{\gamma_E \sqrt{2}} \tag{2}$$

where j is the current density, m is the electron mass, η is the specific resistivity, and γ_E =0.582.

At T_e=1eV a current density of 5-10x10^3 A/cm^2 requires an electric field strength of 200-600 V/cm inside the "bulk" plasma. This field seems rather high because of the following reasons. (a) A high density neutral plasma tends to shield out an externally applied electric field. (b) An electric field of 600 V/cm becomes of the order of the "runaway" field E_c, eq. (1), which means that a Maxwellian distribution cannot be established. Thus it is appropriate to consider a second non-Maxwellian electron group which participates in the current transport process.

B. The Electron Beam Produced in the Cathode Fall

Near the cathode an externally applied electric field cannot be compensated by the available space charge and this region forms the cathode fall . Estimates of the Debye length or space charge limited current flow for a typical BLT plasma parameters lead to a cathode fall width of about 10 μm [7], called the cathode fall sheath in Fig. 1.

The voltage drop at the cathode fall was estimated from (a) preliminary measurements, (b) considerations of the ion current density at the cathode, and (c) calculations performed with an electrical RLC circuit, combining the external circuit with the switch plasma channel, and yield 100-500 V [13],[14] thus the electric field is about 1-5×10^5 V/cm. Since this field is much higher than the "runaway" field E_c, one can assume that electrons which are produced at the cathode will be accelerated to energies which correspond to the voltage drop at this sheath and enter the "bulk" plasma as a monoenergetic beam, i.e. their initial distribution function will be defined in form of a dirac delta function,

$$f_b = n_{e,b} \, \delta(v - v_{e,b}) \tag{3}$$

where $v_{e,b}$ is the velocity component in the direction of the electric field inside the cathode fall and $n_{e,b}$ the electron beam density.

III. Penetration of the "Bulk" Plasma by Injected Electrons

Once fast electrons are injected into the "bulk" plasma, it is of interest to determine how far they penetrate before they lose a significant amount of their energy or become thermalized.

For highly ionized plasmas, which occur during the conduction phase in the BLT, collisions and energy losses occur mainly between charged particles due Coulomb scattering. This process is often described by a small angle scattering and leads to the Fokker-Planck equation [15] which has been evaluated under the assumptions that the electron distribution function is isotropic, electron and ion temperature and density of the "bulk" are equal, beam electrons collide with Maxwellian "bulk" electrons and ions, and the density of electrons characterized by f_e is small, i. e. encounters between the injected particles themselves can be neglected.

Once the distribution function is found the transport momentum equation [8],[16] for those particles can be integrated. The non-thermal electron distribution can be defined by eq. (3) without loosing too much accuracy provided that the considered time scale is smaller than the times in which the distribution becomes too broadened [17]. A detailed description of the solution of the Fokker-Planck and transport equation is given in reference 17. Fig. 2. shows the distribution function of electrons with an initial energy of 100 eV after 3mm penetration of the "bulk" plasma at the anode for a temperature of 1 eV with densities of 1, 2.5, and 5×10^{15} cm^{-3} and different values of the electric field. The electric field of $0.05 \times E_c$ corresponds to 65-325 V/cm for T=1eV and n=1-5×10^{15} cm^{-3}. It can be seen that without external electric field the penetration of the "bulk" plasma is possible without thermalization of the beam electrons if the "bulk" densities are below 5×10^{15} cm^{-3} and the presence of an electric field allows penetration for even higher densities. Since the evaluation of rate coefficients will be carried out in a later chapter of this paper one can conclude from calculations and the slow varying cross sections for energies > 40 eV, that also during the whole penetration of the gap space one doesn't loose to much accuracy in defining the distribution of the injected electrons in form of eq. (3).

Furthermore it was found that for injection energies of 500 eV penetration of the gap and even of the anode space (about 4 cm) is possible provided that an electric field exists. Operation with lower gas pressures of 4 Pa (Argon), i.e. an initial density of 10^{15} cm^{-3}, leads to a "bulk" density of 5×10^{14} cm^{-3} if the ionization degree is assumed to be 0.5. For this case numerical results show that electrons with initial energies of 100-500 eV can easily penetrate the anode space and thus suggest the possibility of using the BLT as a high density electron beam source.

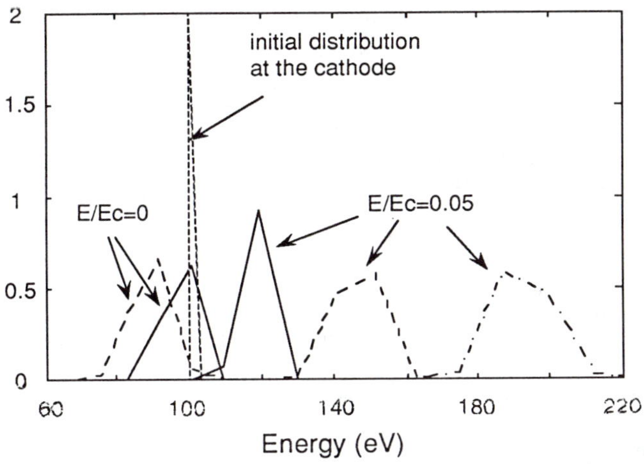

Fig. 2. Distribution function of injected electrons with an initial energy of 100 eV at position of the anode after 3 mm penetration of the "bulk" plasma for different values of the electric field E and "bulk" densities n. E_c is the "runaway field".
———— n = 1x10^{15} cm^{-3} ------- n = 2.5x10^{15} cm^{-3} -.-.-.- n = 5x10^{15} cm^{-3}

In the calculations magnetic field effects are neglected. These can be important at high current and high current densities. The magnetic field at the boundary of a plasma column with homogeneous current density of 10^4 A/cm^2 is about 0.3 T and the corresponding Larmor radius is about 0.1 mm. In this case the injected electrons will gyrate and drift to the anode only due to the electric field in the "bulk" space. Hence the path that the electrons follow to reach the anode will be considerably lengthened, the number of collisions will increase, and electrons at the outer boundary of the column will tend to become thermalized. Closer to the inner boundary the magnetic field will decrease and the results obtained from the Fokker-Planck equation without magnetic field effects become valid. The properties of the electron beam transverse to the axis require additional study.

IV. LINE INTENSITY RATIO AND RATE EQUATIONS

A. Line Intensity Ratio and Rate Equations

In this section we consider modeling of fluorescent emission from hydrogen in order to yield information about "bulk" and beam properties of the plasma and comparison is made between calculated and measured line intensity ratios of atomic radiative bound-bound transitions. Spectroscopic measurements which are compared with calculations are taken from the center of the gap space as indicated in Fig. 1.

The line intensity ratio of two poorly re-absorbed atomic spectral lines for transitions from level k to m and l to m , m<k<l, is

$$\frac{I_{km}}{I_{lm}} = \frac{n_k A_{km} \varepsilon_{km}}{n_l A_{lm} \varepsilon_{lm}} \tag{4}$$

where n_k and n_l are the populations of levels k and l, A_{km} is the spontaneous emission coefficient and ε_{km} the energy difference between two atomic levels k and m.

The population density of the excited levels k and l can be calculated from a set of coupled rate equations. Collisional processes which increase or decrease the population of atomic levels are direct, i.e. electron impact with atomic hydrogen and indirect dissociative processes involving encounters with molecules. Previous calculations concerning the thyratron plasma [5] yield the dissociation degree as about 0.2 which is similar to the value assumed for the BLT plasma. For such high dissociation degrees it was found that the processes, which are important in the production of excited hydrogen, are direct [20] which justifies the omission of atom-molecule and electron-molecule impact. Neglecting also wall processes the collisional and radiative processes taken into account are (a) impact excitation and de-excitation by beam and Maxwellian electrons, (b) impact ionization by beam and Maxwellian electrons, (c) 3-body recombination of hydrogen ions with Maxwellian electrons, and (d) radiative recombination and spontaneous emission.

A collisional radiative plasma model [4], [18]-[22] is extended for two electron groups by including impact of monoenergetic and Maxwellian electrons with atoms. Under the assumption of spatially homogeneous plasma parameters the set of rate equations for production of the atomic state populations can be written as

$$\frac{d}{dt} n_k = n_e n_i [n_e S_{ck} + A_{ck} k_{ck}] + n_e [\sum_{l=1}^{J} n_l S_{lk} - n_k (\sum_{l=1}^{J} S_{kl} + S_{kc})]$$

$$+ \sum_{l>k}^{J;k\neq J} n_l A_{lk} k_{lk} - n_k \sum_{l=1}^{k-1;k\neq 1} A_{kl} k_{kl}$$

$$+ n_{e,b} [\sum_{l=1}^{J} n_l S_{b,lk} - n_k (S_{b,kc} + \sum_{l=1}^{J} S_{b,kl}) - n_i S_{b,ck}] \tag{5}$$

for k = 1, 2, ... J, and the summations have to be carried out for $l \neq k$.

J is the maximum number of atomic levels considered in the calculations, k and l denote atomic levels and c the continuum state, n_l and n_k are the populations of atomic levels l and k, n_e is the "bulk" electron density, $n_{e,b}$ the electron beam density, n_i is the hydrogen ion density, S_{kl} and $S_{b,kl}$ are the rate coefficients for excitation and de-excitation by Maxwellian electron and beam electron impact, S_{kc} and $S_{b,kc}$ are the rate coefficients for ionization by Maxwellian electron and beam electron impact, S_{ck} and A_{ck} are the rate coefficients for 3 body and radiative recombination of Maxwellian electrons, $S_{b,ck}$ is the rate coefficient for capture of an electron under emission of a photon, A_{kl} is the probability for spontaneous radiative emission and κ_{kl} is the radiation escape factor [27].

The equation which describes the production of Maxwellian electrons yields

$$\frac{d}{dt} n_e = n_e [\sum_{l=1}^{J} n_l S_{lc} - n_i (n_e \sum_{l=1}^{J} S_{cl} + \sum_{l=1}^{J} A_{cl} k_{cl})] +$$

$$n_{b,e} [\sum_{l=1}^{J} n_l S_{b,lc} - n_i \sum_{l=1}^{J} S_{b,cl}] \tag{6}$$

where it is assumed that an electron produced by impact ionization of hydrogen by a beam electron becomes Maxwellian because the mean energy of ejected electrons is about 1/4 times the threshold energy [25], which means those electrons become rapidly thermalized. Colliding beam electrons are assumed to remain beam electrons .

B. Rate Coefficients and Radiation Escape factors

In the case of a distribution function defined in eq. (3) the rate coefficients for excitation, de-excitation and ionization reduce to the simple relation $S_{xy} = Q_{xy} (\varepsilon_e) v_e$, where x and y are atomic levels l, k or the continuum state c, ε_e is the impact energy of the electron and v_e the corresponding velocity. The cross sections for excitation and ionization are evaluated from analytical formulae given in reference [23], the cross sections for de-excitation are calculated from the principle of detailed balance [26], and the cross section for capture of an electron by a hydrogen atom with emission of a photon is obtained from reference 24.

The rate coefficients for excitation and ionization by Maxwellian electron impact are evaluated using analytical formulae [23] and detailed balancing yields the rate coefficients for de-excitation and 3-body recombination for impact by a Maxwellian electron [26].

An expression for radiative recombination is given in reference 23 and the A_{kl} coefficients for radiative bound-bound transitions are calculated analytically [26] or can be taken from reference 25.

The radiation escape factors κ_{kl} have been introduced to represent absorption of spectral lines in the plasma [27],[28]. Hence the term $A_{kl} \kappa_{kl}$ in eqs. (6) and (7) includes the effect of spontaneous emission and absorption. In general κ_{lk} increases from zero to its maximum value of 1 with increasing optical depth and the optical depth is proportional to the density of state k. In this paper the escape factors are simply determined by the fact that the values of κ_{lk} correspond to the case of a plasma that is optically thin for a spectral line l->k, $\kappa_{lk} = 1$, or optically thick, $\kappa_{lk} = 0$. This has the result that radiation with wavelength $\lambda = \varepsilon_{lk}/h$ is re-absorbed in the plasma. The evaluation of the radiation escape factors for radiative recombinative processes κ_{ck} can also not be done in a simple way. In the present calculations the radiation escape factors are not evaluated but their values are estimated from basic considerations.

V. STEADY STATE SOLUTION OF THE RATE EQUATIONS FOR THE "BULK" PLASMA

To obtain first estimates of electron temperature and density of the "bulk" plasma it is convenient to solve the rate equations under the assumption of a steady condition assuming that electron density and temperature, density of atomic states, and current density do not change with time. This is often a good approximation at the time of maximum current during the conductive phase but is inappropriate as a description of the ionization rate due to short electron beam pulses as will be shown.

The ionization degree due to beam impact ionization is obtained from eqs. (5) and (6) as a function of time for different pulse durations and densities of the beam. The temperature of the "bulk" plasma is taken to be constant at about 0.5 eV, the beam energy is 100 eV, and the initial neutral density is 10^{16} cm^{-3}. Comparison of time dependent calculations and results obtained from a steady state solution show large differences for beam densities $\leq 10^{13}$ cm^{-3}. Estimates of the ionization rate show that the time which is needed to reach a steady state can be in the order of 50 μs for densities $\leq 10^{11}$ cm^{-3}. Assuming a current cycle half time of about 0.5-1 μs and thus a beam pulse with duration of about 100-500 ns the assumption of a steady state condition is inappropriate and the electron beam will be omitted in the following steady state solutions.

Provided that the temperature of the "bulk" electrons is a given input parameters and no electron beam is present eqs. (5) and (6) reduce to a system of nonlinear equations. Furthermore charge neutrality is assumed to be valid in the "bulk" plasma which means that the density of ions is assumed to be equal to the electron density, $n_i = n_e$.

In addition to the production of excited states and "bulk" electrons an equation for particle conservation is imposed. The total density of electrons, i.e. the sum of free Maxwellian n_e and bound electrons n_k, must be equal to the initial density of atomic hydrogen $n_H(t=0)$:

$$\sum_{k=1}^{J} n_k + n_e = n_H(t=0) \tag{7}$$

The system of equations (5) - (7) is nonlinear in the densities and computations are performed with a modification of Powell's hybrid algorithm [30]. To achieve convergence of the iteration the number of considered atomic levels J is restricted to J=5. Once the electron density is found, eqs. (5) are linear in the atomic level density. The resulting equations can be solved with greater accuracy. Eqs. (5) and eq. (7) then define an overdetermined system of equations which are solved by the method of least square solution [31]. The maximum number of levels taken into account is J=9.

In the temperature range $T_e \leq 1.5$ eV the calculated population of the ground state is still very large, which justifies the assumption of an optically thick Lyman series ($\kappa_{k1}=0$, k=2,...J) because the optical depth of a transition line l->k is proportional to the population density of the state k, whereas the escape factors decrease with increasing optical depth. Because of the low densities of excited states the radiation escape factors for all other lines are taken to be optically thin ($\kappa_{km}=1$, k=m+1,...J).

The following results refer to initial neutral densities of atomic hydrogen of 10^{15} cm^{-3} and 10^{16} cm^{-3}. The value of 10^{16} cm^{-3} corresponds to an dissociation degree of 0.5 for an initial molecular hydrogen density of 2×10^{16} cm^{-3} or pressure of about 40 Pa. For this density the numerical results are plotted for optically thin and thick Lyman series. In addition, values obtained using the Saha equation are added for comparison.

The intensity ratio of the H_α and H_β lines is calculated from eq.. (4) and is shown in Fig. 3 as a function of the electron temperature. Higher temperatures lead to lower line ratios because the lower atomic states will be de-populated. Since the electron temperature increases with increasing current density it is concluded that in the absence of an electron beam the line intensity ratio should decrease in time until the peak current, i.e. maximum ohmic heating, is obtained and increase after time of current maximum. From the results of the steady state solution the minimum of the intensity ratio is therefore expected to occur during the conduction phase. The measured line intensity ratio during the conduction phase at time of decreasing current where no electron beam is expected is 4-6 (6.5 kA peak current) [13] and thus yields the electron temperature to about 1 eV.

The ionization degree due impact ionization by encounters of hydrogen with Maxwellian electrons is plotted in Fig. 4 as a function of electron temperature. In the case of optically thick Lyman series radiation will be strongly re-absorbed which results in a larger population of excited atomic states and stepwize ionization can occur more frequently, hence the ionization degree increases. In case of an optically thin Lyman series ionization has to occur mainly from the ground state with a threshold energy of 13.6 eV and the ionization degree is lower than that obtained with optically thin Lyman lines. Results with optically thick Lyman series are close to that obtained from the Saha equation which can be explained by the fact that for values of $\kappa_{k1}=0$ radiative processes involving the ground state are reduced and the population density is mainly determined by detailed balancing of excitation and de-excitation, a feature of the Saha equation.

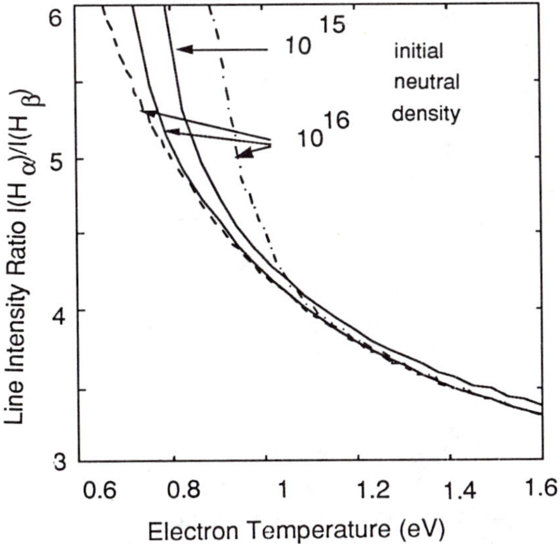

Fig. 3. Calculated line intensity ratio $I(H_\alpha)/I(H_\beta)$ versus electron temperature for different initial densities of atomic hydrogen. No electron beam is included.
——————— optically thick Lyman series
—·—·—·—· optically thin Lyman series ----------- Saha equation

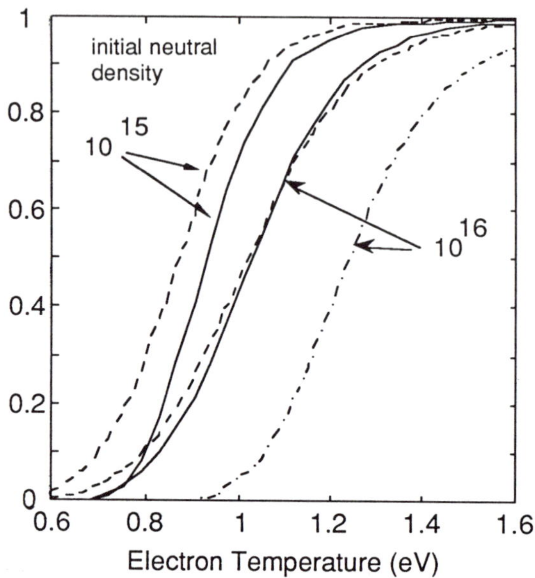

Fig. 4. Calculated ionization degree versus electron temperature for different initial densities of atomic hydrogen. No electron beam is included.
———— optically thick Lyman series
–·–·–·– optically thin Lyman series ---------- Saha equation

For the previously estimated temperature of 1 eV the ionization degree is found to be 0.4 and is comparable to the measured value of 0.3.

VI. TIME DEPENDENT SOLUTION OF THE RATE EQUATIONS

For short current pulses of 0.5-1 μs the assumption of a steady state may be inappropriate and a time dependent solution of the rate equations becomes necessary. Because the rate coefficients are functions of the electron temperature a temporal solution of equations (5) and (6) requires the evaluation of an additional equation which accounts for the energy balance of the "bulk" electrons. The main mechanism for heating is ohmic ($\sim\eta j^2$) and thus the current density has to be specified. Energy losses are found to be mainly due to excitation, ionization, and collisions with neutrals [32]. Processes which account for losses in the electron density are diffusion and recombination at the walls and require the calculation of spatial gradients. The physics of processes which occur at the walls is even not well understood and both types of loss mechanisms will be omitted in the following calculations. The energy balance equation then reduce to an ordinary differential equation and the initial value problem which also includes the rate equations (5) and (6) is solved with a Gear's method [31], suitable for stiff equations.

A parabolic current pulse of 1 μs length is applied and for given initial electron temperature the initial values of electron and excited atomic state densities are obtained from Saha's equation. The model cannot describe the commution phase characterized by low electron densities and high electric field strength in the beginning of the discharge but after a short time of about 200 ns, temperature and densities will quickly change their initial values and adjust themselves according to the given non-equilibrium conditions and the current density. If a current density of 10^4 A/cm^2 is specified the temperature is found to increase to high values ≥ 5 eV which can be due to the omission of diffusion and wall processes. In this case fully ionization of the hydrogen is calculated. Thus the current density is chosen artificially to match numerical and the measured value of the electron density of 2-3x10^{15} cm^{-3}. The temperature of neutrals is taken to be constant at 0.2 eV, the initial population of the ground state is 10^{16} cm^{-3}, and the Lyman series is optically thick. Calculations are also performed including an monoenergetic electron beam with energy of 100 eV and an parabolic pulse with 500 ns length which starts 250 ns after the beginning of the discharge.

Electron temperature and population of the first excited atomic level are plotted in Fig. 5 as a function of time. Due ohmic heating the maximum temperature is obtained at the time of maximum current (0.5 μs). The plasma then becomes colder and at the end of the current pulse the after pulse behaviour [32], i.e. an increase in the population density of excited states, can be seen . If collisions with monoenergetic electrons are included Fig. 5 shows that the atomic state density increases and lower electron temperatures are obtained because of higher ionization degrees.

The line intensity ratio I(H$_\alpha$)/I(H$_\beta$) was calculated according to eq. (4) and is shown in Fig. 6 as a function of time. Since the density of excited states increases during the

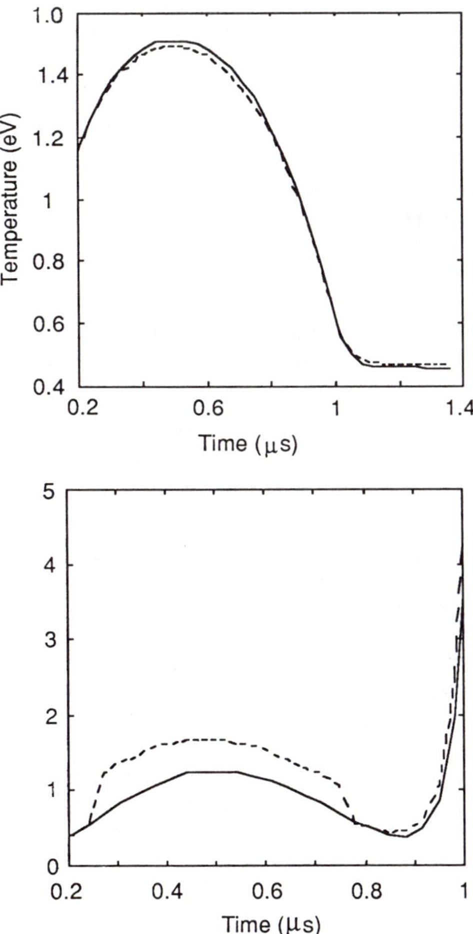

Fig. 5. Electron temperature and density of the first excited atomic level versus time.

———— no electron beam, ------ electron beam with density of 10^{12} cm^{-3}

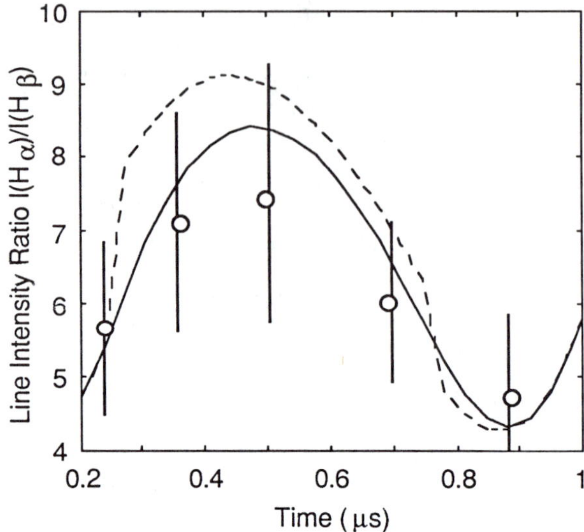

Fig. 6. Line intensity ratio I(H$_\alpha$)/I(H$_\beta$) versus time.
——— no electron beam, ------ electron beam with density of 10^{12} cm
Circles and bars indicate the measured line ratios including the experimental error.

conduction phase the line ratio also has its maximum value at the time of the current maximum, which could not be explained with the steady state solution. The measured line intensity ratios [13] (6.5 kA peak current), indicated by circles and error bars, correspond well to numerical results provided that the electron beam density is $\leq 10^{13}$ cm^{-3}.

Electron beam densities $\geq 10^{14}$ cm^{-3} result in line ratios of about 15 which are not confirmed by experiment, hence the upper limit of the beam density can be estimated to about 10^{13} cm^{-3}.

VII. CONCLUSIONS

In this paper several aspects of a cathode fall produced electron beam during the conduction phase of a high current hydrogen pseudospark or BLT are analyzed using a model that considers two groups within the electron distribution function. The cathode fall - produced beam is considered important because of the small separation between anode and cathode, 2 - 3 mm. A Maxwellian plasma, without a second, higher energy component, is shown to require a high electric field of about 200-600 V/cm in order to provide the conductivity necessary to achieve the observed high current densities of 10^4 Acm^{-2}.

Temporal behaviour and values of measured Balmer line intensity ratios correspond to results obtained from a time dependent solution of the rate equations. Comparison of calculations and measurements yield a "bulk" electron temperature of 1-1.5 eV and an ionization degree of 0.3 if impact ionization is mainly due by Maxwellian electrons (beam density $\leq 10^{12}$ cm^{-3}), resp. ≥ 0.5 if the electron beam density is $\geq 10^{13}$ cm^{-3} (beam pulse of 500 ns).

For a BLT with neutral density of 10^{16} cm^{-3}, peak current ≈ 10 kA, current pulse length ≈ 1 μs, and charging voltage of 10 kV the plasma parameters are estimated as follows:

314

a) The "bulk" plasma has an electron density of about 3×10^{15} cm^{-3}, electron temperature of 1-1.5 eV, and a Maxwellian velocity distribution.

b) An electron beam with density of 10^{12}-10^{13} cm^{-3} participates in the transport processes with current density of 10^2 -5×10^3 A/cm^2, energy of about 100-500 eV, and pulse duration of about 100-500 ns during the conduction phase.

First time dependent calculations shows that under the described conditions a steady state condition is inappropriate for times of about 1 μs (current pulse length) but give reasonable estimates for larger times $\geq$ 10 μs. Furthermore it was found that diffusion and wall processes such as recombination may be important loss mechanisms for the electrons especially in the case of the small spatial dimension of 3mm in the BLT gap.

Detailed knowledge of thermalization times and penetration depths is necessary to justify the form of the velocity distribution function and energy of injected particles, and improved knowledge of the electric field inside the cathode fall plasma sheath is also needed. Lower electric field strengths will lead to lower injection energies and decrease the thermalization times and penetration depths [17]. If cathode fall generated electrons become thermalized in distances which are shorter than the gap space ($\approx$ 3 mm) they will become part of the Maxwellian "bulk" plasma, but will be negligible because of their low density.

For lower gas pressures, e.g. below 4 Pa (30 mtorr), these results and calculations of the penetration depth [17] suggest the possibility of extracting the conductive phase electron beam and using it as a high density electron source.

ACKNOWLEDGEMENT

This research was supported by the Air Force Office of Scientific Research and the Army Research Office. We would like to thank Dr. W. Hartmann, Prof. M.J. Kushner, Dr. P. Choi, Prof. J. Christiansen, and Prof. J. Kunc for valuable conversations.

REFERENCES

[1]   K. Frank, E. Boggasch, J. Christiansen, A. Goertler, W. Hartmann, C. Kozlik, G. Kirkman, C. Braun, V. Dominic, M. A. Gundersen, H.Riege and G. Mechtersheimer, "High Power Pseudospark and BLT Switches," IEEE Trans. Plasma Sci. **6**, 317 (1988).

[2]   P. Chen, J. Dawson, R. W. Huff, and T. Katsouleas, "Acceleration of Electrons by the Interaction of a Bunched Electron Beam with a Plasma," Phys. Rev. Lett. **54**, 693 (1985).

[3]   G. F. Kirkman, H. Figueroa and M. A. Gundersen, "A plasma lens with highly stable properties," Proceedings of the 1989 Workshop on Advanced Accelerator Concepts, to appear.

[4]   J. A. Kunc, S. Guha and M. A. Gundersen, "A fundamental theory of high power thyratrons, I: The electron temperature," Laser and Particle Beams **1**, 395 (1983).

[5]   J. A. Kunc and M. A. Gundersen,"A fundamental theory of high power thyratrons, II: The production of hydrogen and positive ions," Laser and Particle Beams **1**, 407 (1983).

[6]   J. A. Kunc, D. E. Shemansky and M. A. Gundersen, "A fundamental theory of high power thyratrons for high power lasers and beam applications III: The production of radiation," Laser and Particle Beams **2**, 129 (1984).

[7] W. Hartmann, M. A. Gundersen, "Origin of Anomalous Emission on Superdense Glow Discharge," Phys. Rev. Lett. **60**, 2371 (1988).

[8] I. P. Shkarofsky, T. W. Johnston, M. P. Bachynski,*The Particle Kinetics of Plasmas*, Addison-Wesley Publishing Company (1966).

[9] H. Dreicer, "Electron and Ion Runaway in Fully Ionized Gases. I," Phys.Rev. **115**, 238 (1959).

[10] H. Dreicer, "Electron and Ion Runaway in Fully Ionized Gases. II," Phys. Rev. **117**, 329 (1960).

[11] L. Spitzer, *Physics of Fully Ionized Gases*, Interscience Publishers, Inc., New York (1962).

[12] L. Spitzer and R. Haerm, "Transport Phenomena in a Completely Ionized Gas," Phys. Rev. **89**, 977 (1953).

[13] G. Kirkman and M. A. Gundersen, "Spectroscopic analysis of the BLT plasma," Proceedings of the Seventh IEEE Pulsed Power Conference, Monterey (1989).

[14] H. Bauer, G. Kirkman, and M. A. Gundersen, "Modeling of the discharge plasma in a Back Lighted Thyratron during the conduction phase, " Proceedings of the Seventh IEEE Pulsed Power Conference, Monterey (1989).

[15] M. N. Rosenbluth, W. M. Mc Donald and D. L. Judd, "Fokker-Planck Equation for an vol. Inverse Square Force," Phys. Rev. **107**, 1 (1957).

[16] N. A. Krall, A. W. Trivelpiece, *Principles of Plasma Physics*, McGraw-Hill Book Company, NY (1973).

[17] H. Bauer and M. A. Gundersen, " Penetration and Equilibration of Injected Electrons into a High Current Hydrogen Pseudospark-type Plasma," to be published.

[18] D. R. Bates, A. E. Kingston and R. W. P. McWhirter, Proc. Roy. Soc. **A267**, 297 (1962).

[19] D. R. Bates, A. E. Kingston and R. W. P. McWhirter, Proc. Roy. Soc. **A270**, 155 (1962).

[20] J. A. Kunc, "Stepwise Ionization in a Non-Equilibrium Steady-State Hydrogen Plasma," J. Quant. Spectr. Radiat. Trans. **32**, 311 (1984).

[21] J. F. Shaw, M. Mitchner and C. H. Kruger, "Effect of Nonelastic Collisions in Partially Ionized Gases, I . Analytical Solution and Results," Phys. Fluids **13**, 325 (1970).

[22] J. F. Shaw, M. Mitchner and C. H. Kruger, "Effect of Nonelastic Collisions in Partially Ionized Gases, I . Numerical Solution and Results," Phys. Fluids **13**, 339 (1970).

[23] L. C. Johnson, "Approximations for collisional and radiative Transition Rates in atomic hydrogen," Astrophys. Journal **174**, 227 (1972).

[24] Y. B. Zel'dovich and Y. P. Raizer, *Physics of Shock Waves and High-Temperature Phenomena*, Academic Press, New York (1966).

[25] R. K. Janev, W. D. Langer, K. Evans, D. E. Post, Jr., *Elementary Processes in Hydrogen and Helium Plasmas - Cross Sections and Reaction Rate Coefficients*, Springer Verlag, Berlin-Heidelberg (1987).

[26] M. Mitchner, C. Kruger, Jr., *Partially Ionized Gases*, John Wiley & Sons, New York (1973).

[27] T. Holstein, "Imprisonment of Resonance Radiation in Gases II," Phys. Rev. **83**, 1159 (1951).

[28] R. W. P. McWirther, *Plasma Diagnostic Techniques*, Academic Press, New York (1965).

[29] J. More, B. Garbow and K. Hillstrom, "Use guide for Minipack-1," Argonne National Laboratory Report, ANL-80-74, Argonne, Illinois (1980).

[30] G. Dahlquist, A. Bjoerk and N. Anderson, *Numerical Methods*, Prentice- Hall, Inc., NJ (1974).

[31] C. W. Gear, *Numerical Initial Value Problem in Ordinary Differential Equations*, Prentice-Hall, Inc., Englewood Cliffs, NJ (1971).

[32] C. D. Braun, D. A. Erwin, and M. A. Gundersen, "Fundamental processes affecting recovery in hydrogen thyratrons," Appl. Phys. Lett. **50**, 1325 (1987).

ELECTRON IONIZATION RATE COEFFICIENTS AT VERY HIGH E/N

L.C. Pitchford

Centre de Physique Atomique de Toulouse
CNRS Laboratoire Associeé URA N° 277
Université Paul Sabatier, 118, route de Narbonne
31062 Toulouse, CEDEX, France

INTRODUCTION

The growth of ionization in gas discharges in regions of very high values of E/N, the ratio of the electric field strength to the neutral density, can be the result of several different collisional processes including electron impact ionization, ion impact ionization, ionization by fast atoms and multi-step or cumulative ionization. Recently Phelps and his colleagues have published a series of experimental and computational results (Jelenkovic and Phelps, 1987, Phelps et al, 1987, Phelps and Jelenkovic, 1988, Gylys et al, 1989) which emphasize the importance of heavy particle contributions to the growth of ionization at very high E/N. In this article we are primarily interested in ionization by electron impact at high values of E/N where until recently very little information existed on the values of the ionization rate coefficients. Moreover, even the existence of well-defined values of the rate coefficients had not been definitely established.

In regions of low to moderate E/N, the electron ensemble or "electron swarm" (by analogy with a drifting, spreading swarm of bees) is well described by electron swarm parameters consisting of transport and rate coefficients which are functions of E/N and the background gas composition (Kumar et al, 1980). This functional dependence arises because, as the electrons are pulled through the neutral background gas under the influence of the electric field, they undergo many collisions with the neutrals during their transit between electrodes. The electrons gain energy in falling through the electric field and lose energy in the collisions with the neutrals, and the net effect is the establishment of a local equilibrium between the electron energy distribution and the local value of E/N. Since the electron swarm parameters are averages over the electron energy distribution function, they can thus be simply and conveniently expressed as functions of E/N and the neutral gas composition. Numerous data exist for swarm parameters as functions of E/N and gas composition, and several extensive compilations of the electron swarm data have been published (Dutton, 1975, Gallagher et al, 1983).

As E/N increases, however, the average energy of the electron swarm increases and eventually attains values such that the efficiency of the collisional energy and momentum loss processes begins to decrease as a result of the decreasing magnitude and increasing forward component of the electron scattering cross sections with increasing electron energy. This point occurs in most molecular gases between 100 and 200 eV. Thus, the electrons at high values of E/N have a tendency to gain energy from the field without limit. For such conditions where any individual electron can "runaway", there is no longer a local equilibrium between the energy gained by the electrons from the field and that lost in collisions with the background gas. Under these conditions, questions as to the existence of a steady-state electron energy distribution function and hence a unique, well-defined value of the electron impact ionization rate coefficient have been raised and partially addressed in several publications as outlined by Phelps, et al in 1987.

In the initial phases of pseudo-spark discharges, extremely high values of E/N can exist in the main discharge gap, and models of the high field regions in such discharge devices require some way of treating the electron transport and rate coefficients. The model reported by Dr. Mittag in these proceedings makes use of Monte Carlo simulations in this high E/N region precisely in order to avoid uncertainties associated with values of ionization rate coefficients and other transport coefficients at high E/N. In this article we briefly summarize a recent experiment designed to measure the electron ionization rate coefficients at high values of E/N and we present a simple, intuitive model which has successfully described the experimental results and which agrees well with more sophisticated Monte Carlo simulations which are also described below.

SUMMARY OF EXPERIMENTAL RESULTS

We have previously described a technique for measuring ionization and excitation rate coefficients at very high values of E/n, the ratio of the electric field strength to the neutral density, which eliminates uncertainties in the interpretation of the experimental data due to the presence of physical boundaries and their inevitable secondary electron emission (Hays, et al, 1987). This technique, which is based on the equivalence of electron heating rates by a microwave field at cyclotron frequency in a low pressure weakly ionized gas and by a DC electric field, was used to measure ionization rate coefficients in nitrogen up to 10 kTd (1 kTd = 10^3 Td = 10^{-14} V cm^2) and, more recently, in hydrogen up to 15 kTd (Verdeyen et al, 1988) by monitoring the time-dependent growth of the electron density after application of an electric field.

The experiment, carried out at Sandia National Laboratories in Albuquerque, New Mexico, was designed to measure the time-dependence growth of the electron density, and hence the ionization rate coefficient, after periodic application of a fast rising and high electric field. The electron density was monitored continuously with a microwave interferometer. The heating pulse was a microwave with a rise time of less than 10 ns, and the entire apparatus was embedded in the coils of a solenoid which provided a constant magnetic field. The heating pulse frequency was chosen to be the cyclotron frequency.

The conditions of the experiment are such that the electrons make several oscillations in the microwave field per collision with the gas molecules. According to Allis (Allis, 1956), the energy distribution function of electrons in crossed microwave and constant magnetic fields is equivalent to the one generated by an effective DC electric field, E_{eff}, given by

$$E_{eff}^2 = 1/4 \ E_o^2 \left(\frac{1}{1 + \{(\omega - \omega_b)/\nu_m(v)\}^2} + \frac{1}{1 + \{(\omega + \omega_b)/\nu_m(v)\}^2} \right) \quad 1)$$

where E_o and ω are the amplitude and angular frequency of the applied microwave, ω_b is the cyclotron resonance frequency and $\nu_m(v)$ is the momentum transfer collision frequency which depends on the electron speed, v. Two limits are obvious from eq. 1:

$$E_{eff} = \left(\begin{array}{ll} E_o/2 & \text{if } \omega = \omega_b \gg \nu_m \\[2ex] E_o/\sqrt{2} & \text{if } \omega = \omega_b \ll \nu_m \end{array} \right. \quad 2)$$

The experimental conditions are always such that the first limit is valid.

The utility of this experimental technique and in particular the exploitation of the equivalence of electron heating rates in a microwave field at cyclotron frequency with an effective DC field are limited by the onset of a relativistic mass increase for the high energy electrons and a corresponding shift in the resonance frequency for those electrons. A practical, but somewhat arbitrary, upper limit to E/N can be taken as that for which the calculated average electron energy is 10 keV, corresponding to about 15 kTd in hydrogen, for example. Although the high energy tail (>50 keV) of the electron energy distribution function starts to deviate from that rigorously equivalent to a DC distribution for E/N above 15 kTd in hydrogen, the deviation affects the measured value of the ionization rate coefficient only slightly because collisions between the high energy electrons and the neutral molecules are infrequent and inefficient for producing secondary electrons.

Some considerable effort on the theoretical side was devoted to checking the effective field concept by performing Monte Carlo simulations using both the predicted effective field and the time varying electric field and constant magnetic field (Li and Pitchford, 1987). In addition, on the experimental side, checks were performed to verify that the gas composition did not change during the course of the experiment due to power loading effects. Scaling with E/N and with Nt (the product of the neutral density and the time) were also confirmed.

CALCULATIONAL MODELS

The ionization rate coefficient as a function of time after application of a rapidly rising electric field for the conditions of the experiment can be immediately calculated from the electron energy distribution function. Solutions of the Boltzmann equation or Monte Carlo simulations are the two usual ways of determining the electron energy distribution function. The determination of the electron energy distribution function at high E/N from a numerical solution of the Boltzmann equation is complicated because it is not known a priori what boundary conditions or numerical approximations are valid. Our initial calculations for comparison with experiment relied on "two-term" solutions of the Boltzmann equation, a common solution technique in which the angular dependence of the velocity is expanded in Legendre functions and truncated after the first two terms. This approximation of near isotropy in the velocity distribution function is of uncertain validity at high E/N.

Although these initial calculations seemed to yield good agreement with the experiment in nitrogen, we prefer to avoid potential uncertainties in the approximate Boltzmann approach used in our earlier work and to develop alternate computational techniques which also provide more physical insight. The computational results described here are based on an extension of the multi-beam model described by Muller (1962) and more recently by Phelps, et al (1987) and on Monte Carlo simulations which we take as our benchmark computational results. The multi-beam model provides an intuitive picture of electron transport phenomena at very high E/n where an individual electron gains much more energy between collisions than it can lose in collisions. The multi-beam model predictions become increasingly accurate at high E/n as more and more of the electron energy loss is in the production of new secondary electrons.

Multi-beam Model
================

This model provides a simple computational recipe and an intuitive picture of the phenomena involved in electron transport at very high field strengths, field strengths high enough that individual electrons can gain energy without limit in the field and runaway. The model considers first a primary electron beam starting from rest whose momentum and energy evolve in time according to the applied field and collisional "friction" terms which represents the momentum and energy loss in collisions. The equations determining the time evolution of the momentum and energy are

$$\frac{dv_z}{dNt} = \frac{eE}{mN} - \frac{\nu_m}{N} \, v_z \qquad\qquad\qquad 3)$$

$$\frac{d\varepsilon}{dNt} = \frac{E}{N} \, v_z - \frac{\nu_u}{N} \, \varepsilon \qquad\qquad\qquad 4)$$

where v_z is the component of the beam velocity in the field direction, ε is the beam energy, ν_m/N and ν_u/N represent momentum and energy friction terms which will be described below, and e and m are the electron charge and mass, respectively.

Secondary beams are produced by the primaries (and the subsequent secondaries) in ionization events at a rate given by the product of the ionization cross section and the primary electron speed. Secondary beams are born at zero energy and their energy and momentum evolve in time exactly as that of their parents. Thus there is only one possible trajectory in velocity space and the primary and all subsequent secondaries exist at some point along that trajectory. It is a matter of bookkeeping to write the beam (or electron) number as a function of time, $n_b(t)$,

$$n_b(t) = \left(1. + \int_0^t n_b(t-t') \, Q_i(t')v(t') \, dt' \right) n_b(t=0), \quad 5)$$

where $Q_i(t')$ is the cross section for electron impact ionization at energy $\varepsilon(t')$ given by eq. above, and the velocity v is

$$v(t') = (\varepsilon(t')2e/m)^{1/2}. \qquad\qquad\qquad 6)$$

The ionization rate coefficient, $k_i = \nu_i/N$, is related to the growth of the electron density as

$$\frac{dn_B(t)}{dNt} = \nu_i/N \; n_b(t) . \qquad 7)$$

Note that if we were also considering a spatial dependence of the electron density, eq. 5 would contain additional terms.

Once eqs. 3-5 have been solved, it is straightforward to calculate all the average properties of the electron swarm. It is simple to show that any quantity (electron energy, velocity or the excitation or ionization frequency) averaged over all the beams, $X(t)$, can be determined from a knowledge of the same quantity, $X_b(t)$, for the primary electron and from $n_b(t)$, and can be expressed as

$$X(t) = X_b(t) + \int_0^t X_b(t-t') \frac{dn_b(t')}{dNt'} dNt' \qquad 8)$$

Input to this model consists of electron neutral scattering cross sections which define the collisional friction terms. The energy loss collision frequency, ν_u/N, depends on the total (angle integrated) cross sections for inelastic scattering, Q_k^0, and the threshold energy, ε_k, for each of the k inelastic processes. The elastic recoil energy loss is also included and this depends on the elastic momentum transfer cross section, Q_{el}^m, and the ratio of the electron to the neutral mass, m/M:

$$\frac{\nu_u}{N} = \sum_{k>0} Q_k^0(\varepsilon) \; \varepsilon^{-1/2} \varepsilon_k \left(\frac{2e}{m}\right)^{1/2}$$

$$+ \quad \frac{m}{M} \; \varepsilon^{1/2} Q_{el}^m \left(\frac{2e}{m}\right)^{1/2} \qquad 9)$$

Ionization is included as another inelastic process in the sum in eq. 9. The only effect of anisotropic electron scattering is in the momentum loss friction term. For isotropic scattering, this term depends on the total cross sections for elastic, Q_{el}^0 and for inelastic scattering,

$$\frac{\nu_m}{N} = Q_{el}^0(\varepsilon) + \sum_k Q_k^0(\varepsilon) \; \varepsilon^{1/2} \left(\frac{2e}{m}\right)^{1/2} \qquad 10)$$

In the limit of forward scattering and in the Born approximation, the friction to momentum gain can be expressed in terms of the energy loss function, and the momentum exchange frequency becomes (Phelps and Pitchford, 1985)

$$\frac{v_m}{N} \Rightarrow \left(\frac{L(\varepsilon)}{2\varepsilon} + Q_{el}^m \right) \varepsilon^{1/2} \left(\frac{2e}{m} \right)^{1/2} \qquad 11)$$

where

$$L(\varepsilon) = \sum_{k>0} Q_k^0(\varepsilon)\ \varepsilon_k \qquad 12)$$

We present results below for isotropic scattering and for forward scattering in the Born approximation.

The cross sections adopted for the calculations reported below in nitrogen were those described by Phelps and Pitchford (1985b). This cross section set include three effective inelastic channels (vibrational, singlet and triplet excitation) and elastic scattering as well as ionization. The total cross sections were extrapolated to energies greater than the values (usually 1 keV) tabulated in ref. using a log(energy)/energy scaling appropriate to allowed transitions and log(energy)/energy^2 for the elastic momentum transfer cross section.

Because collisions are treated in effective friction terms, this model is similar in spirit to solutions of the Boltzmann equation in the continuous slowing down approximation. Dispersion is introduced in the electron energy distribution not by collisions but by the birth of new electrons. At high E/N where most of the electron energy loss goes into ionization, this description becomes increasingly accurate. In the limit of collisionless, multiplying beams where there is no collisional friction, but which allows the creation of new secondary electrons, this description is a solution of the Boltzmann equation.

Monte Carlo Simulations

Our simulation algorithm is based on a simulation code developed by Reid (1979), but it has been extensively modified. Several time saving features have been introduced in addition to the null collision method (Skullerud, 1968) for facilitating the calculation of the collision time. In particular, sampling before collisions following Friedland (1977) and artificial attachment rescaling (Li, et al, 1989) in order to maintain a roughly constant number of particles in the simulation in spite of the large ionization rate coefficients have been implemented in the computer code. In addition, we monitor the steady-state electron energy balance in order to confirm that the simulations are not biased by the finite maximum energy allowed for the electrons. The calculated energy balance in steady-state was always better than 1% indicating that the simulations accounted properly for the electron energy gain and loss mechanisms even in the regime where electron runaway is possible. Results from a typical simulation are shown in Li, et al, 1989.

The obvious deficiency in results from Monte Carlo simulations is the difficulty in distinguishing small physical effects from statistical fluctuations. This is a particularly important when the issues under investigation are the physical phenomena themselves rather than questions as to the magnitude of phenomena known to exist. Thus, our conclusions as to the existence of steady-state at all values of E/N investigated do not rest on Monte Carlo simulations alone.

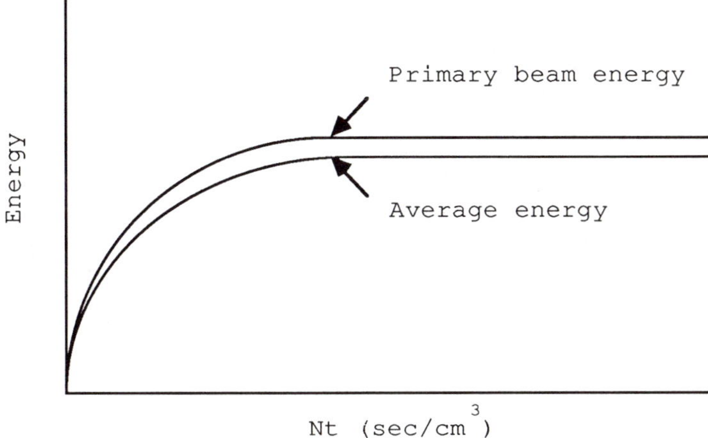

Fig. 1. Schematic of single, primary beam energy (from eq. 3) and total swarm average energy (from eq. 8) as functions of reduced time, Nt, for a moderate value of E/N.

RESULTS AND DISCUSSION

We first present schematically results from the multi-beam model to compare the behavior of the electron swarm at low or moderate with that at high E/N. Monte Carlo results are shown for comparison, and finally a comparison with the experiment in nitrogen will be presented.

<u>Existence of Steady-State Swarm Parameters at High E/N</u>

Figure 1 shows schematically the solution of eq. 2 for the energy of a beam at a constant value of E/N where an equilibrium is eventually established between the energy lost by the electrons in collisions and the energy gained from the field. We have assumed an E/N high enough that appreciable ionization takes place. Also shown in the figure is the

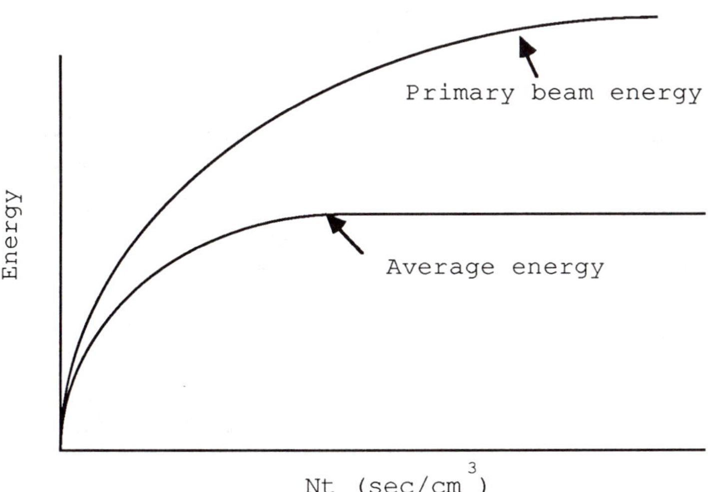

Fig. 2. Schematic of single, primary beam energy (from eq. 3) and total swarm average energy (from eq. 8) as functions of reduced time, Nt, for a high value of E/N.

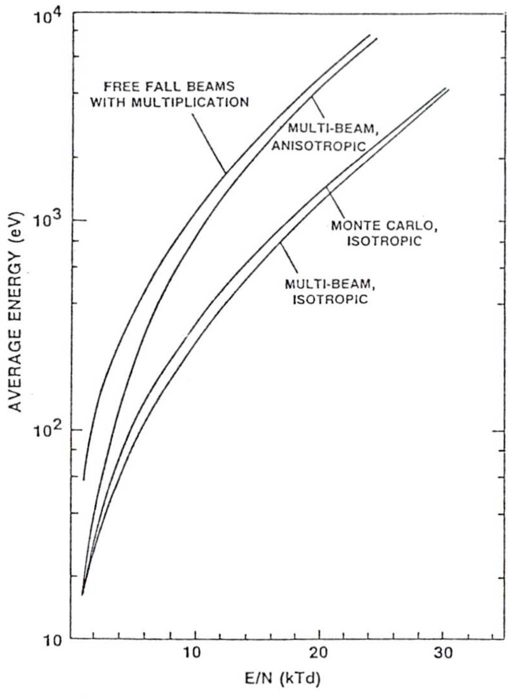

Fig. 3. Steady-state, swarm average energy as a function of E/N in N_2 calculated from multi-beam model with three different assumptions forms for the energy and momentum loss friction terms and Monte Carlo simulations.

average energy of the electron swarm as a function of the reduced time for the same case. The initial conditions were zero energy and velocity, and after a short time, both the primary electron energy from the solution of eq. 2 and the swarm average energy from eq. 6 reach a steady-state. The average energy of the swarm, however, is slightly lower than that of the primary because at any time a finite number of secondary electrons exist which have not yet attained their steady-state energy.

As E/N is increased, a point is reached where the energy gained from the field cannot be balanced by the energy lost in collisions because the electron energy is in the region where the energy and momentum loss friction are decreasing functions of energy. Thus, the higher the electron energy, the less resistance the electron feels to further gains in energy and momentum. Figure 2 illustrates schematically the solution of eq. 2 for these conditions where the energy of the primary electron increases without limit. *Because of the increasingly large number of secondary electrons which exist in the low energy region, however, the swarm average energy reaches a steady-state even if the individual electron energies do not.* This behavior is typical of all the swarm average quantities including the ionization rate coefficient as will be illustrated below.

Comparison of Multi-beam and Monte Carlo Results

Figure 3 shows the calculated, steady-state swarm average energy as a function of E/N in nitrogen from 2 to 25 kTd. As mentioned above, our benchmark calculation is the Monte Carlo simulation which is shown in Fig. 3 along with the multi-beam results for three different assumptions as to the angular scattering. The first feature to note in the figure is the excellent agreement between the multi-beam and the Monte Carlo simulations for isotropic scattering over the entire range of E/N considered, with an increasingly good agreement apparent at the highest E/N. Equally good agreement between the multi-beam and Monte Carlo

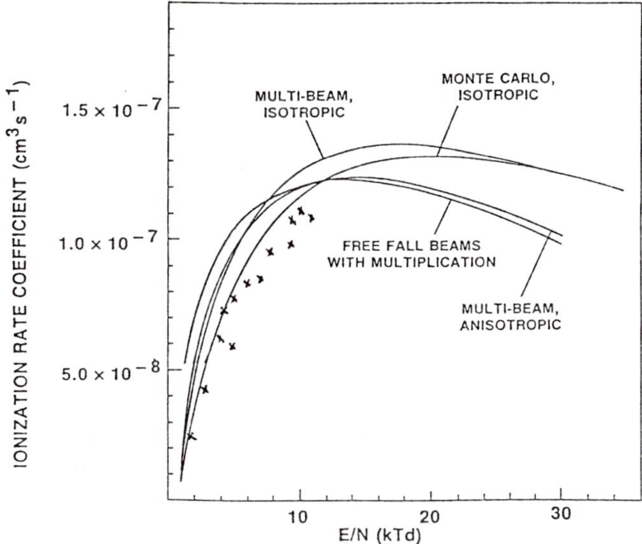

Fig. 4.   Steady-state, swarm ionization rate coefficient as a function
of E/N in N_2 calculated from multi-beam model with three
different assumptions forms for the energy and momentum loss
friction terms and Monte Carlo simulations. The experimental
points are shown by x's for comparison.

results have been obtained for other conditions as well; i.e., in
hydrogen, for anisotropic scattering, and during the transient approach
to steady-state. We have thus developed a good deal of confidence in
the quantitative predictions as well as in the overall physical
description provided by the multi-beam model.

The effect of different assumed angular scattering distributions is
also illustrated in Fig. 3. The curve labeled "free fall beams" results
from a calculation in which the energy and momentum loss frequencies
were set equal to zero. Thus, the only effect of collisions was
therefore the production of a secondary electron at zero energy at the
time of the collision, the primary electron exiting the collision event
unperturbed. This yields an upper limit to the average energy, and the
average energy calculated assuming forward scattering in the Born
approximation is seen to approach this limit at very high E/N. The
isotropic scattering calculation should represent a lower limit to the
average electron energy, and, as expected, the Born approximation
approaches the isotropic results at the lowest values of E/N considered.
For the entire range between about 4 kTd and 20 kTD in nitrogen, these
results indicate that it is important to consider the anisotropy in the
scattering cross sections.

<u>Comparisons of Calculations and Experiment</u>

A comparison of the multi-beam calculations for the same three
cases as in Fig. 3, the Monte Carlo simulations and the experimentally
measured ionization rate coefficients in nitrogen from 2 to 12 kTd
(Hays, et al 1987) are shown in Figure 4. The agreement among the
various calculations shown in the figure is consistent with the comments
in Fig. 3 above. The experimental points, indicated by the x's fall
slightly lower than the calculations. It is difficult to assign error
bars to the experimental points, although ± 25% would be a conservative
guess, and this brings the experiment into agreement with the

calculations. A theoretical maximum of the ionization rate coefficient, the maximum value of the product of the ionization cross section and the velocity, is $1.96 \ 10^7 \ cm^3 \ sec^{-1}$ using the cross sections of Rapp and Englander-Golden. (1965) Such a high value for the ionization rate coefficient could only be realized if the electron energy distribution function were monoenergetic at 250 eV, the energy at the maximum of the product of the ionization cross section and the velocity in nitrogen.

The calculations predict a maximum in the ionization rate coefficient followed by a slow decrease with increasing E/N past about 15 kTd. The experimental points do not go higher than 12 kTd so the decrease at high E/N in nitrogen has not been observed. Recent experiments in hydrogen, however, have confirmed a decrease in V_i/N with E/N for E/N greater than about 5 kTd (Verdeyen, et al, 1988). The maximum value of the ionization rate coefficient as a function of E/N occurs at a lower value of E/N in hydrogen than in nitrogen because of the smaller energy losses in hydrogen and consequent larger average energies as a function of E/N. Other recent experimental observations consistent with multi-beam beam and Monte Carlo predictions include the transient behavior of V_i/N, and in particular, an overshoot in the transient ionization rate coefficient in hydrogen for E/N greater than about 5 kTd.

CONCLUSIONS

The main conclusions to be drawn from these experimental and computational results is that swarm parameters at high E/N do attain steady-state values, at least for value of E/N up to 10 or 20 kTd, due to the large density of newly-born, secondary electrons which are always present at low energies. Thus, it is completely reasonable to develop models at high E/N which make use of swarm data. The multi-beam model presented briefly above provides a convenient framework for qualitative as well as quantitative descriptions of electron transport in this high E/N regime.

The implications of these results in the context of pseudo-sparks are several:

1) Values of electron impact ionization rate coefficients at very
high fields in the common gases are available from experiment and calculations,
2) Relatively simple computations are sufficient for the electron transport in the high field regions provided they are constant fields,
and,
3) Calculations at high fields must include effects of anisotropic scattering.

The issues remaining to be considered in the formulations of the multi-beam for the context of pseudo-spark discharges are:

1) These is an analogous multi-beam description of space-dependent, steady-state electron transport. Can this be extended to include space _and_ time dependence?
and
2) Is the extension to nonuniform fields useful?

It is also important to remark that ionization by heavy particle (fast neutral or ion) impact can play an important role in the charged particle kinetics in regions of high E/N if the time scales of interest are at least on the order of an ion transit time.

ACKNOWLEDGEMENTS

The experiments described above were carried out by Drs. G.N. Hays and J.B. Gerardo of Sandia National Laboratories in Albuquerque, New Mexico, and by Prof. J.T. Verdeyen of the University of Illinois, and the computational results were performed in collaboration with Dr. Y.M. Li of GTE Laboratories in Waltham, Massachusetts.

REFERENCES

Allis W.P., 1956, in Handbuch der Physik, edited by S. Flugge (Springer, Berlin) Vol. 21, PP. 383-444.
Dutton J, 1975, J. Phys. Chem. Ref. Data, 4, 577.
Friedland L., 1977, Phys. Fluids 20, 1461.
Gallagher J.W., Beaty E.C., Dutton J. and Pitchford L.C., 1983, J. Phys. Chem. Ref. Data, 12, 109.
Gylys V.T., Jelenkovic B.M. and Phelps A.V., 1989, J. Appl. Phys. 65, 3369.
Hays G.N., Pitchford L.C., Gerardo J.B., Verdeyen J.T. and Li Y.M., 1987, Phys. Rev. A 36, 2031.
Jelenkovic B.M. and Phelps A.V., 1987, Phys. Rev. A 36, 5310.
Kumar K, Skullerud H.R. and Robson R.E., 1980, Aust. J. Phys. 33, 343.
Li Y.M. and Pitchford L.C., 1987, Proc. of IEEE Int. Conf. on Plasma Science, Washington, DC.
Li Y.M., Pitchford L.C. and Moratz T.J., 1989, Appl. Phys. Lett. 54, 1403.
Muller K.G., 1962, Z. Phys. 169, 432.
Phelps A.V. and Pitchford L.C., 1985a, Phys. Rev. A 131, 2932.
Phelps A.V. and Pitchford L.C., 1985b, JILA Inforlation Center Report No. 26, unpublished.
Phelps A.V. and Jelenkovic B.M.,1988, Phys. Rev. A 38 2975.
Phelps A.V., Jelenkovic B.M. and Pitchford L.C., 1987, Phys. Rev. A 36, 5327.
Rapp D. and Englander-Golden P., 1965, J. Chem. Phys. 43, 1464.
Reid I.D., 1979, Aust. J. Phys. 32, 231.
Verdeyen J.T., Pitchford L.C., Li Y.M., Gerardo J.B. and Hays G.N., 1988, 41st Gaseous Electronics Conference, paper N-3.

PLASMA-BASED DEVICE CONCEPTS BASED ON THE

PSEUDOSPARK AND BLT

Martin A. Gundersen

Department of Electrical Engineering-Electrophysics
University of Southern California
Los Angeles CA 90089-0484

ABSTRACT

This chapter discusses several new approaches to the development of practical plasma based devices needed to test and develop concepts for plasma based accelerators, electromagnetic sources, and related devices such as plasma lenses. Although there has been considerable simulation effort, there are very few practical devices, and the realization of such devices is one of the most important limitations in the further development of plasma-based devices for applications. These concepts are based on recent results wherein a uniform, high repetition rate, pulse repeatable, high density plasma source has been developed, and a very high emission cathode has been demonstrated in the hollow cathode pseudospark-type device. Applications to electron beam devices, accelerators, and electromagnetic wave sources are considered.

INTRODUCTION

The development of new, innovative devices for plasma based accelerator research and technology, and for quantum electronics devices, will provide an important element for major advances in these technologies. This chapter considers several new approaches that are based on the plasma that is produced in the pseudospark-type device. The important properties of the device include the cathode, with current density of 10,000 A/cm^2, and the formation of a pulse-repeatable super-dense glow plasma.

Although plasma based devices are proposed to form the means of advancing accelerator technology beyond accelerators such as the SSC, there are as of this writing very few approaches to the development of these devices that actually address the device technology. At present it is very difficult to produce a high density, uniform, pulse repeatable plasma. The approach described is based on the recent development of high density plasma sources, including the pseudospark and the back-lighted thyratron (BLT). In the following section the background research is described briefly.

PSEUDOSPARKS AND BLTS

Accelerators, pulsed discharge lasers, such as CO_2 and the excimer families, and laser systems such as pulsed free electron lasers, rail guns, microwave sources, and EMP generators require a pulsed power conditioning system. For an accelerator such as the Stanford Linear Accelerator, pulse shaping, peak current, repetition rate, and current rate of rise are significant issues, and pulse modulators are required for the klystrons and the

Physics and Applications of Pseudosparks
Edited by M. A. Gundersen and G. Schaefer
Plenum Press, New York, 1990

switching magnets (kickers). These systems, however, are more complex than might be anticipated from a simple consideration of the required components -- a DC power supply, a switch, an energy storage capacitor, and two electrodes. The design and implementation of the system impact the quality, energy, timing, performance and the reliability of the specific device. At the heart of these devices is a switch. Switches in the pulsed power area include spark gaps, semiconductor switches such as the thyristor, the thyratron, and the new pseudospark and BLT. For an accelerator that must operate at a repetition rate of the order of or greater than 100 Hz, the switch most commonly chosen at this time is the thyratron. Normally, repetition rate limitations, and the need for gas handling systems limit the applications of spark gaps.

Since the development of the ceramic envelope thyratron in the 1950s there have been a number of innovations, with possibly among the most important being the grounded grid thyratron (EG&G, ≈ 1970) and the hollow anode thyratron (EEV, ~ 1980). The grounded grid thyratron led to improved current rate of rise, and the hollow anode improved reverse current handling capability.

Recently considerable interest has developed in a new class of thyratrons, called the pseudo-spark, and a related, optically switched thyratron (Figure 1) called the back-lighted thyratron (BLT). Research and development of pseudo-sparks has occurred mainly in Europe, with the discovery of the psuedo-spark occurring in the laboratory of Prof. J. Christiansen of the University of Erlangen[1]. The BLT is optically triggered, has a hollow cathode and a hollow electrode. Optical triggering has been demonstrated with ultraviolet radiation from a variety of sources including 1) a spark, 2) flashlamp, and 3) laser (Kirkman, 1986, 1988). The required optical power is low; it is wavelength dependent, of the order of 1 mJ at 300 nm, and 10 µJ at 220 nm. The conductive phase is initiated by photoemission from the back of the cathode.

The pseudospark is a closely related device with the main difference being that triggering is electrical, analogous to a conventional thyratron. With the exception of performance characteristics that are specifically related to triggering (such as optical isolation, or jitter associated with a particular triggering design) other characteristics remain the same. Thus pseudospark data is useful in characterizing the expected performance of the switch, particularly in terms of high power, high current, and high repetition rate applications. The isolation possible with the BLT is important for the conceptually new modulator, but is not as critical a consideration as the externally unheated cathode, which is common to both the BLT and pseudospark. Attractive features of these new switches include improved current rise, improved peak current, improved reverse current, and glow discharge operation with a cold cathode.

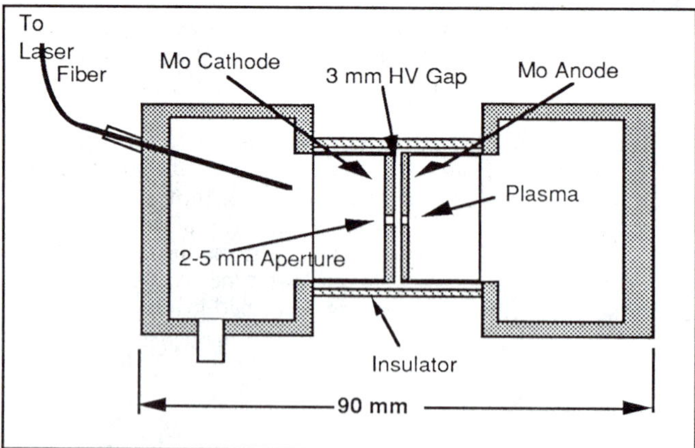

Figure 1. Structure of a BLT switch with optical triggering through an optical fiber. The switch is filled with low pressure (0.01-0.3 Torr) gas and is triggered by unfocused UV light incident on the back of the cathode surface.

The two electrode concept implicit in the BLT is a key element of the device. Further investigative work is needed to determine the differences in comparison with more standard devices such as thyratrons.

A fiber optic waveguide is a simple and practical way to distribute the light and can be designed so that the light is delivered exactly to the edge of the cathode hole - which leads to lower delay and jitter and reduced optical energies required for triggering. At a wavelength of 308 nm (XeCl excimer laser), *subnanosecond* jitter triggering was found (Braun et al 1988) with as little as 1.5 mJ of laser energy incident in a 10 ns pulse on the electrode. The fiber (either a 1mm diameter PCS (Plastic Coated Silica) good to approximately 250 nm or an all silica fiber 1.5 mm diameter good to 190 nm) was sealed in a glass tube to form a vacuum feedthrough and was positioned relative to the cathode electrode by a small, unobtrusive molybdenum mount. Typically, the light from the fiber is directed immediately around the electrode hole (some light goes through to the anode but this has no significant effect).

The investigation of the cathode emission process is important. Based on existing experimental data on pseudospark performance, scaling to extremely high currents should be fairly straightforward. It should be possible to ultimately produce a device that can switch a MA (10^6 A).

The anomalously high cold cathode emission (>>hot cathode) and current densities (40,000 A/cm^2, >>normal thyratron plasmas), obtained in a non-arcing mode and with a simple structure, strongly encourage further basic study of plasma devices. The physics of the cathode emission and high current density require study, and further device applications should be sought. In summary, versions of these switches under varying operating conditions have switched over 100 kA, operate in a glow discharge mode without electrode degradation from arcing, have demonstrated over $2 \cdot 10^{12}$ A/sec dI/dt, and sub-nanosecond jitter, using a cold cathode.

The study of the plasma in the device was key in the development of the switch, and one of the most important aspects is the uniform glow mode at high plasma density. The results suggest that it will be possible to extend the plasma to new applications wherein a high density plasma is required.

The cathode emission mechanism, described in the next section, is responsible for the high plasma density. This mechanism was resolved recently (Hartmann and Gundersen, 1988). A fundamental aspect of the cathode is the current density, which is $\approx$ 10,000 A/cm^2, and extends over a macroscopic area ($\approx$ 1 cm^2). The cathode, and the uniform plasma that the cathode emission properties support, encourage consideration of a variety of new applications.

THE SUPER-EMISSIVE CATHODE

The BLT and Pseudospark have been shown to operate with a self heated super emissive cathode[2] producing electron emission current densities >10kA/cm^2, suggesting the use of this cathode to produce high current electron beams in high vacuum and plasma for accelerator applications. The emission is observed to occur over a surface area $\approx$ 1 cm^2, and the cathode operates without forming an arc, and hence is useful for thyratrons and other applications. This emission is readily achieved in a simple, externally unheated configuration. It appears feasible to extend performance to devices requiring peak currents over 100,000 A. This is an anomalously high -- super-emissive -- regime of operation for a cold cathode operating in a glow discharge plasma. In spite of the considerable phenomenological understanding of gas discharge technology, these data were not predicted, and these results apparently will have important device applications. Although in the past higher but localized current densities have been achieved through the formation of filamentary arcs, devices (e.g. spark gaps) tend to be limited melting, sputtering and cratering of electrode material, and addition of electrode material to the arc plasma. This new cathode supports currents that formerly required arc-type devices, such as spark gaps.

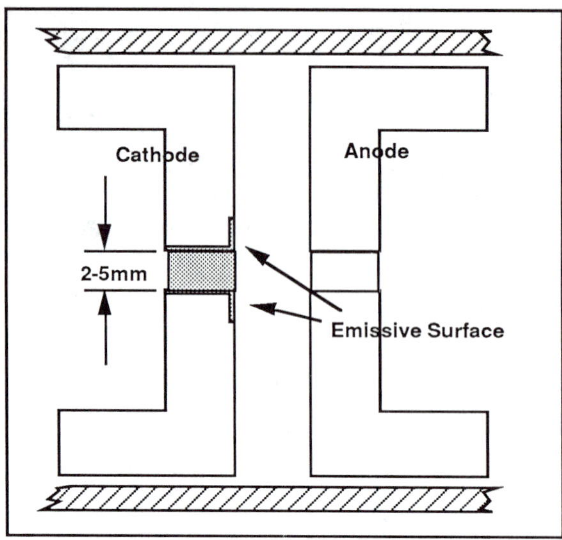

Figure 2. Electrode structure indicating the region of the highly emissive hot cathode surface. The super-emissive area is approximately 1 cm^2.

The cathode materials used are pure refractory metals. Materials such as Mo and pure W are actually among the best thermionic emitting materials simply because they melt at high temperatures, and thermionic emission is an especially violent function of temperature. These materials by themselves are ordinarily not used for high current emission in devices such as thyratrons, however; typically alkali-oxide coatings, or special preparation of a W matrix that contains Ba, are used, which have much lower work functions and good emission characteristics at temperatures of the order of 800 K.

The cathode emission process is characteristic of the geometry of the device and is distinct from others reported, such as liquid metal plasma cathode or spark gaps. The experimental structure consists of a hollow cathode and anode as shown in Figure 2. The longest path in this device is in the region of the central holes thus the discharge will begin here when the gap is overvolted or triggered by a pulse of electrons in the hollow space behind the cathode.

ELECTRON BEAM AND ION BEAM GENERATION

The highly emissive cathode may be useful to produce an electron beam for additional accelerator applications. Pulsed electron and ion beams have been produced using this structure in low pressure gas. During the initial phase of discharge plasma formation there is an intense electron beam emitted through the anode aperture and an ion beam is emitted through the cathode aperture. The energy of these beams are about equal to the initially applied voltage. Table I shows the electron beam results from several laboratories.

Ion beams have also been measured at the University of Erlangen[3] with peak currents of 35A and current density 10^4A/cm^2 produced at a voltage of 50kV in low pressure hydrogen. At the KFK Karlsruhe, West Germany ion beams of 10^6A/cm^2 have been observed in a pseudospark structure operating at over 200kV.

Table I. Electron beam results from several laboratories.

GROUP	peak current	current density	pulse length	energy	gas pressure
	A	A/cm2	ns	keV	
U. Erlangen	10^3	10^5	20	35	< 100 Pa
U. Maryland	10^2	----	20	30	$\approx$ 10 Pa
Dusseldorf	10^3	10^5	0.7-100	50	< 100Pa
USC	10^2	$>10^3$	>20	15	$\approx$ 10-20 Pa

Electron beams can be produced in both single and multi-gap structures. The single gap structures consist of a cathode and anode electrode separated by about 3mm and operate at voltages of 10 to 50 kV.Multi-gap structures consist of several electrode discs with central holes separated by glass, quartz or ceramic insulators and can operate at voltages greater than 200kV. A high voltage pulse of 10 to 50 kV is applied to the single gap structure. When the voltage across the gap nears its maximum UV light from a flashlamp or unfocused laser is incident on the back surface of the cathode electrode releasing electrons through photoemission. The electrons released are accelerated through the gap and anode hole into the drift tube. Some of these electrons collide with neutral gas atoms creating ions that bombard the cathode producing electrons by secondary emission and heating the cathode to a temperature where thermionic emission is possible.

These electron emission processes produce a high current of electrons that are accelerated to the applied voltage and pass through the anode into the drift tube. This pulse of electrons lasts for 10 to 100ns after which a plasma is formed in the anode cathode gap. After formation of this plasma the voltage on the device is 100-200V allowing only low energy electrons to be produced.

In addition to the pulsed beam during discharge formation there can be a high current low energy beam during the steady state phase of the discharge after plasma formation. The heated surface of the cathode is present for many microseconds. The inter-electrode plasma will form a cathode sheath layer at the surface of the cathode with a potential drop of 100 - 500 V. Electrons emitted are accelerated in this cathode sheath before entering the bulk plasma. The mean free path between collisions with electrons and ions of the bulk plasma for 100ev electrons is greater than the typical 3mm electrode spacing, hence many electrons emitted can pass through the anode aperture undisturbed by collisions. The current in this low energy electron beam may be a large fraction of the total discharge current.

An experimental apparatus for the study of the pulsed beam is shown in Figure 3. The BLT anode was modified by drilling a three millimeter diameter hole to allow the electron beam to propagate out of the device. The beam was collected by a Faraday cup and the beam current measured with a Rogowski coil. These results suggest the possibility of extracting and accelerating the *low* energy beam. With the addition of an extraction grid and an accelerating gap it should be possible to extract the beam and accelerate it to high energy. Beam generation in the BLT can be studied by observing damage to the metal structures.

This device produces beams in low pressure gas and is therefore especially suited for operation in accelerators that may be gas filled for plasma beam guiding or acceleration. This cathode does not suffer from the poisoning problems associated with thermionic cathodes. In applications where high vacuum is required there should be no trouble maintaining high vacuum in the accelerator by differential pumping as the amount of gas required for efficient cathode operation is very small.

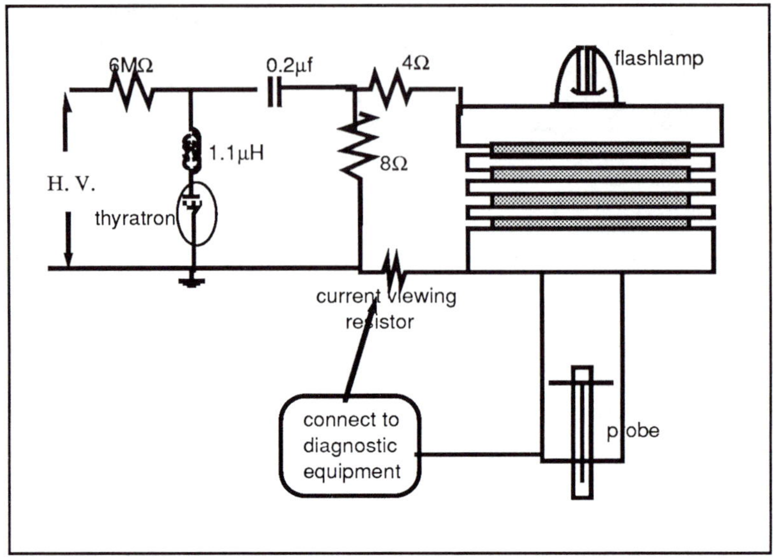

Figure 3. Experimental schematic for initial probe studies of electron beam. Shown is a flashlamp switched BLT. The BLT shown has three gaps in order to achieve higher stand-off voltage relative to a single gap device.

PLASMA BASED ACCELERATORS

Plasma-based systems are under consideration as new accelerators for high energy physics because a plasma can support very large electric fields which scale as SQRT(n) where n is the plasma density. For a density of 10^{16} cm^{-3}, the maximum longitudinal electric field will be on the order of 10 GeV/m , which exceeds by several orders of magnitude the current accelerating gradients. Acceleration due to one of the proposed plasma based accelerator schemes, the plasma wake field accelerator[4], was demonstrated[5] using a 50 cm long hollow cathode chamber that contained a plasma with a density of 3×10^{13} cm^{-3}. Partly as a result of the low density, accelerating gradients of 1 MeV/m were obtained. Although there will be trade-off issues related to background collisional processes that have yet to be quantitatively determined, it appears that in order for a plasma based accelerator to be realizable, long plasma sources of several meters in length and densities higher than 10^{16} cm^{-3} will have to be developed. These plasmas will likely be pulsed. A high pulse to pulse reproducibility is required for extended periods of time -- requiring the formation of a glow-type plasma with high density and pulse to pulse repeatability of the order of 1%. It is clear from present research efforts in related areas that the plasma *quality* is a key problem.

The pseudospark is a promising high density hydrogen plasma source that satisfies the stability requirements mentioned above to an extent not previously possible. This high quality plasma source, based on the pseudospark or BLT device plasma, occurs because of the very high emission properties of the cathode. Measurements of the plasma electron density and temperature show a very uniform, pulse to pulse repeatable distribution in the transverse and longitudinal directions. The distribution also persists for several microseconds following the current pulse. The electron density has been determined by spatially and temporally resolved measurements of the Stark broadened profile of the H$_\beta$ and other Balmer emission using several spectroscopic methods and a streak camera. Within the limits of the experimental apparatus, no shot-to-shot variation was detected.

This approach allows one to produce plasmas *less* dense than most presently proposed schemes and simulations[6], but *more* dense than can now be produced while satisfying the constraints listed in the previous paragraph, and with a *low* background. The plasma density is also more appropriate if one is to reduce background from scattering such as that which produces copious amounts of synchrotron radiation, which will be excessive for beam bunch densities above 10^{10} and at energies in the multi-GeV range.

The densities that have been achieved in reproducible experiments are readily applicable to wakefield experiments -- in order to test concepts, and develop technologically useful devices. Approaches that have *demonstrated* pulse to pulse reproducibility and spatial uniformity at the densities have not been achieved to date. There are excellent prospects for scaling to higher plasma densities, using the same device principles.

Approaches that have been suggested previously include use of a laser for photoionization of a small gas volume, use of a wall stabilized arc, various Θ and Z pinch schemes, and ionization and plasma formation using an electron beam. The use of a laser for multiphoton ionization of hydrogen gas is very inefficient at best, requiring an extremely expensive laser with very high maintenance requirements that will produce less than full ionization for gases such as hydrogen. Although laser assisted photoemission is a technique to be investigated in order to obtain properly shaped pulses for wakefield generation[7], the multiphoton ionization cross sections of hydrogen are small, and the process requires a laser with very high intensity at a wavelength that is as short as possible. There is considerable information in the literature describing both the physical processes and the hardware required[8]. Thus even with an expensive and complex laser system, the production of a fully ionized hydrogen plasma will be a difficult task.

The current emission mechanism in the cathode of the pseudospark or BLT is an important intrinsic advantage over the conventional Z or Θ type plasma source, in that it allows production of a uniform glow, because it has a large and highly emissive cathode area. As this plasma is scaled to higher densities, it should be possible to achieve uniform, cold plasmas with densities comparable to those achieved in the pinch device -- without the turbulence, and with reproducible high repetition rate capability.

The use of an electron beam for ionization and plasma formation may be an important approach. However, without using the electron beam from a large accelerator, there are certain limitations. On the basis of data obtained at USC on electron beam generation in the BLT, we conclude that the BLT overs a significant simplification in this approach as well.

Alternate approaches to the development of the plasma for wakefield generation also include the successful work of Rosenzweig et.al., which found experimental evidence for wakefield densities in plasmas of the order 10^{13} cm^{-3}. Thus the pseudospark or BLT plasma density is already over two orders of magnitude higher than that used by Rosenzweig et. al., and is more readily implemented in a pulse mode, with simpler support requirements.

By using a 10^{15} cm^{-3} plasma density column one will be able to increase the accelerating fields reported by Rosenzweig et al. by a factor of 10, operate with a pulse repeatable, simple device, and therefore, scale the dynamics of the beam-plasma interaction to high plasma densities. An added advantage is that the magnetic field lines created by the discharge will tend to confine the injected electron beam to the central region rather than disperse it as is the case with the theta-pinch-type discharge.

PLASMA LENSES FOR FINAL FOCUS

One of the technological issues that the future accelerators are facing, is that in order to obtain luminosities higher than 10^{33} cm^{-2} sec^{-1}, final focusing forces much stronger than the ones provided even with today's superconducting magnets need to be developed.

Plasmas can play an important role in this field too, since a relativistic electron beam traveling through a plasma can excite transverse focusing fields of comparable strength to the longitudinal fields used to accelerate a trailing beam. Thus, a beam going through a dense enough plasma can self focus to a very small radius. The characteristics of the focused beam as well as the lens focal length are a strong function of the plasma density.

The next generation of light particle accelerators will include linear colliders capable of accelerating particles to energies of ≈ 1 TeV. Such high energies require luminosities on the order of 10^{33} cm^{-2} sec^{-1}, which are to be achieved by increasing the machine repetition rate, shaping the beams, increasing the number of particles per bunch, and by developing extremely tight focusing lenses[9]. At present, focusing field gradients on the order of 10 kG/cm can be achieved with superconducting magnets. However, in light of the ability for plasmas to support very large electromagnetic fields, it is possible to create in a plasma medium focusing forces that will exceed those of a superconducting magnet by several orders of magnitude. Such large focusing fields can be created by exciting in the plasma radial wake fields[10]. Therefore, among the candidates for tight focussing lenses are a new class based on plasma engineering.

A plasma lens, however, developed for a linear collider will have to meet the stringent requirements set by the inherent characteristics of the accelerator. In addition to the high plasma density, these include high repetition rate, high reproducibility from pulse to pulse, device reliability with low maintenance requirements, and low jitter. These constraints place very strong demands on the performance of repetitively pulsed high density plasma sources, which are characteristically unreliable. As described above, the pseudospark and BLT already satisfies most of the requirements listed above.

Measurements of the electron density and temperature of the plasma show a very uniform distribution in the radial direction (figure 4.), which also lasts for several microseconds following the current pulse -- even when the source is an impedance matched, square current pulse. The electron temperature was studied with a spectroscopic investigation of the H_α and H_β emission, and using both a non-equilibrium model and a local thermal equilibrium model for the ratio of the line intensities, which gave a 'low', uniform plasma temperature of the order of one to a few eV. Related physical processes are discussed below. The electron density was spatially and temporally resolved by measuring the Stark broadened profile of the H_β emission. Effects due to Doppler broadening were found to be negligible for hydrogen temperatures of ≈ 1 eV and electron densities $> 1 \times 10^{15}$ cm^{-3}. These data, and streak camera recordings demonstrate that initially during the rising part of the current pulse the electron density is centered around the cathode hole, with a diameter comparable with that of the hole. In the configuration that has been investigated, and not optimized as a plasma lens, the electron density reaches its maximum of 3×10^{15} with a diameter of 1.74 cm. At 300 ns after the peak of the current pulse the plasma density has decayed to 2.8×10^{15} cm^{-3}, reaching radial density scale-lengths of ≈ 1 cm. A plot of the radial plasma density near the peak of the discharge is shown in Fig. 4.

The hollow structure with an axial bore, together with a strong axial current and a high plasma density, make the BLT a very attractive device for focusing of charged particles. First, the axial bore of the BLT allows for an easy injection of the beam into the lens. Second, the high axial current of 5 to 15 kA uniformly distributed over 5 mm in radius, generates radial magnetic field gradients of 4 to 12 kG/cm with field strengths on the order of 2 to 6 kG at a distance of 5 mm from the axis. Such high field gradients can be used to focus relativistic beams on the order of the MeV's to very small spot sizes, and also to perform fine alignment. Third, dense particle beams injected through the cathode bore can create strong focusing forces as a result of the radial wake fields excited in the plasma. These strong focusing forces can focus highly energetic beams of energies 50 GeV and higher to very small spot sizes as needed in order to keep the luminosity high. These two focusing mechanisms make the BLT a versatile lens capable of performing effectively under a variety of experimental conditions.

For example, focussing can be shown to occur assuming a normalized emittance E_n of 8.8×10^{-4} m. rad., relativistic gamma = 40, beam radius at lens location of 0.5 cm and a lens thickness of 1.5 cm. β is the Twiss parameter defined by

$$\frac{E_n}{\gamma} \beta(z) = a^2(z),$$

where a is the beam radius and z is the spatial coordinate along the beam axis[11]. These parameters correspond to the beam of the AATF at Argonne National Laboratory. In this case the focusing forces are determined by the lens geometry and are independent of the characteristics of the beam.

For a beam with a radius less than the plasma skin depth l_s ($l_s = \lambda_p/2\pi$), the focusing forces from the wake fields become the dominant mechanism. This corresponds to the beam characteristics of SLAC and also the Advanced Accelerator Test Facility of Argonne National Laboratory. For the 50 GeV (SLAC) case, we have undertaken calculations[12] assuming a long beam limit (beam length ~ 2 mm) in order to estimate the focusing force from a 3×10^{15} cm^{-3} dense plasma by using the linear theory approximation. Linear and transverse parabolic beam profiles were assumed, and a lens thickness of 1.5 cm., a normalized emittance of 5×10^{-5} m.rad. and 5×10^{9} particles in a beam of radius 50 µm are assumed. Under such conditions, a beam of originally 50 µm in radius will be focussed to a spot of 4 µm in radius at its midpoint in a distance of 38 cm. In this case the BLT acts as thin lens. In the case of energies of 20 MeV (Argonne) a normalized emittance of 8.8×10^{-4} m.rad., lens thickness of 1.5 cm and also 10^{11} particles in a focused beam of radius 100 µm and 5 mm long are assumed. In this case the beam focuses within the plasma lens, since the focusing strength becomes very large. Nonlinear effects must be considered resulting in higher field gradients than estimated from the linear theory. Under these conditions the plasma acts as a thick lens and a simple calculation of the beam spot size at focus is not possible. The fact that the BLT behaves as a thin lens at 50 GeV and as a thick lens at 20MeV makes it possible to study both configurations with this device.

Wake field focusing experiments using the Argonne Test Facility will require prefocusing to increase the beam density. This prefocusing of the 20 MeV beam can be realized by taking advantage of the magnetic fields generated by the high current discharge of the BLT. The final focusing can thus be achieved by placing a second BLT at the focal point of the first one. A BLT can be used for prefocusing whether the final focusing lens is a BLT or other device.

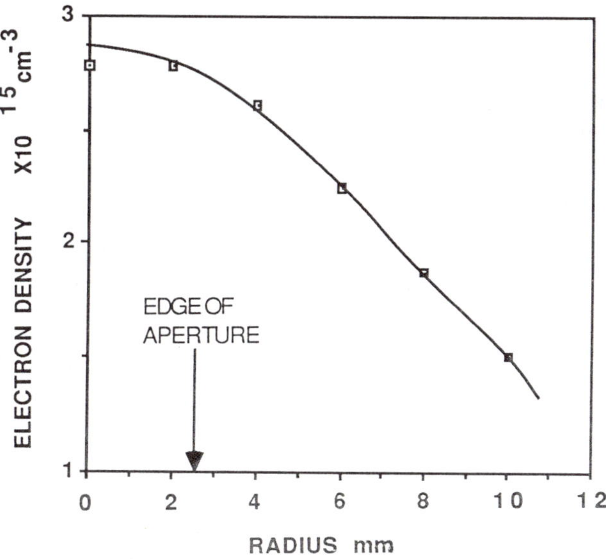

Figure 4. Electron density as function of radius during pulse.

ELECTROMAGNETIC SOURCES

The propagating electron beam in a pseudospark is known to produce microwaves[13]. Reasonable transmission of microwaves occurs through pyrex glass and alumina, being essentially lossless at microwave frequencies. Detection can be carried out using diode detectors and cutoff waveguides in various bands (L-Band, 1-2 GHz , X-Band, 8-12 GHz,and Ka Band, 27-40 GHz, for example).

Plasmas with a periodic density variation present forbidden bands of frequencies in the electromagnetic dispersion relation[14]. We are studying the possibility of applying these forbidden bands to the design of efficient "plasma mirrors" of controllable reflectivity in the millimeter wavelength region. Preliminary calculations indicate that it is possible to obtain reflectivities higher than 95% with only 5 plasma slabs of thickness 5/4 l and densities of 0.6nc each, separated from each other by vacuum regions of the same thickness. Here, l is the wavelength of e.m. wave one wants to reflect, and nc is its associated critical density. The absorption levels are less than 1%. Thus, for example, a mirror composed of 5 plasma slabs of thickness 1.25 mm each, and densities of 6×10^{14} cm^{-3}, separated from each other by vacuum regions of thickness 1.25 mm, will reflect 97% of a 1 mm e.m. wave incident perpendicular to the plasma slab faces. These calculations, however, represent ideal cases in which the plasma layers are considered to be perfect slabs of sharp boundaries. More realistic calculations that will allow for diffused boundaries will need to be made. A particular case will be that of a sinusoidal density variation. At present, we are designing an experiment consisting of parallel BLT discharges to test these ideas.

The structure of the multigap device described above has dimensions appropriate for an interaction between an electron beam that can produce mm waves. Injection of a beam into the multigap structure will provide a means for testing beam interactions in a plasma loaded system. This allows higher current densities than previously achievable, because of the space charge neutralization resulting from the (uniform, glow) plasma.

CONCLUSIONS

Successful development of the pseudospark device will have applications to electromagnetic wave generation in several ways, including plasma loaded free electron lasers, generation of microwave radiation for fusion heating, and millimeter wave generation, through the interaction of the electron beam with the plasma. The electron and ion beams may be expected to impact microwave, accelerator, and other technology, such as X-ray generation. The prospects for these experiments to produce exciting results are strongly encouraged by recent results -- particularly the remarkable plasma and cathode properties. Research into the physical properties will have important applications, providing a basis for an entirely new technology in plasma devices.

ACKNOWLEDGEMENTS

The work presented here has been supported by the Air Force Office of Scientific Research and the Army Research Office. The plasma lens calculations and plasma mirror concept are due to Dr. Humberto Figueroa. The author would like to acknowledge the work of Dr. C. Braun, Prof. D. Erwin, Dr. G. Kirkman, Mr. H. Bauer and Dr. W. Hartmann, as well as valuable conversations with Prof. A. McCurdy, Prof. J. Christiansen, Dr. K. Frank, Dr. R. Temkin, and Dr. H. Riege in developing these ideas.

REFERENCES

1.   G. Kirkman and M. A. Gundersen, "A Light Initiated Glow Discharge Switch for High Power Applications," Appl. Phys. Lett. **49** (9) 494, (1986). For comprehensive reviews see "High power hollow electrode thyratron-type switches," K. Frank, E. Boggasch, J. Christiansen, A. Goertler, W. Hartmann, C. Kozlik, G. Kirkman, C. G. Braun, V. Dominic, H. Riege and M.A. Gundersen,

Proceedings, Sixth IEEE Pulsed Power Conference, June 1987, and "High power hollow electrode thyratron-type switches," K. Frank, E. Boggasch, J. Christiansen, A. Goertler, W. Hartmann, C. Kozlik, G. Kirkman, C. G. Braun, V. Dominic, M.A. Gundersen, H. Riege and G. Mechtersheimer, IEEE Trans Plasma Science **PS-16**, 317, (1988).

2. W. Hartmann and M. A. Gundersen "Origin of Anomalous Emission in Superdense Glow Discharge", Phys Rev Lett., **60**, 2371 (1988).

3. See "Workshop on the Fundamentals and Applications of the Pseudospark Discharge", Bad Honnef, West Germany, October 1987.

4. J.B. Rosenzweig, D.B. Cline, B. Cole, H. Figueroa, W. Gai, R. Konecny, J. Norem, P. Schoessow, and J. Simpson, Phys. Rev. Lett., **61**, 98 (1988).

5. T. Tajima and J. Dawson, "Laser electron accelerator", Phys. Rev. Lett. **43**, 267 (1979).

6. P. Chen, J.M. Dawson, R.W. Huff and T. Katsouleas, "Acceleration of electrons by the interaction of a bunched electron beam with a plasma", Phys. Rev. Lett. **54**, 693 (1985).

7. T. Katsouleas, J.J. Su, W.B. Mori, C. Joshi and J.M. Dawson, "A compact 100 MEV accelerator based on plasma wakefields", presented at OE /LASE, Los Angeles, Jan. 1989.

8. See, for example A.D. Bandrauk, Ed., "Atomic and Molecular Processes with Short Intense Laser Pulses", NATO ASI Series B: Physics, Plenu Press (1987). This provides a current review of multiphoton ionization problems, and several articles specifically address multiphoton ionization in hydrogen.

9. See, for example, S. Turner, Ed., "Proceedings of the workshop on new developments in particle acceleration techniques", Orsay, Fr, 29 June - 4 July 1987, available from CERN, Publication CERN 87-11, ECFA 87/110 (1987).

10. E.B. Forsyth, L.M. Lederman and J. Sunderland, "The Brookhaven-Columbia plasma lens", IEEE Trans. Nucl. Sci. **12**, 872 (1965). See also P. Chen, "A possible final focusing mechanism for linear collider", Particle Accelerators **20**, 171 (1982), P. Chen, J.J. Su, T. Katsouleas and J.M. Dawson, "Plasma focusing for high-energy beams", IEEE Trans. Plasma Sci. **PS-15** (2), 218 (1987).

11. H. Figueroa, W. Gai, R. Konecny, J. Norem, A. Ruggiero, P. Schoessow, and J. Simpson, "Direct measurement of beam-induced fields in accelerating structures," Phys. Rev. Lett. **60**, 2144 (1988).

12. H. Figueroa, G. Kirkman, and M.A. Gundersen, "A plasma lens candidate with highly stable properties," Proceedings of the 1989 Workshop on Advanced Accelerator Concepts, Lake Arrowhead, California, Jan. 9-13, 1989. in press.

13. F. Muller, Reported at the "Workshop on the Fundamentals and Applications of the Pseudospark Discharge", Bad Honnef, West Germany, October 1987.

14. H. Figueroa and C. Joshi, Laser Interaction and Related Plasma Phenomena, **7**, 241, Plenum Press (1986).

EMITTANCE MEASUREMENT OF A PSEUDOSPARK-PRODUCED

ELECTRON BEAM

M. J. Rhee and E. Boggasch

Laboratory for Plasma Research
University of Maryland, College Park, Maryland 20742 U.S.A.

ABSTRACT

We report the first measurement of the rms emittance of a pseudospark-produced electron beam. A six gap pseudospark chamber filled with argon gas was operated at $\sim$25 kV, producing a $\sim$10 Hz repetitive pulse train of electron beams. Typically, a beam of average energy 20 keV, peak current $\sim$50 A and pulse duration $\sim$10 ns FWHM was extracted from the chamber. The rms emittance was evaluated by using the slit-hole method. A typical value of rms emittance was found to be $\epsilon \simeq 55$ mm-mrad, yielding a normalized emittance of $\epsilon_n \simeq 15$ mm-mrad. The normalized brightness of the beam was then estimated as $B_n = I/\epsilon_n^2 \simeq 2 \times 10^{11} \mathrm{A}/(\mathrm{m}^2\mathrm{rad}^2)$.

I. INTRODUCTION

Interest in high quality, high current electron beams has been stimulated by their application with regard to such field as advanced accelerators and free electron lasers.[1,2] Pseudospark discharge phenomenon with interesting charged particle emission characteristics was reported by Christiansen and Schultheiss[3] in 1979. At the University of Maryland, an experiment was performed to measure for the first time the emittance of an electron beam produced by a pseudospark discharge.

In this paper, we shall describe the operation of a six gap pseudospark chamber which is characteristically similar to the devices reported by other laboratories.[3-7] Repetitively pulsed ($\sim$10 Hz) electron beams were extracted through an anode hole into a drift region. The beam currents were measured by a Rogowski coil at the anode and a Faraday cup placed downstream. A simple emittance meter[8] consisting of a series of thin slits and detector film was placed downstream of the anode. Angular distributions of sheet beamlets formed after passing through the slits were recorded on radiachromic film.[9] The optical density distribution of the film after exposure to the beam provide us with the distribution of the transverse components of electron velocities from which the rms emittance is evaluated. Subsequently, we describe the emittance analysis and estimation of brightness of the beam. Finally, the experimental results are discussed.

Physics and Applications of Pseudosparks
Edited by M. A. Gundersen and G. Schaefer
Plenum Press, New York, 1990

II. EXPERIMENTAL ARRANGEMENT

The experimental setup is shown in Fig. 1. The discharge chamber consists of a hollow cathode, five modules of intermediate electrodes and insulators, and an anode with a center hole for electron beam extraction. Intermediate electrodes made of 3.2 mm thick brass have outer diameter of 6.35 cm and center hole diameter of 3.2 mm. The 3.2 mm thick plexiglas insulator washers have outer and inner diameters of 7 cm and 2.54 cm respectively. In addition to the chamber capacitance of 11 pF between anode and cathode, a low inductive external capacitor of 420 pF was added for the present work. A 50 cm long drift chamber is attached to the anode side to accommodate diagnostics such as the emittance meter or the Faraday cup. A capacitance-manometer type vacuum gauge was used to measure the gas fill pressure which was almost statically balanced by a needle valve and a throttle vacuum valve as shown in Fig. 1. Two Rogowski coils were molded into axisymmetric grooves milled into both side of the anode flange so that the azimuthal component of the magnetic field, which arises from the axisymmetric current, is predominantly supported and other components (noises) are suppressed. The rise time of the Rogowski coil system may be approximately given by L/Z_0, where L is the inductance of the coil and Z_0 is the characteristic impedance of the transmission line used, and is found to be less than 0.5 ns. A Faraday cup consisting of a 3 cm diam graphite beam collector and a 10 mΩ current viewing resistor placed on axis of the downstream drift chamber was employed to measure the beam current at various axial positions. The chamber voltage was measured by a homemade high impedance resistive voltage probe of division ratio 1:20,000 into a 50 Ω load. The response of the probe was greatly improved by a proper compensation for the stray capacitance in the carbon resistors used. The resulting rise time is less than 0.5 ns, and the RC droop time constant due to the blocking capacitor is $\sim$120 ns. For emittance measurement, a slit-hole type emittance meter was employed and was placed in the drift chamber 9 cm downstream of the anode. The emittance meter consists of a series of parallel thin slits of 200 μm width and 2 mm spacing constructed from 0.6 mm thick stainless steel plate; 2 mil thick radiachromic film, used as a beam detector was placed 12 mm downstream of the slit plane.

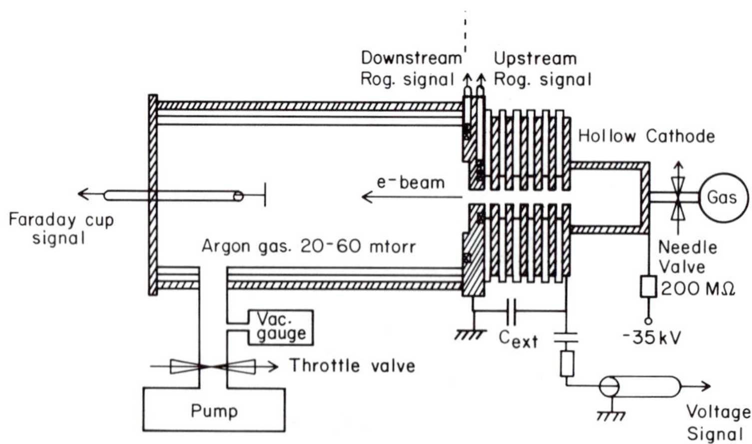

Fig. 1. Experimental setup.

The system was operated typically at 30 kV, with an argon gas pressure of 30 mTorr.

For the emittance measurement, exposures of approximately five consecutive beam pulses to the radiachromic film through the slits in the emittance meter were enough to produce an appropriate density profile, with the peak value of the optical density not exceeding 0.5. This ensure that the measured optical density distribution is linear to the beam intensity.[9] The density profile was obtained by scanning the film using an optical microdensitometer, and is shown in Fig. 2.

III. ROOT-MEAN-SQUARE EMITTANCE ANALYSIS

The root-mean-square (rms) emittance[10] (or effective emittance[11]), which has been widely used as a measure of beam quality, was defined as

$$\epsilon = 4\left\{ <x^2><x'^2> - <xx'>^2 \right\}^{1/2}, \tag{1}$$

where x' is the gradient of the particle trajectory given by $x' = dx/dz = p_x/p_z$, and the angular brackets denote average values over the two-dimensional trace space as

$$<\phi> = \int \phi\rho(x,x')dxdx', \tag{2}$$

where ρ is the projected density in two-dimensional trace space, and is assumed to be normalized, i.e., $\int \rho dxdx' = 1$.

It is very reasonable to assume that as in Ref. 8, the beam produced in this experiment is axisymmetric and of Maxwellian transverse velocity distribution. This allows us to use the simple slit-hole type emittance meter mentioned above, whose results can be easily analyzed.[8] We find empirical functions $\alpha(r)$, $\beta(r)$, and $\sigma(r)$ as functions of radial position r from the density profile shown in Fig. 2, where $\alpha(r)$ is the mean diverging angle, $\beta(r)$ represent the peak values, and $\sigma(r)$ is the rms width of the individual distributions. Numerical integrations were then performed using the empirical functions α, β, and σ to find ρ and the moments $<x^2>$, $<x'^2>$, and $<xx'>$ (see Ref. 8 for details). Several isodensity contours of the resultant $\rho(x,x')$ were constructed in $x - x'$ space (known as an emittance plot) as shown

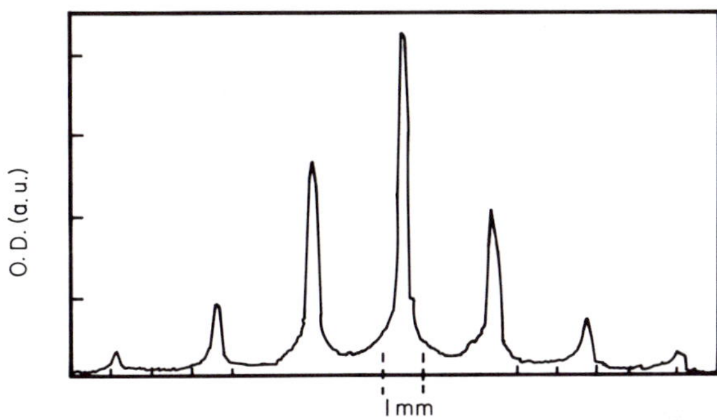

Fig. 2. Optical density distribution on the radiachromic
film after exposure to the beamlets.

in Fig. 3. The measured rms emittance, given by Eq. (1) with the obtained moments, was found to be

$$\epsilon \simeq 55\text{mm} - \text{mrad}. \tag{3}$$

It is interesting to note that the emittance plot in Fig. 3 shows a slightly S-shaped contour; this is indicative of the existence of nonlinear focusing(defocusing) in the beam indicating the possible existence of space charge field of substantial strength. The normalized rms emittance $\epsilon_n = \beta\sigma\epsilon$, which is invariant[11,12] when there is acceleration in the axial direction, is conveniently used for comparing beam qualities at different energies. With an average beam energy of 20 keV inferred from the voltage probe signal, the normalized emittance is estimated to be $\epsilon_n \approx 15$ mm-mrad. Another useful invariant of the beam associated with the emittance is the brightness.[11] We use here a simpler mathematical expression of normalized rms brightness $B_n = I/\epsilon_n^2$, so as to compare with other results by other laboratories using different definitions.[13] With a measured current of $\sim$50 A, we obtained the brightness

$$B_n = I/\epsilon_n^2 \simeq 2 \times 10^{11} \text{A/m}^2\text{rad}^2, \tag{4}$$

which is an order of magnitude higher than that of other high brightness sources.[13] It should be noted here that the results in this work are all time integrated over 5 shots.

IV. CONCLUSIONS

We have measured, for the first time, the emittance of a pseudospark-produced electron beam. A six gap pseudospark chamber with argon gas was operated at $\sim$ 25 kV and produced a $\sim$ 10 Hz repetitive pulse train of electron beams. Typically, a beam of average energy 20 keV, peak current $\sim$50 A and pulse duration $\sim$10 ns FWHM was extracted from the chamber. The rms emittance was evaluated by using the slit-hole method. The typical value of rms emittance was found to be $\epsilon \simeq$ 55 mm-mrad, yielding a normalized emittance of $\epsilon_n \simeq$ 15 mm-mrad. These values

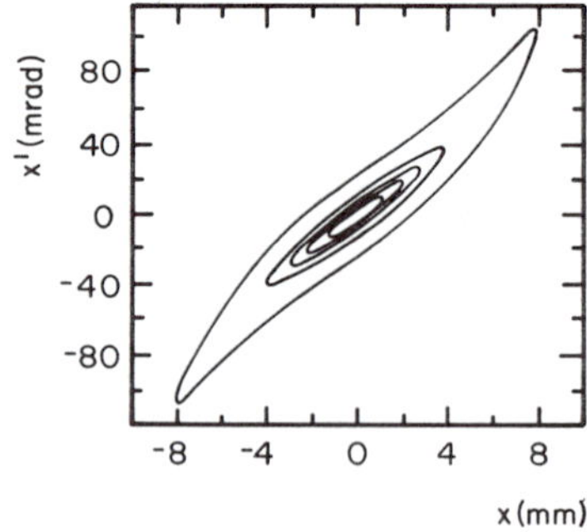

Fig. 3. The emittance plot in x-x' trace space.
The isodensity contours correspond to 0.1, 20,
40, 60, and 80% of the peak value.

are time integrated over 5 shots; thus they represent upper limits. The normalized rms brightness of the beam was then estimated as $B_n = I/\epsilon_n^2 \simeq 2\times10^{11}\,\mathrm{A}/(\mathrm{m^2 rad^2})$, which is an order of magnitude higher than existing high brightness electron beam sources.

ACKNOWLEDGMENTS

We have benefitted from valuable discussions with Dr. M. Reiser. This work was supported by the Air Force Office of Scientific Research and the U. S. Department of Energy.

REFERENCES

1. R. Stiening, "The status of the Stanford Linear Collider," Proc. 1987 IEEE Particle Accelerator Conference (Edited by E.R. Lindstrom and L.S. Taylor, 1987), p. 1.

2. IEEE J. Quantum. Electron. QE-21, (1985), pp. 804-1119, special issue papers therein.

3. J. Christiansen and C. Schultheiss, "Production of high current particle beams by low pressure spark discharges," Z. Phys. A, 290, 35 (1979).

4. X.L. Jiang, K.F. Chen, S.C. Jiang, and Y.B. Piao, "Neutralization and focusing of pulsed beams with high current densities," IEEE Trans. Nucl. Sci. NS-32, 1814 (1985).

5. P. Choi, H.H. Chuaqui, M. Favre, and E.S. Wyndham, "An observation of energetic electron beams in low pressure linear discharges," IEEE Trans. Plasma Sci., PS-15, 428 (1987).

6. C. Schultheiss, "Insulator with reduced E-field on surface," Nuclear Instruments and Methods A 254, 398 (1987).

7. H. Gundel, H. Riege, J. Handerek, and K. Zioutas, "Low pressure hollow cathode switch triggered by a pulsed electron beam emitted from ferroelectrics," Appl. Phys. Lett. 54, 2071 (1989).

8. M.J. Rhee and R.F. Schneider, "The root-mean-square emittance of an axisymmetric beam with a Maxwellian velocity distribution," Particle Accelerators, 20, 133 (1986).

9. W.L. McLaughlin, R.M. Uribe, and A. Miller, "Megaray dosimetry," Radiat. Phys. Chem., 22, 333 (1983).

10. P.M. Lapostolle, "Possible emittance increase through filamentation due to space charge in continuous beams," IEEE Trans. Nucl. Sci., NS-18, 1101 (1971).

11. J.D. Lawson, The Physics of Charged-Particle Beams (2nd ed., Clarendon, Oxford, U.K. 1988).

12. M.J. Rhee, "Invariance properties of the root-mean-square emittance in a linear system," Phys. Fluids, 29, 3495 (1986).

13. D.A. Kirkpatrick, R.E. Shefer, and G. Bekefi, "High brightness electrostatically focused field emission electron gun for free electron laser applications," J. Appl. Phys. 57, 5011 (1985), references therein.

NEW WAYS OF ELECTRON EMISSION FOR

POWER SWITCHING AND ELECTRON BEAM GENERATION

H. Riege

CERN
Geneva, Switzerland

ABSTRACT

Strong electron emission from a plasma or a solid surface is the basis of many applications such as high-power switching or intense electron beam generation. The electron emission in pseudospark geometries, from the interior of a hollow cathode and from the outer cathode surface, is discussed. These types of emission are compared with other emission methods by taking into account the electron current density, the spatial and the energy distribution, the beam quality, and the mechanisms, which allow each process to be controlled. A combination of different types of emitters may be the best solution for efficient power switching and for intense electron beam production. As examples, ferroelectric emitters in pseudospark-like switches and in electron beam sources are described.

1. INTRODUCTION

In many technical applications there is a steadily growing need for energy concentration in space and time. This is true not only in the field of high-energy particle accelerators but also for other high-power devices, such as lasers and free-electron lasers (FEL), inertial confinement fusion (ICF) drivers, material processing systems, and space applications. The increasing resistance against gigantism motivates the creation of very compact systems, which are energized for preferably only a limited time, during which they produce energetic output.

The problems arising from such systems are related mainly to the efficiency of conversion of primary energy into output energy. Low efficiency implies necessarily high losses, which in smaller and smaller volumes lead to material fatigue and destruction. A method or technology that does not provide very efficient energy conversion *cannot be considered for high-power devices*.

An example of such a method is the apparent weakness of high-pressure spark gaps in the field of high-frequency, high-power switching. Even at low repetition rates the performance is limited by severe erosion of the electrodes owing to the transition of the current from a narrow spark channel to the electrode surfaces. A considerable part of the switched energy is dissipated in the spark channel and on the areas of impact on the electrodes.

In the domain of particle accelerators, a high electric accelerating field $\vec{E}_{acc}$ is often claimed as the most important feature. A strong $\vec{E}_{acc}$-field is, however, not sufficient to accelerate an *intense, high-brightness* beam to high energy. The primary energy has to be transferred *efficiently* from the accelerating structure to the beam. The high-intensity, high-energy plasma beat-wave accelerator is therefore an example of a technically impracticable device, because of the very poor overall energy conversion efficiency.

In this paper, selected practical methods of high current density power switching and of high-brightness electron beam generation are reviewed and proposed. There are certainly other schemes that give similar results. However, the methods and devices that are dealt with here have been partly

Physics and Applications of Pseudosparks
Edited by M. A. Gundersen and G. Schaefer
Plenum Press, New York, 1990

proven in experiments and in technical operation. A vast potential for further performance upgrading should render them interesting for a large range of technical switching and electron beam applications.

2. HIGH-CURRENT EMISSION AND SWITCHING APPLICATIONS

In vacuum and gas switches, the charge transfer from one electrode to the other is mainly governed by the electrode surface properties and by the immediate interface with the interelectrode medium. Generally, even extremely high current densities can be obtained, e.g. 10^9 A/cm^2 have been measured in the hot spots of vacuum arc switches. The choice of current amplitude, current density, and discharge geometry for a specific high-power application is, however, challenging when the erosion of the electrode material should be kept low. The task of building an efficient, compact, low-inductance power switch, having a long lifetime, includes finding the optimum compromise between current and current density, discharge channel diameter, electrode surface area, and electrode material.

High-pressure spark gaps, thyratrons, ignitrons, and triggered vacuum-gap switches are nowadays the most widely used gas switches in high-power applications. Spark gaps are able to switch the highest peak powers and pulse energies, but at the expense of strong electrode erosion and complicated trigger systems.[1] Hence their use is restricted to low repetition rates. Careful design and choice of material can nevertheless lead to impressively long lifetimes under hard conditions, as has been shown by the CERN spark gaps[2] that were developed for the fast ejection system of the Serphukov IHEP accelerator[*).

Ignitrons are widely used in high-current switching applications that are not too demanding in terms of switching precision and low inductance. They also suffer from a long recovery time and a decay of voltage hold-off capability under high pulse-energy conditions.

Thyratrons are probably the most common high-power precision gas switches used in power lasers, radar systems, pulsed magnet circuits, and many other pulsed power systems. Thyratrons can be operated at repetition rates above 10 kHz, voltages above 100 kV, and currents above 50 kA, however, not all at the same time and only for short pulse lengths. The current densities in thyratrons are generally limited to less than 10^3 A/cm^2 and high-current amplitudes are only feasible with large, heated current-emitting cathodes. The hold-off capabilities of thyratrons at high charging voltage are diminished by the steady presence of a d.c. discharge in the tube, which is necessary for the precise triggering.

In 1981 a new type of switch was developed at CERN in collaboration with the University of Erlangen.[3] These switches, based on the pseudospark phenomenon,[4] combine some of the advantages of spark gaps, ignitrons, and thyratrons in a single unit.[5] Their special geometry permits the distribution of the current over a relatively large electrode surface area, although the discharge is not so homogeneous as in a thyratron. With a cold cathode, current densities far beyond 10^4 A/cm^2 have been obtained in the CERN high-current switches[6] for the ACOL plasma lens test generator. This can be considered as a major progress in switching technology, since destructive effects are much smaller than in high-pressure spark gaps and only slightly worse than in thyratrons. Pseudospark switches can be triggered at least as precisely and efficiently as thyratrons and are therefore very suitable for high repetition rates.[5,7] Pseudospark switches also withstand current reversals up to 100% without any problems.

The operation of all high-power gas discharge switches, including vacuum gaps, requires a large number of initial charge carriers at the time of breakdown, and efficient physical electron emission mechanisms that tolerate the high current density during the conduction phase. Both requirements are generally fulfilled with electrons emitted from a cathode surface by one of the following four conventional processes, or by a combination of them:

 i) heating of a cathode surface *(thermionic emission)*;
 ii) *secondary emission* from a cathode by bombardment with other particles (ions, neutrals);
 iii) extraction by a high electric field *(field emission)*;
 iv) illumination with photons *(photoemission)*.

All four mechanisms can be significantly enhanced by the proper choice of a cathode material with low work function. High-power gas switches are generally ignited by the processes of field emission and photoemission (laser triggering). Often the initial charge carriers are produced in the plasma of an auxiliary discharge.[5] In thyratrons, an auxiliary discharge is maintained all the time and is routed into the main discharge gap at the desired moment of breakdown.

The emission processes during the conduction phase of all high-power gas switches are more difficult to identify. Common to all types of switches is the presence of a discharge plasma in contact

---

*) These spark gaps have been operated for 17 years in pulse generators of 80 kV, 8 kA, and 5 μs with the *same* high-quality tungsten electrodes, and at current densities of 10^7 A/cm^2 in the spark channels!

with the cathode surface. The slightly positively charged plasma fulfils three important functions for the current conduction:
   i) Whereas the bulk plasma is at constant potential, there is a sharp decay of potential across the Debye layer of width:

$$\lambda_D = \sqrt{\frac{\epsilon_0 k T_e}{e^2 n_e}} \quad . \tag{1}$$

Here ϵ_0 is the dielectric constant, k is the Boltzmann constant, T_e is the electron temperature, e is the elementary charge, and n_e is the plasma density. For $n_e = 10^{17}$ 1/cm^3 and a plasma potential of $kT_e = 50$ eV, we obtain an electric field of the order of MV/cm! This strong field is present just at the electrode surface, where it is most important for the emission.

  ii) Space-charge forces, which tend to push the electrons back to the emitting surface, are partly reduced when the electrons enter the plasma region.
 iii) The plasma is a rich source of ions which after being accelerated through the Debye layer, bombard the cathode surface for a short time and heat it up to the melting temperature. Hence thermionic emission, being a steeply rising function of temperature, will be substantially enhanced.[8] The current density of ions with mass m_i passing through the Debye layer towards the electrode is given by Bohm's equation (with a $\approx$ 1):

$$j_i = a e n_e \sqrt{\frac{k T_e}{m_i}} \quad , \tag{2}$$

which can greatly exceed the normal Langmuir–Child space-charge limit.

In all the gas switches under discussion, the combined thermionic and field emission, together with space-charge neutralization by the plasma and surface heating by ion bombardment, may result in an almost unlimited current density as observed in the hot spots of vacuum arcs. Such high densities can never be obtained in a pure vacuum.

Pseudospark switches profit from a favourable geometry, which leads to reduced electrode erosion while tolerating very high current densities ranging between those of spark gaps and thyratrons. Up to now, only the usual trigger methods of pseudospark switches[5] present certain disadvantages. The long-term performance of the so-called surface-discharge trigger is very poor. The charge injection trigger,[9] on the other hand, is a source of premature discharge, as in thyratrons, especially in high-voltage switching applications. These disadvantages can be overcome with a new type of switch, which is triggered by a radically new method of electron emission.

Contrary to the four classical emission processes listed above, the proposed method[10] is based upon the fast reversal of the spontaneous polarization P_s of a *ferroelectric* sample. In this type of emission the reservoir of available charge on the sample surface is limited to maximum values of the order of $\Delta q/\Delta F = P_s = 0.5$ C/m^2, where Δq is the charge, ΔF is the active surface area and P_s is the spontaneous polarization of the ferroelectric material. The amplitude of the emitted electron current density, $j_e = \Delta P_s/\Delta t$ is determined by the speed of the polarization reversal. The current amplitude is given by the area of the emitting surface, $I = j_e \times \Delta F$. The method is ideal for the preionization of large volumes of arbitrary shape. If the polarization change ΔP_s is induced quickly enough, the main gap of a high-power switch can be fired directly with the emitted electrons, without additional charge-carrier multiplication. A prototype switch of this type has been described elsewhere.[11]

Here we propose a multistage high-voltage low-pressure gas switch with a hollow, cold cathode, which is triggered by a cold ferroelectric trigger in the absence of any d.c. high-voltage potential (Fig. 1).

The switch geometry resembles that of a pseudospark structure, but in principle the electron beam cross-section can be shaped into any desired form following the layout of the electrodes, e.g. into multiple beam bundles with multihole electrodes, or into hollow beams of large internal diameter with ring-shaped electrodes. In fact, a charge multiplication process, such as in a pseudospark discharge, is not required for breakdown initialization owing to the abundant electron emission from the ferroelectric trigger FE. The e$^-$ beam triggers all stages of the switch simultaneously, thus reducing breakdown delay and current rise-time. The FE sample (generally made from a ceramic lead–lanthanum–zirconium–titanate, known as PLZT) is subjected to a fast HV pulse from a transistor-switched electronic power circuit. The sample electrode facing the HV stages is perforated to allow for e$^-$ emission from the bare surface. The auxiliary grid AG shown in Fig. 1 can be used either for investigating the emitted e$^-$ beam or for increasing the hold-off voltage of the main switch by applying a potential of a few hundred volts.[3] In this case, some gas amplification will further enhance the trigger-beam intensity. The main switch stages must be graded potential-wise as for thyratrons.

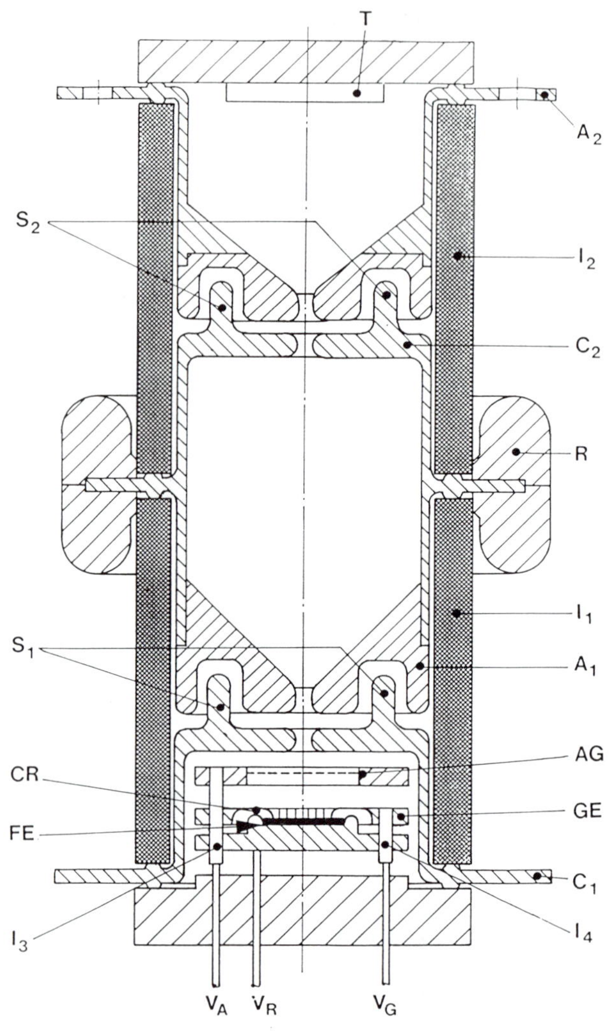

A$_{1,2}$ = anodes
C$_{1,2}$ = cathodes
I$_{1-4}$ = insulators
S$_{1,2}$ = electrode screens
R = grading ring
T = electron beam stopper
AG = auxiliary grid
FE = ferroelectric disk
CR = contact ring
GE = grid electrode
V$_A$ = potential of AG
V$_R$ = pulse voltage on the back of the FE sample
V$_G$ = potential on GE

Figure 1. Scheme of a hollow cathode, low-pressure, two-stage, high-voltage switch triggered by an electron beam emitted from the surface of a ferroelectric sample (FE)

We can summarize the advantages of the new HV switch as follows:

i) The ferroelectric trigger system does not contain any heated element (cold electrodes, ceramic insulators and ferroelectrics) or any permanent discharge.

ii) The e$^-$ trigger beam is so dense that gas amplification is not needed and direct firing of the main switch becomes possible. The switching delay is of the order of several tens of nanoseconds. In Ref. [11] less than 5 ns jitter were measured with a non-optimized FE trigger system in combination with a pseudospark switch structure. Delay and jitter are determined mainly by the characteristics of the electronic trigger circuit, by the distance of the FE sample from the main gap, and by the type of ferroelectric material.

iii) Unlike surface discharge triggers with organic glide-spark insulators, the ferroelectric emitter does not develop organic vapours or other contaminating products. The auxiliary grid AG protects the emitting FE surface from contamination by deposits coming from the main discharge region and against intensive contact with the discharge plasma.

iv) The main switch is built with a cold cathode. Owing to the absence of hot elements and of permanent glow discharges inside the switch, a better hold-off voltage can be achieved than with thyratrons. The excellent trigger efficiency allows the switch to be run far away from the breakdown voltage so that, compared with high-pressure spark gaps, a gain in prefiring safety can be obtained.

v) As in pseudospark switches, high main-current densities of 10^4 to 10^5 A/cm^2 and currents up to 100 kA can be combined with low electrode erosion rates.

vi) The strong e$^-$ beam passing through the main switch after breakdown can be dumped on a thoriated-tungsten target or on a LaB$_6$ target. The e$^-$ beam can also be diffused inside the hollow anode by means of an insulating ceramic target.

The full use of the favourable features listed above should make it possible to build compact, low-inductance, single-unit switches for voltages above 100 kV and 100 kA.

3. ELECTRON EMISSION AND BEAM SOURCES

In high-power switching, it is the current density that is of main interest, and it is of less importance, how the beam current is emitted from the surface and transferred into the adjacent gap medium. However, the realization of a particle beam source depends strongly on the way in which the beam is emitted and then transported to its destination. Generally, several problems have to be solved, when the electrons produced by any of the five emission methods described in the previous section, shall be transformed into a technically manageable beam.

In most of today's e$^-$ beam sources, electrons are injected thermionically into vacuum with the assistance of a high d.c. or r.f. electric field. Recently, laser-illuminated photocathodes have been successfully implemented in direct combination with r.f. field extraction into vacuum.[12] With thermionic sources, hardly more than 10 A/cm^2 beam-current density can be extracted from an emitter, whereas with laser photocathode sources more than 500 A/cm^2 have been achieved on a picosecond time-scale. Beam pulse-shaping in time and geometry is the big advantage of this method. However, when injecting into vacuum, both methods have to fight with space-charge forces whenever very high beam-current densities are envisaged. The Langmuir–Child law is a natural limit for the emitted current density, and the extraction field cannot be infinitely increased. The thermal material resistance is, on the other hand, the limiting factor for heating the emitter electrically or by means of an intense laser pulse. Hence the classical ways of raising the emitted current density—i.e. by increasing the surface temperature, and/or the laser intensity, and the extraction field strength—become practically inefficient when one requires emitted beam-current densities that are one or more orders of magnitude higher than are feasible today.

The method of e$^-$ emission is also a decisive factor for the *beam quality* of the source. For high-energy accelerators and FELs, low normalized emittance and high brightness are essential beam parameters. The normalized emittance ϵ_n is derived from the r.m.s. emittance ϵ [13] by

$$\epsilon_n = \beta\gamma\epsilon \, , \tag{3}$$

where β is the ratio of electron velocity v_e to the velocity c of light, and $\gamma = (1 - \beta^2)^{-1/2}$. The normalized emittance of an emitter of absolute surface temperature T and of radius r_{em} can be written as

$$\epsilon_n = 2\pi r_{em} \sqrt{\frac{kT}{m_0 c^2}} \, , \tag{4}$$

with m_0 being the electron rest mass measured in units of mrad. The normalized emittance ϵ_n can also be expressed in terms of current I and current density j_e, with $I = \pi r_{em}^2 j_e$, as

$$\epsilon_n = 2 \sqrt{\frac{\pi kTI}{m_0 c^2 j_e}} \tag{5}$$

with I in A and j_e in A/m^2 $^{*)}$.

The normalized brightness B_n of the beam is given by

$$B_n = \frac{I}{\epsilon_n^2} \, . \tag{6}$$

For a given current requirement, the low emittance and high brightness are obtained with the *high current density and low temperature* of the emitting medium. The equations for emittance and brightness are only theoretical limits. In practice, one generally achieves performances that are worse by one or two orders of magnitude. A thermionic emitter of 1100 K surface temperature (kT = 0.1 eV) is able to emit up to 20 A/cm^2. For a total current of 100 A it should, theoretically, produce a

---

*) Note, by this definition ϵ_n contains π.

beam with a minimum emittance of $\epsilon_n = 3.5 \times 10^{-5}$ mrad and at best a brightness of $B_n = 10^{11}$ A/(mrad)^2.

In photocathodes the equivalent temperature is considerably higher than in thermionic emitters. However, much higher current densities of several hundred A/cm^2 have been obtained by laser illumination, and with a special low-work-function photocathode material the effective surface temperature can also be shifted below $kT = 1$ eV. With a current of 100 A, a photon energy of $kT = 0.5$ eV, and a current density of 500 A/cm^2, a normalized emittance of 1.6×10^{-5} mrad and a brightness of 4×10^{11} A/(mrad)^2 are, theoretically, feasible.

The electron emission from ferroelectrics into vacuum has been studied in PLZT ceramics with HV-pulse polarization reversal.[14] In a rather slow 50 Ω cable pulser, emitted current densities greater than 10 A/cm^2 have been measured. Taking into account again a total current of 100 A and $kT = 0.025$ eV (room temperature!), the emittance from a ferroelectric surface is calculated to be 2.5×10^{-5} mrad and the brightness 1.6×10^{11} A/(mrad)^2. This is better than a thermionic emitter, but worse than a laser-illuminated photoemitter. With faster and more powerful electronic pulsers, considerably faster polarization reversal—and consequently higher current densities—can be expected from ferroelectrics.

In vacuum, all emitters suffer from strong space-charge fields induced by the electrons that have just left the surface, and from the irregularities of the emitting surface, which lead to strong field distortions near the surface. Both effects are responsible for the emittance degradation and the blow-up of the ejected beam, even when strong extraction fields are applied. Magnetic focusing could counteract the beam blow-up, but it is inefficient owing to the small velocity of the electrons near the surface.

Now we may ask why electron beam sources should not profit from the favourable plasma surface interaction, as in the case of high-power switches. First we will discuss some data on electron emission from a plasma, using the example of pseudospark electron beam sources, which have been studied for several years. A striking feature is the current density of these sources, which beat all other emitters by orders of magnitude.[4] In Ref. [15] measurements of total current and of beam diameter were used to deduce the emittance data. It should be pointed out that the pseudospark beam is very complex, since the current density, the beam diameter, and the kinetic energy of the electrons vary strongly along the beam pulse. Hence, also the emittance and the brightness are not constant. Recently, the *average* normalized emittance and brightness of a pseudospark beam have been measured experimentally to be 5×10^{-5} mrad and 2×10^{11} A/(mrad)^2, respectively.[16] Parts of the beam certainly have smaller emittances. With a beam current density of 10^5 A/cm^2, a total current $I = 100$ A as above, and a plasma temperature of $kT = 50$ eV, we can calculate an emittance of 1.2×10^{-5} mrad and a brightness of 7×10^{11} A/(mrad)^2. These data depend strongly on the estimated plasma temperature. With a 10 eV plasma, 5×10^{-6} mrad and 4×10^{12} A/(mrad)^2 become possible.

Interesting measurements have also been done with ferroelectrics in a plasma environment.[17] A plasma was produced by a pseudospark electron beam. A second beam was generated by polarization reversal from a ferroelectric sample and then sent directly into the plasma. The second beam travelled for 10 cm through the plasma before being partially ejected into vacuum. Beam-current densities of more than 1000 A/cm^2 have been measured. Assuming again a total current of 100 A for such an emitter, which is at room temperature, an emittance of 2.5×10^{-6} mrad and a brightness of 1.6×10^{13} A/(mrad)^2 can be calculated. There are good chances that the current density emitted into a plasma from a ferroelectric sample can be further increased by choosing faster emitting material and polarization reversal circuits. One can then expect theoretically an emittance below 10^{-6} mrad and a brightness above 10^{14} A/(mrad)^2.

The significantly higher emission into the plasma can—like the current emission in high-power switches—be explained by the action of the strong electric field in the Debye layer just on the surface of the emitter. After having left the emitter surface, the beam is much less subject to space-charge forces and emittance blow-up than a beam emitted into vacuum. Even if the plasma density is less than the beam density, there is a strong reduction of the beam space-charge field. It has been shown in many experiments that the beam can travel rather long distances without being focused through a plasma, or even, with some ionization losses, through a low-pressure gas.

Whilst electron emission and transport are relatively easy to handle in a plasma, acceleration is a more difficult problem. The plasma tends to cancel the radial electric field E_r of the beam by a space-charge neutralization factor f_e,

$$E_r = 2\pi e n_0 r_b (1 - f_e) , \tag{7}$$

and the magnetic self-field B_θ by a current neutralization factor f_m,

$$B_\theta = 2\pi e n_0 r_b \beta (1 - f_m) , \tag{8}$$

where n_0 is the plasma density and r_b is the beam radius. Then the beam envelope can be approximately described by[18]

$$\ddot{r}_b - \frac{2eI}{\gamma m_0 \beta c r_b} \, [1 - f_e - \beta^2(1 - f_m)] = 0 \ . \tag{9}$$

The plasma screening not only reduces the beam self-fields but also prevents external accelerating fields from penetrating through the plasma towards the beam. The only possible way around this obstacle is to decrease the plasma density to below the beam density until the penetration depth of the external field exceeds the beam diameter r_b ('under-dense plasma'). Induction acceleration with nanosecond pulses seems the best way of raising the kinetic energy of plasma–embedded beams.

The use of an under-dense plasma may force the reintroduction of a focusing field. A simple, passive, electrostatic focusing method has been tested on a pseudospark beam by H. Gundel as reported in Ref. [19]. A pseudospark electron beam was injected into an insulator tube of 3 mm inner diameter and 60 cm length. The mean beam energy was 35 keV. The beam current was measured by two beam-current transformers, one at the entrance and one at the exit of the tube (Fig. 2). The radial focusing was so strong that a significant longitudinal distortion of the beam was observed; the front electrons were accelerated and the trailing ones decelerated to a few hundred eV. The resulting, mean current neutralization factor f_m was about 0.7. The violent compression effect on the beam electrons, which is caused by the electrons expelled from the plasma towards the insulator wall, will be smoother if the beam is short and longitudinally more uniform, and if the tube is filled with plasma before injection.

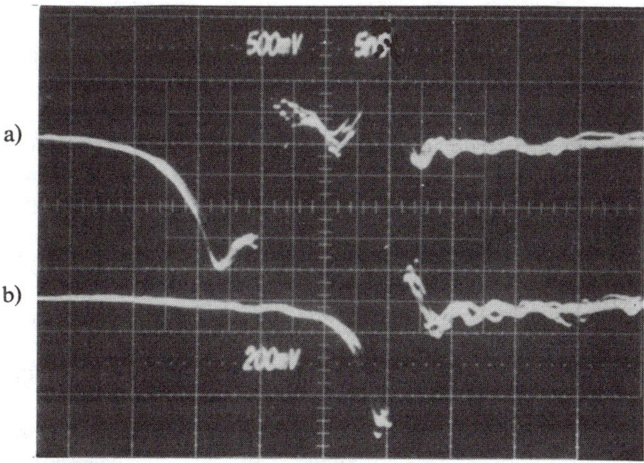

Figure 2.   Electron beam current a) in front of and b) behind an insulating tube, which acts as strong, transverse focusing element while increasing the energy spread of the beam. Each trace contains five pulses: a) 5 A/V, 10 A peak current; b) 5 A/V, 4 A peak current.

A proper choice of tube diameter, plasma diameter, and plasma density will mean that the strength of focusing and the beam distortion in the expansion direction can be controlled, while offering the possibility of simultaneous induction acceleration.

A plasma electron beam source is now proposed that combines all the beneficial effects described above, in order to produce beam intensities that are an order of magnitude better than those generated by existing sources. At the same time, beam emittances that are much lower than those from conventional sources may be achieved. Figure 3 shows a source consisting of a pseudospark section, which generates an e^- beam for plasma production in the chamber and along the adjacent beam transfer tube (insulator). During the breakdown phase of the pseudospark chamber, a ferroelectric emitter is activated by rapid polarization reversal, and a short e^- pulse is emitted into the pseudospark plasma, as described in Ref. [17]. It is immediately focused by the insulators and the floating electrodes of the chamber, and later by a tapered tube along which the density of the plasma decreases owing to differential pumping. After acceleration to relativistic energy by several induction units, the beam is injected into vacuum through a small hole. The beam energy is now sufficiently sensitive to magnetic focusing and is ready for the application of classical bunch compression and r.f. accelerating techniques. The decoupling of the source and the conventional accelerator part allows also efficient synchronization of the beam pulse with the r.f. phase.

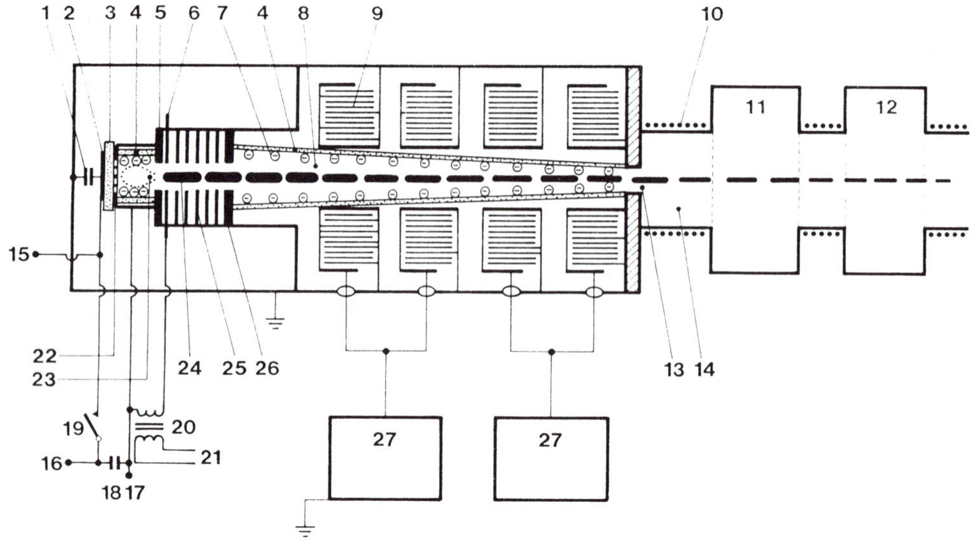

<table>
<tr><td>

1: coupling capacitor
2: RE (rear electrode of ferroelectric sample)
3: FE (ferroelectric sample)
4: ceramic tubes for electrostatic focusing
5: cathode of pseudospark chamber
6: trigger electrode
7: focusing charges expelled from plasma
8: high-density plasma region
9: induction field coils
10: solenoid coils
11: conventional bunch compression and r.f.

</td><td>

feedback systems for synchronization with standard r.f. vacuum cavity acceleration
12: r.f. cavity
13: exit hole to vacuum
14: vacuum region
15: negative d.c. charging voltage HV_1 for RE of FE
16: negative d.c. charging voltage HV_2 for polarization reversal circuit
17: negative d.c. charging voltage HV_3 for pseudospark chamber ($HV_2 < HV_3 < HV_1$)

</td><td>

18: storage capacitor for FE-reversal circuit
19: fast transistor switch
20: coupling transformer for pseudospark trigger pulse
21: trigger pulse
22: grid electrode GE of ferroelectric sample FE
23: hollow cathode plasma
24: electron beams from pseudospark and FE
25: multidisk pseudospark chamber
26: anode
27: PFN for induction acceleration modules.

</td></tr>
</table>

Figure 3. Schema of an e^- beam source with e^- emission from a ferroelectric surface with a differentially pumped plasma tube, with strong, passive electrostatic focusing, with inductive acceleration, and with transmission into vacuum

It is not absolutely necessary to have a ferroelectric medium as emitter in this particular source; it is just convenient. Any other method that generates short, intense charge pulses at the pseudospark plasma–emitter interface may give good results. Short beam pulses have been obtained with a simple capacitor discharge on the rear side of the pseudospark chamber. If plasma–resistant photocathodes can be found, a laser could be used for beam *and* plasma production. The pseudospark is just a simple and cheap way of doing the same job.

4.   CONCLUSIONS

The plasma-electrode interaction is the basis for high current density generation in high-power gas switches. The combination of the high extraction field in the Debye layer with the heating of the electrode surface by ion bombardment is the reason for the abundant currents observed since a long time in most types of gas switches. The new type of high-voltage switch triggered by a ferroelectric trigger, which has been described here, does not contain any hot element, discharge, or high d.c. potential. Such switches may be used in future in the fast-pulsed beam deflection and beam-dump systems of high-energy particle accelerators as well as for the modulators of radiofrequency and radar systems. They should be able to commute substantially higher power at higher current and voltage levels than can the conventional thyratrons.

The beneficial effects of a plasma are also interesting for intense electron beam sources. Apart from very high beam-current densities, one can expect excellent beam quality in terms of emittance and brightness. An easy-to-build electron beam source is described. The behaviour of most of its elements has been studied sucessfully in separate experiments. The whole assembly is believed to give one order of magnitude higher beam current density and a substantially lower emittance than can be achieved with conventional sources. The source may be used as an injector for linear accelerators, for FELs, for gas-laser preionization, for electron-beam-controlled switches, and for electron guns applied to machining and to X-ray or microwave generation.

Acknowledgements

Most people who have actively participated in the experiments on which this paper is based are mentionned in the bibliography. Especially, I have to thank J. Handerek and H. Gundel for their experimental work on electron emission from ferroelectrics, H. Kuhn for the mechanical design of the HV switch and for setting up the test pulse generator, D. Boimond for assisting in the trigger and electron source experiments, and B. Hadorn and G. Knott for preparing the drawings for this paper.

REFERENCES

[1] D.R. Humphreys, K.J. Penn, J.S. Cap, R.G. Adams, J.F. Seamen and B.N. Turman, Rimfire: A six megavolt laser-triggered gas-filled switch for PBFA II, *Proc. 5th IEEE Pulsed Power Conf., Arlington, 1985* (85C 2121-2 of IEEE Catalog, New York, 1985), p. 262.

[2] L. Caris and E.M. Williams, The spark-gap switches of the new fast beam extraction system at the CERN Proton Synchrotron, *Report CERN 69-13* (1969).

[3] D. Bloess, I. Kamber, H. Riege, G. Bittner, V. Brückner, J. Christiansen, K. Frank, W. Hartmann, N. Lieser, C. Schultheiss, R. Seeböck and W. Steudtner, The triggered pseudospark chamber as a fast switch and as a high-intensity beam source, *Nucl. Instrum. Methods* 205:173 (1983).

[4] J. Christiansen and Ch. Schultheiss, Production of high-current particle beams by low-pressure spark discharges, *Z. Phys.* A290:35 (1979).

[5] P. Billault, H. Riege, M. van Gulik, E. Boggasch, K. Frank and R. Seeböck, Pseudospark switches, *Report CERN 87-13* (1987).

[6] E. Boggasch, V. Brückner and H. Riege, A 400 kA pulse generator with pseudospark switches, *Proc. 5th Pulsed Power Conf., Arlington, 1985* (85C 2121-2 of IEEE Catalog, New York, 1985), p. 820.

[7] G.F. Kirkman and M.A. Gundersen, Low pressure, light initiated, glow discharge switch for high-power applications, *Appl. Phys. Lett.* 49 (9):494 (1986).

[8] W. Hartmann and M.A. Gundersen, Origin of anomalous emission in superdense glow discharge, *Phys. Rev. Lett.* 60 (23):2371 (1988).

[9] G. Mechtersheimer, R. Kohler, T. Lasser and R. Meyer, High repetition rate, fast current rise, pseudospark switch, *J. Phys.* E19:466 (1986).

[10] H. Gundel, H. Riege, E.J.N. Wilson, J. Handerek and K. Zioutas, Fast polarization changes in ferroelectrics and their applications in accelerators, *Nucl. Instrum. Methods* A280:1 (1989).

[11] H. Gundel, H. Riege, J. Handerek and K. Zioutas, A low-pressure hollow cathode switch triggered by a pulsed electron beam emitted from ferroelectrics, *Appl. Phys. Lett.* 54 (21):2071 (1989).

[12] R.L. Sheffield, E.R. Gray and J.S. Fraser, *The Los Alamos photoinjector program,* LA-UR-87-3036 (1987).

[13] P. Lapostolle, Possible emittance increase through filamentation due to space charge in continuous beams, *IEEE Trans. Nucl. Sci.* 18 (3):1101 (1971).

[14] H. Gundel, J. Handerek, H. Riege, E.J.N. Wilson and K. Zioutas, Copious electron emission from PLZT ceramics with a high zirconium concentration, *Ferroelectrics,* 100, 1 (1989).

[15] H. Riege, Approach towards fully plasma loaded electron-positron colliders, CERN/PS/86-8 (AR) (1986) and CERN/CLIC Note 14 (1986).

[16] E. Boggasch and M.J. Rhee, Experimental study of pseudospark-produced electron beams, Paper presented at the *2nd Int. Conf. on High Density Pinches, Laguna Beach, 1989.*

[17] H. Gundel, H. Riege, J. Handerek, and K. Zioutas, Pulsed electron emission from ferroelectrics, CERN PS/88-66 (AR) (1988) and CERN/CLIC Note 82 (1988).

[18] R.B. Miller, *An introduction to the physics of intense charged particle beams* (Plenum Press, New York, 1982), p. 34.

[19] H. Riege, Acceleration of high-density electron beams in an inductive gap, Talk given at the *Frühjahrstagung der Deutschen Physikalischen Gesellschaft, Göttingen, 1987*, Vortragsnr. III, K-16 [Verhandl. DPG (VI) 22 (1987)].

INDEX

Annular glow, 26
Arc ignition potential, 46

Back-of-the cathode lighted thyratron
        (BLT), 10, 155-165, 219
    cathode emission, 159-161
    cathode surface
        emission from, 160
        temperature, 160
    discharge character, 159-161
    discharge formation, 161
        hollow cathode phase, 163
        superdense glow discharge phase, 163
    high current operation, 159
    laser window cleaning, 158-159
    optical triggering
        fiber-optic triggering, 158
        flashlamp triggering, 157
        laser triggering, 157
    reverse current handling, 159
    structure and operation, 155-156
Bartels' model, 139-141
Beam-bulk model
    BLT performance, 221-224, 227-232
Beam quality, 353
Beam sources, 353
Bohm's formula, 6
Boltzmann's equation, 110
Boltzmann's law
    in LTE plasma, 134
Brightness, 353
Bulk plasma
    electron temperature and density of, 309
    of pseudospark and BLT, 304, 305

Cathode cavity
    inversion of, 18-19
Cathode emission mechanisms, 350
Cathode erosion, 85-87
Cathode fall
    in a hollow cathode, 58, 59
    ion transport through, 63
    potential distribution, 62
    thickness of, 26
Cathode fall region, 109
Cathode fall sheath, 305
Cathode fall width, 82
Charge carrier multiplication, 3

Chemical sputtering
    as erosion mechanism, 85
Continuity equations
    finite difference method, 203
    high resolution schemes, 205-215
        Flux Corrected Transport (FCT), 205-206
         incorporation of source terms in, 210
        Monotonic Upstream-Centered
          Scheme for Conservation Law
          (MUSCL), 206-210
        MUSCL with source terms, 211
    hybrid method of characteristics, 199
    numerical solution of, 198
Convective scheme
    cathode fall simulation, 110
    model of electrons, 122-130
Cowan-Dieke's model, 141-151
    deviation from excitation equilibrium, 148
    population temperatures in non-LTE Arcs
        Arcs, 146
    pressure determination in LTE Arcs, 145
    temperature determination in LTE Arcs, 145
Current balance
    at the cathode surface, 111-115
Current conduction of discharge plasma, 351
Current neutralization, 354
Current quenching, 84

Dark space, 15, 16
Delay
    of discharge in hydrogen, 10
Discharge(s)
    high voltage, low pressure, 22-24
    hollow cathode, 28
    hydrogen, 22-24
    hindered, 22
    self-sustained, 1
    silent, 22
Discharge formation time
    of hollow-cathode spark gaps, 42
Double layers(DLs), 277
    collisionless, 279
    solutions of Vlasov-Poisson
        system, 279-280
    strong, 279
        in collisional plasma, 284
DREICER, 32

Electrode
    auxiliary, 27
    phenomena, 27
Electrode material, 95
    physical properties of, 96
Electrode surface
    of gas-filled switch, 95
Electron beam(s)
    glow-discharge, 34
    produced in the cathode fall, 305
Electron energy distribution, 61, 219-221,
        321
Electron emission, 353
Electron emission processes
    at the cathode, 79, 80-85
Electron guns
    hollow-cathode, 34
Electron penetration
    into bulk plasma, 306
Electron swarm, 319
    steady-state parameters, 325-326
Electron temperature and density
    in LTE plasma, 134-135
Emittance, 353
Eulerian fluid, 197
Evaporation
    as erosion mechanism, 85, 86-87

Fast-electron current, 32
Ferroelectric trigger
    for pseudospark switches, 351
Field emission of electrons, 82
Fluid model(s), 221
    hollow cathode region, 235-243
        basic equations, 235-236
        boundary condition, 236-238
    pseudospark onset time, 238
    results of the simulations, 243-252
        buildup of charges, 245-252
        current density growth rate, 243-245
Fluorescent emission from hydrogen
    modeling of, 307-309
Fokker-Plank equation, 306

Gas density
    in the cathode fall, 114
Glow discharge
    high-current hollow cathode, 27-32
    hollow cathode, 19-20
    normal, 15-16
    pre-ignition, 39
    superdense, 20, 24-27
    transient behavior, 271-274
    types of, 15-33
Glow-to-arc transition, 16, 46

High-current plasma commutator, 35
Hollow-anode
    configurations, 33
    gap-discharge, 33
    thyratrons, 33

Hollow-cathode
    cold, 16
    geometry, 67
    mechanisms, 58
    system, 27
Hollow-cathode discharge, 3-5, 32
    applications, 55-57
    cold, 16
    main discharge, 294
    negative glow, 295
    pendulum motion of electrons, 295
    predischarge, 293
    pulsed, 70
    superdense glow discharge, 295
Hollow-cathode effect, 17-18, 47, 55,
        267-268
    mechanisms, 58
Hollow-cathode switch, 56
Hollow cold cathode
    for high-current switching, 35
Hybrid method of characteristics, 235
Hydrodynamic equilibrium approximation, 110

Intense particle beams
    production of, 34
Ion current density
    limitation of, 83
Ion velocity
    in hydrodynamic equilibrium, 112
Ionization rate coefficient, 321

Jitter
    of discharge in hydrogen, 10

Lagrangian methods, 197
Laser-Induced Fluorescence (LIF), 168-171
    density of neutral and singly ionized
        tungsten, 177
    density of singly ionized chromium, 179
    influence of collisional transfer on, 175
    vacuum arcs, 171
Line fluorescence, 168
Local Field Approximation (LFA), 221
Local thermodynamic equilibrium (LTE), 133
Low energy electrons
    in negative glow, 120
LTE plasma
    inhomogeneous, optically thin, 137

Magnetic Field
    influence of, 68-70
        in cathode fall region, 68-70
Magnetization
    self-fields, 32
Metastables
    balance equations of, 119
    role on hollow cathode discharge, 64
Metastable density measurement
    in negative glow, 118
Modeling of DC and transient glow
        discharges

numerical methods, 261
  physical basis, 256-258
  two electron group model, 258-260
    beam electrons, 258
    boundary condition, 259
    bulk electrons, 258
    reduced ionization rate, 260
Models of DC glow discharges
  1D model, 261-263
    field reversal, 261
  2D model, 264-267
Molybdenum electrode, 97
  PVD-coated, 100
Monte Carlo simulations, 110, 115-117, 324
Multi-beam model, 322-324
  implications in pseudosparks, 328
  secondary beams, 322
  primary electron beam, 322
Multichannel pseudospark switch
  (MUPS), 10

Negative glow, 58, 109, 117-122
  low energy electrons in, 120
Nonlinear plasma effects, 32

Optogalvanic detection of Rydberg atoms, 111

Paschen curve, 23
  equations, 296
Paschen law, 1, 9, 11
  violation of, 20
Pendel-electron effect, 296
  modeling of, 296-300
Pendulum effect
  in hollow cathode discharges, 267-271
Pendular electrons, 58, 60-61
Photo effect, 81
Photons,
  role on hollow cathode discharge, 64
Physical sputtering
  as erosion mechanism, 85, 86
Plasma, 7-8
  decay of the, 8
  inhomogeneous, optically thick, 138
  recovery, 8
  resistance of the, 7
Plasma based accelerators, 336-337
Plasma-electrode interaction
  of pseudospark, 97
Plasma in LTE-state, 134
Plasma lens, 337-339
Pseudospark
  as electron beam source, 34
  cathode emission mechanism,, 95
  electron beam from, 4
Pseudosparks and BLTs, 331-333
  as electromagnetic sources, 340
  beam generations, 334-335
  super-emissive cathode, 333-334
Pseudospark discharge, 233
  ignition of, 2

properties of, 1-13
  self-sustained, 1
  temporal evolution of, 2-8
Pseudospark switch, 350
  compared to thyratrons and spark gaps, 12
  electrode effects, 92
  high current, 37
  high power, 37
  medium power, 43
  non-conducting of, 45
  properties of, 9
  trigger mechanisms, 9

Radial breakdown, 294
Rate equations, 307
  steady state solution, 309
  time dependent solution, 312
Rectifying devices
  high power, 46-50
Resonance fluorescence, 168
Root-mean-square emittance, 345
  pseudospark produced electron beam,
    345-346
Rydberg-atom experiment, 117

Saha's law
  in LTE plasma, 134
Schottky emission, 6, 8
Secondary electron emission, 81
  at high current densities, 95
  at low current densities, 95
Space-charge neutralization, 354
Spark gap(s), 96
  low-pressure hollow-cathode, 37
Spectral line emission
  inhomogeneous light source, 134-135
Sputtering, 65-67
Streamer(s), 185
  emission, 191
  free-electron density, 192
  gap current, 192
  resistance, 192
  variation with charging voltage, 189
Superdense glow, 79
Superdense glow discharge, 6-7, 21
  applications of, 15
  review of, 15-54
Switch(es)
  high power closing, 35
  high-power opening, 44-46
  opening, 11-12

Temperature
  cathode surface, 82-83
Thermionic electron emission
  field-enhanced, 82-83
Townsend-α coefficient, 3
Townsend discharge, 78
Trigger delay time
  of hollow-cathode spark gaps, 40
Trigger mechanisms, 9

Triggering,
  of hollow-cathode spark gaps, 39
Two-electron-group plasma model
  for pseudospark and BLT, 304-306

Virtual Anode, 2, 225
Voltage drop
  of cathode fall, 82